MOLECULES TO MAN

Molecules to Man

SIR GEORGE PORTER, F.R.S.
Director, The Royal Institution of Great Britain, London

RICHARD J. HARRISON
Professor of Anatomy, University of Cambridge

PAUL R. EHRLICH
Professor of Biology, Stanford University, California

D. C. PHILLIPS, F.R.S.
Professor of Molecular Biophysics, University of Oxford

G. J. V. NOSSAL
Director, Walter and Eliza Hall Institute of Medical Research, Melbourne

CHAPMAN PINCHER
Science Editor, "Daily Express", London

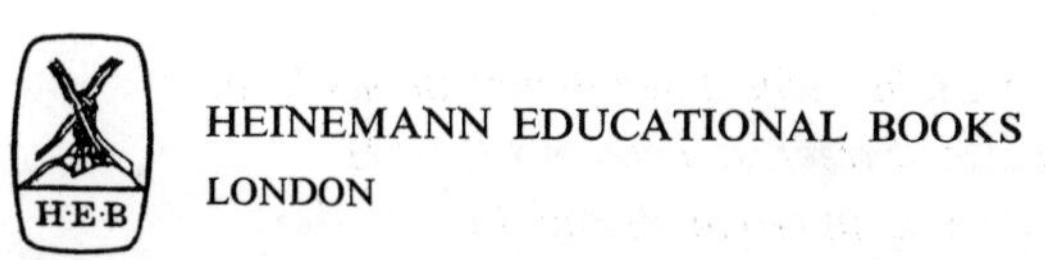

HEINEMANN EDUCATIONAL BOOKS
LONDON

Heinemann Educational Books Ltd

LONDON EDINBURGH MELBOURNE TORONTO
AUCKLAND JOHANNESBURG SINGAPORE KUALA LUMPUR
HONG KONG IBADAN NAIROBI NEW DELHI

IN THE SAME SERIES
Nuclear Energy Today and Tomorrow
Pioneering in Outer Space

ISBN 0 435 60235 7

First published in Great Britain 1972

Published by Heinemann Educational Books Ltd
48 Charles Street, London W1X 8AH
Printed in Great Britain by
Biddles Ltd., Guildford, Surrey

Preface

The contributions comprising this book were delivered in 1971 at the fourteenth International Science School organized annually by the Science Foundation for Physics within the University of Sydney. These Schools are attended by students selected from Australia and New Zealand in addition to twenty students, chosen for their outstanding ability, from America, Britain, and Japan. The aim is to stimulate and develop science consciousness in Australia and throughout the world.

On behalf of the Foundation we wish to take this opportunity of thanking Professor Paul R. Ehrlich, Professor Richard J. Harrison, Professor G. J. V. Nossal, Professor D. C. Phillips, F.R.S., Professor Sir George Porter, F.R.S. and Mr. Chapman Pincher, for having given so generously of their time and effort.

Sydney, August, 1971 H. MESSEL and S. T. BUTLER

THE SPONSORS

The Science Foundation for Physics within the University of Sydney gratefully acknowledges the generous financial assistance given by the following group of individual philanthropists and companies, without whose help the 1971 International Science School for High School Students and the production of this book would not have been possible.

Full Sponsors

The James N. Kirby Foundation

The Nell and Hermon Slade Trust

The Sydney County Council

Part Sponsors

Ampol Petroleum Limited

A. Boden, Esq.

Mobil Oil Australia Ltd.

Philips Industries Holdings Ltd.

The Shortland County Council

Standard Telephones & Cables Pty. Ltd.

Contents

Molecules to Man

by

Sir George Porter

CHAPTER ONE

In The Beginning, The Molecules

Of all the things which are made up of molecules, the most remarkable is man and, to be quite impartial, woman. In making this statement we may appear to be a little egocentric but, since we can claim no credit for the way we are made, we are really being modest about our own efforts in chemistry compared with those of nature. Quite apart from what we may think of man's achievement and his civilisation, looked upon purely as a chemical product of evolution, he is still the most wonderful and the most highly organised system of molecules in the universe we know.

We have to believe in man because he is there. But it is perhaps asking more than our imagination is capable of, to believe at the same time that this great organism arose by chance out of a disorganised, chaotic, universe and subsequently out of a soup-like sea or a slimy shore. Yet the alternatives don't take us very far . . . an earlier organisation or being, a God, would solve the problem except for the difficulty which arises in answering the question "Who Made God"?

Perhaps we are being arrogant and presumptuous in supposing that we can ever answer questions of this magnitude, perhaps we should not even ask them. But man has asked them, ever since he was able to write and probably before that; it seems natural that the first words of Genesis should be concerned with this theme and that, of the six days of Creation, three days are occupied with the creation of life. We cannot stop ourselves asking these questions and we should be very half-hearted about the problem if we did not apply to it the most powerful methods of reasoning which are now available to us, the methods we call science.

There have always been people, and there is still probably a majority, who believe that there are phenomena in our universe and ourselves which are in some way different from the material world of science and which are beyond the powers of our reason ever to understand. Such people may be right but we shall never find unless we seek. Time and time again mysterious forces, spirits and demons have been shown, by those courageous enough to investigate them, to yield to careful scientific investigation and to be capable of being understood as clearly as we understand anything. So, as long as we are capable of asking questions and applying our powers of reasoning to the best of our ability in trying to answer them, I hope most of us will continue to do so, hopefully but patiently. We must not expect to answer all the great questions at the beginning of our studies; this is what the older kinds of philosophy attempted and consequently they have progressed hardly at all. In the context of these lectures, and indeed in the present state of our knowledge, if the proper study of mankind is man, then the proper study of man is physics, chemistry and biology.

Molecules

Chemistry is a description of the world at the atomic and molecular level. It covers an immense field but it is concerned with only two things, the structure of chemical substances and the way in which one substance or structure changes into another; what we may call chemical statics and chemical dynamics.

This division into statics and dynamics is one which applies not only to chemistry but to most sciences, as well as to descriptions of life in the everyday world. For example, a fat man slips on a banana skin; that is life, part statics and part dynamics. Our descriptions of the fat man standing up or lying down give the static structure of the phenomenon before and after the event but the essence of the story is that the man moves, probably in an interesting way, from one structure to another; that is dynamics. And just as this story begins with a description of a man who was fat, so our study of the molecular world naturally begins with a description of the shape and structure of the molecules themselves.

Chemistry, more than most sciences, owes its phenomenal success to the use of models. Only in this way can the mind grasp the full import of what is known about a complex structure of many atoms. The sophistication of the model depends on the problem and the most detailed model is not always the most revealing.

Some who are not chemists are doubtless already contemplating this bewildering array of molecular models with a sense of resignation. Yet practically the whole of this great science is based on a few concepts, developed one hundred years ago, which are about as complex as the rules of a simple parlour game. For substances containing only carbon, nitrogen, oxygen and hydrogen, which comprise a major part of all known organic substances, the basic rules of the game of molecule building can be stated in a sentence. Hydrogen has a valency or combining power of one, oxygen of two, nitrogen of three and carbon of four; the four bonds in carbon are disposed in a tetrahedral arrangement, whilst the bonds in oxygen and nitrogen are separated by angles somewhat greater than a right angle. Simple as they appear to us to-day these rules were discovered painfully slowly, no Einstein or Newton appeared who alone could give a clear lead through the tangle of prejudice and confusion which was chemistry. Rather the truth slowly emerged, as a result of the accumulated evidence of many men, being forced upon a generation who seemed obstinately reluctant to formulate the obvious and indispensable theory of valency.

The basic idea on which chemistry rests, the idea of the "compound atom" or molecule, with its fixed numbers of atoms of one or more kinds, was first clearly proposed by John Dalton in 1803. Dalton's compound atoms look remarkably like the molecular structures which we write to-day, they imply that a molecule has not only a definite number of atoms but that these atoms are arranged in a particular way, the molecule has a shape. In fact these drawings implied more than Dalton wanted them to imply and neither he nor his successors carried the idea through to its logical conclusion of atoms in a special arrangement; they were content to regard their atoms and compound atoms as a numerical shorthand.

The main point that Dalton missed was the idea that each kind of atom had its own distinctive combining power or valency and it was to be another half century before this was realised. Frankland made the first clear statement to this effect in 1852 but did not follow it up as fully as he should have done and another ten years elapsed before Couper, Crum-Brown and Kekulé used formulae similar in most respects to those that we use to-day. But in one respect the picture was still not complete and this is very well illustrated by a lecture given by Hofmann at the Royal Institution in the year 1865 before H.R.H. The Prince of Wales. In this lecture for the first time he transferred the new formulae of chemistry from the printed page to models made of croquet balls and wire. But what he did not do illustrates how difficult it is for man's mind to advance more than one step at a time and to rid itself of preconceptions. Figure 1.1 shows a drawing of the models from Hofmann's paper and you will see that Hofmann had literally translated his molecules from the printed page so that his models also were two dimensional.

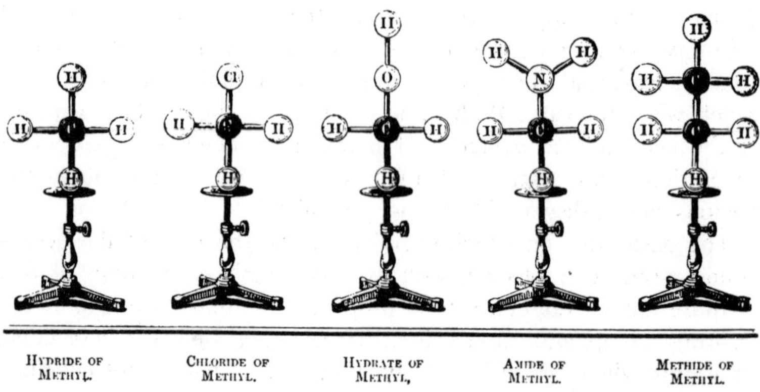

Figure 1-1. Illustration from the discourse by Hofmann, 1865.

The final phase of molecular geometry came in the last quarter of the century when chemistry left the planarity of the printed page. If the four bonds in carbon lay in a plane then, if all four of the atoms attached to carbon were different, there would be three different isomers with different chemical properties. In fact there were no such isomers, only one was known in such cases as far as chemical properties were concerned, though there were two varieties in another sense, the behaviour towards polarised light. It had been known for some years that light could be polarised and that the orientated molecules in a crystalline lattice affected this in many remarkable ways.

But even more remarkable, and more relevant to our present purpose, is that some molecules rotate the plane of polarised light even when dissolved in a liquid, in which case their orientations must be entirely disordered and random. A solution of cane sugar rotates the plane of polarisation to the right, different colours being rotated to different extents. When sucrose is broken down into its monosaccharide components, one of them, glucose, rotates the plane to the right and the other, fructose, rotates it to the left.

All this, as well as other forms of isomerism, was explained by Van't Hoff in 1874, (and independently by Le Bel). First, he pointed out that all then-known molecules showing optical activity, as this effect was termed, had at least one carbon atom in which all four bonds were attached to different atoms. He pointed out, a fact which had escaped Pasteur, that if the bonds are arranged tetrahedrally, even in a perfectly regular tetrahedron, there are just two forms which should be indistinguishable chemically, but which are mirror images of each other and retain this mirror image relationship whatever the orientation, just like a bag of left-handed or right-handed screws. This was the final vindication of the structure theory. It was clearly no longer possible to describe chemical compounds without it and indeed it was now clear that the finest differences in three dimensional structure were accessible to, and revealed by, experimental observations on the bulk substance. It was clear, as well, that chemists must take account of these differences between left-handed and right-handed forms because, even if they were not distinguishable by the chemist, they were of great concern to living things which usually produced one form exclusively. Above all, the tetrahedral arrangement of carbon's four bonds was established.

Thus, by 1874, the essential ideas of the chemical bond had been formulated in sufficient detail to make them, simple as they were, one of the most powerful, intellectual tools of the natural sciences. Even to-day it is for most purposes entirely satisfactory to represent a chemical bond as a bit of wire connecting two wooden balls, provided the proper number and angles are used. But what does this bit of wire represent, what are the forces which connect the atoms?

The answer to these questions can only be given when we know something of the internal structure of the atoms themselves and the successful solution of this problem fell to the physicists. They were a quite separate breed for whom the whole process of education into a belief in the reality of atoms had to begin all over again. That there is some regularity in the atoms' structure had been shown by Mendeleev's discovery that, when arranged in a certain way in order of atomic weights, the valencies and other properties of the elements varied in a periodic way. By 1913, Rutherford had developed the planetary theory of the atom and Bohr, shortly afterwards, explained the periodicity of properties in terms of this atom. But this model was unable to account for the simplest chemical bond.

Only in terms of quantum or wave mechanics is it possible to understand how atoms can form stable bonds and the complete description is highly mathematical and complex. For molecules, even as simple as water, complete solutions of the mathematical equations are far too difficult to be possible. This does not sound promising for the chemist who is more likely to be interested in a molecule with many more atoms than water. Nevertheless, in his characteristic chemical way, the scientist interested in big molecules has absorbed as much from modern wave mechanics as can be usefully handled and has learned to make models of molecules embracing these new ideas, which have in the last few years invaded most modern chemical texts.

The essential characteristic of these new models is that they embrace the idea of indeterminacy in the electron's position, its wave nature. The electron, whether in an atom or molecule must be represented in probability density terms, rather like a population density map which tells only the probability of finding a certain person at a place and nothing more. These probability density distributions for the electron are represented mathematically as three dimensional standing waves with the square of the amplitude of the wave giving the probability of finding the electron at that place.

If the equation is solved for one electron, we find a number of possible distributions. The spherically symmetrical one is called an s orbital, the double egg-shaped distribution with a node between the "eggs" a p orbital, and so on.

Now, to see what the electron distribution in any particular atom will be, we simply fit the electrons into the lowest possible energy arrangement. In doing this we must use a new principle, that of Pauli, which says that no two electrons can be exactly the same. Since the only way in which two electrons can differ, apart from their position, is in the way they spin, and since there are only two possible ways of spinning, no more than two electrons can go into any one orbital. With this restriction in mind we build up all the atoms of the periodic table, obtaining the electron distributions or shapes as we do so. Immediately, many things become clear, particularly the reason for periodicity in properties.

Now we have the shapes and structures of the atoms, can we understand valency and the chemical bond? We have to let the electron move in the field of two or more nuclei in a molecular orbital, with the same restriction as before, that no more than two electrons can occupy the same molecular orbital. When these two singly occupied atomic orbitals overlap, whilst the electron is in the overlap region it will be attracted by both nuclei, the energy will be lowered and the united atoms will be more stable than the separate ones. This is the chemical bond—nothing more mysterious than an electrostatic attraction between oppositely charged particles, but one which must be described in the spirit of the uncertainty principle and modern quantum mechanics. In this way, we can see why H, O and N form one, two and three bonds in the way they do. For carbon, things are a little more difficult. We would expect only two bonds but find four. What happens is this. One of the s electrons in the carbon atom is promoted to this vacant p orbital so that we now have four valencies. The energy required to do this is a good investment and is more than recovered from the energy of the two extra bonds which can be formed. However, to do this as efficiently as possible, the eight electrons have to get as far away from each other as possible and the best way of doing this is to take up the tetrahedral arrangement.

So, in this way, we obtain a satisfactory account of the chemical bonds, the bits of wire which the chemist used so successfully and so long. The new theories help us to explain many subtleties and exceptions to the rules of valency which have arisen, especially in the more complicated field of inorganic chemistry. But in full justice to the experimental chemist we must remember that the explanations, even today, are usually given retrospectively after the anomaly has already been discovered.

There is a great deal of difference between a simple formula written on paper or made up of ball and wire models and the "space filling models" which take some account of the size of the electron clouds around the molecules. There is an even greater difference between these and the models which try to show the actual shape of the orbitals or the mathematical equations from which these are derived. Which is the best model for us to use? This depends very much on the problem we are trying to solve and during these lectures, and others like them, you will find that the type of model used is often changed as the problem changes. For example, the orbital model of water tells us a lot about the electron distribution and is helpful in understanding the electrical polar properties of the molecule and why two molecules of water attract each other. If we now wish to see how many molecules of water are bound together in a crystal of ice we turn to a ball and wire or similar model because even a space filling model would block our view of the interior structure of the crystal.

In a similar way, a mathematical model is useful in providing numbers and quantitative information which often can be manipulated to calculate other properties not apparent from the models. On the other hand, if the molecule is complex, the amount of numerical information necessary to describe it, even as fully as it is described by a ball and wire model, may be so great as to be beyond our comprehension. For example, the coordinates of each of the atoms in a relatively simple protein molecule if simply printed out in several pages of a book would be meaningless and the coordinates of all the atoms in a microscopic biological object would fill all the books in our library. As we progress from simple molecules to complex biological systems, we must therefore choose models which are less and less detailed, but which still represent as accurately as possible the features under examination and relevant to the problem in hand.

Organic Molecules

Organic chemistry was originally the chemistry of substances made by living things. When it was discovered that these substances were all based on the element carbon, the science of organic chemistry was broadened to mean the chemistry of all carbon compounds including those synthesised by man but not found naturally. Since man is part of the biosphere, it is still true that all organic compounds arise from life.

The biosphere, including both animals and plants, is composed principally of four elements which make up 98% of its mass as follows,

Oxygen 73%. Carbon 14%. Hydrogen 9%. Nitrogen 2%.

The rest is made up of 1% calcium, and traces of some forty other elements of which the most common and important are sulphur and phosphorus, though we certainly could not live without the others such as the metals potassium, magnesium, iron and sodium. All the elements necessary for life are found in sea water, which is not surprising because life almost certainly started there. Although it is a long time since we crawled out of the sea, life continues in a watery environment which comprises 65% of the human body, and 90% of that of marine animals.

Dissolved in the water are salts and a number of quite small organic molecules but by far the greater part of living matter, after water, is made up of very large molecules, macromolecules, linked together so as to create a structure which is no longer liquid, since it has shape and form, though it cannot be called a solid in the crystalline sense. Our understanding of the chemistry of very large molecules and polymers is relatively recent, having begun about 1920 when the word macromolecule was given to these large organic structures by Staudinger. New methods of studying them had to be developed since traditional chemical methods usually led to nothing more illuminating than a sticky charred mass at the bottom of the vessel. These were physical methods such as spectroscopy, X-ray diffraction, osmometry and ultra-centrifugation. During the 1939-1945 war, methods for their synthesis became highly developed and led to the plastics and synthetic fibre industry.

Macromolecules

Most of the man-made polymers are composed of repeating units of the same group of atoms. The simplest is polyethylene in which the repeating unit is a $-CH_2-$ group:

$$-CH_2-CH_2-CH_2- \ldots\ldots CH_2-CH_2-$$

It is made, as its name implies, from many "monomer" ethylene molecules $CH_2{=}CH_2$.

There are many derivatives which can be made from substituted ethylenes $CH_2{=}CHX$ so as to give the structure

$$\begin{array}{l} -CH_2-\underset{\displaystyle X}{\underset{|}{CH}}-CH_2-\underset{\displaystyle X}{\underset{|}{CH}} \ldots\ldots CH_2-\underset{\displaystyle X}{\underset{|}{CH}}-CH_2-\underset{\displaystyle X}{\underset{|}{CH}}- \end{array}$$

for example

polypropylene	X = methyl, CH_3
polystyrene	X = phenyl, C_6H_5
polyvinyl chloride	X = chlorine, Cl

These polymers do not occur naturally, although smaller "oligomers" of polyethylene $(-CH_2-)_n$ occur in petroleum hydrocarbons. Candle wax is an example, with n about 30.

Another hydrocarbon polymer which occurs naturally is natural rubber which has a repeating unit of isoprene:

$$-CH_2-CH{=}\underset{\displaystyle CH_3}{\underset{|}{C}}-CH_2-$$

This is distinguished by its double bond which has an important effect on the chemical reactivity. Natural rubber is sticky and tends to flow when heated but if it is reacted with sulphur, links are formed across the molecules, holding them together, this is the process of vulcanisation discovered by Goodyear in 1838. The more sulphur, the greater the number of cross links formed, until the substance becomes the hard material known as ebonite. The ability of sulphur to form cross links between polymer molecules is also used by nature in forming cross links between parts of a protein molecule.

It is not necessary that all the units in a polymer molecule should be the same and many polymers are made by linking together two types of molecule so as to give the structure

$$-A-B-A-B-A-B-$$

The method of condensation polymerisation, a reaction in which a small molecule is eliminated by reaction of the end groups of two different molecules, was developed by Carothers and led to the manufacture of nylon. There are many different nylons, depending on the length of the molecules used but all are based on reactions of the following type:

$$NH_2\text{–}(CH_2)_n\text{–}NH_2 + \begin{matrix} O \\ \| \\ HO\text{–}C\text{–}(CH_2)_m\text{–}C\text{–}OH \\ \quad\quad\quad\quad \| \\ \quad\quad\quad\quad O \end{matrix} + NH_2\text{–}(CH_2)_n\text{–}NH_2$$

$$\downarrow$$

$$NH_2\text{–}(CH_2)_n\text{–}NH\text{–}\overset{\displaystyle O}{\overset{\|}{C}}\text{–}(CH_2)_m\text{–}\overset{\displaystyle O}{\overset{\|}{C}}\text{–}NH\text{–}(CH_2)_n\text{–}NH_2$$

$$+ H_2O \qquad + H_2O$$

Nylon was discovered as a result of a systematic attempt to make, in the laboratory, polymer materials which were similar to natural silk and wool. Silk and wool are proteins and have the repeating unit:

$$\text{–}NH\text{–}\underset{\displaystyle X}{\underset{|}{CH}}\text{–}CO\text{–}$$

where the group X can be any one of a number (about 20) of amino acid units.

One of the very important properties of proteins and similar molecules is that they form excellent strong fibres, in which, as one would expect, the long chain molecules are lined up alongside each other.

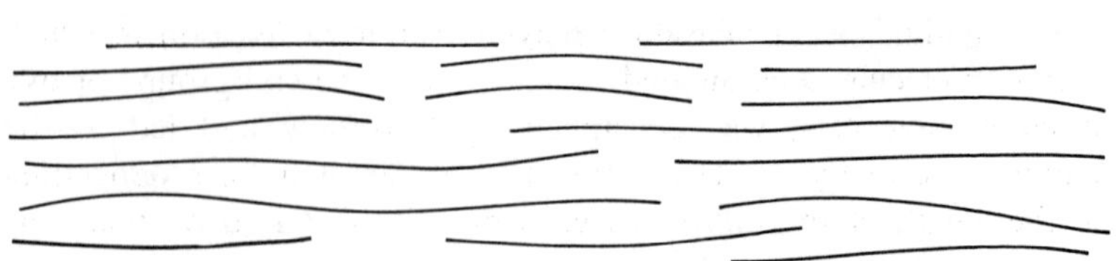

But since there are no ordinary chemical bonds cross linking the molecules, why do they not fall apart? The explanation is to be found in terms of a special kind of bond which is of very great importance in molecules of this kind and particularly in biological polymers, the hydrogen bond.

The Hydrogen Bond

The hydrogen atoms in a water molecule are held to the oxygen atom by covalent bonds in which the two electrons spend a good deal of their time between the nuclei of oxygen and hydrogen and so hold the atoms together by electrostatic attraction. But the bonding electrons are not attracted equally by these two nuclei but are on the average closer to the nucleus, so that there is a residual negative charge on the oxygen atom and a corresponding positive charge on the hydrogen atoms.

```
      Oδ-
     /  \
  δ+H    Hδ+
```

In liquid or solid water the molecules will tend to line up so that these partial charges neutralise each other, in the configuration with lowest energy

```
   H
    \
     O
     |
     H
     :
     O          H
    / \        /
   H   H . . .O
               \
                H
```

One of the reasons why hydrogen forms this type of bond much more effectively than other atoms, which may, of course, also be in polar structures, is that there are no inner electrons in the hydrogen atom and the second atom with which it forms the hydrogen bond may, therefore, approach closely without encountering repulsion.

The electrons on oxygen which form the hydrogen bond will be principally the lone pair electrons on the "backside" of the water molecule. A hydrogen atom in the polar groups O–H, N–H, F–H,

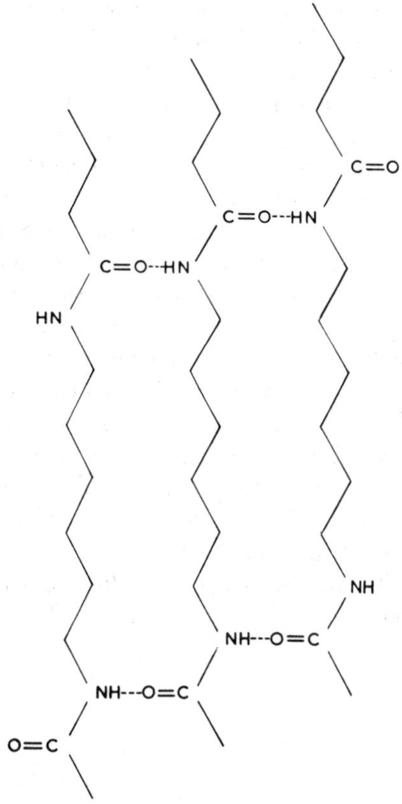

Figure 1-2. Arrangement of neighbouring chains in crystal of nylon 66.

and S–H will form hydrogen bonds with the lone pair electrons of oxygen or nitrogen atoms in this way

The strength of a hydrogen bond is only about one twentieth of that of a coordinate bond such as O–H or C–H but, between two long polymer molecules lying in a favourable position alongside each other, many such bonds can be formed so that the molecules are very strongly bound together. The molecules in a nylon filament which has been stretched so that the molecules are ordered in a partially crystalline arrangement will be held together by hydrogen bonds in the manner shown in (Figure 1.2).

Such hydrogen bonds can also be formed between groups within the same molecule and so hold the molecule in one particular configuration out of the many which it could take up as a result of the free rotation which is possible about the single bonds.

So the atoms form molecules, the molecules build up into macromolecules composed of repeating units and the macromolecules can be linked together by hydrogen bonds into still more complex organised structures. How these structures themselves build up into the specific enzymes, code-carrying D.N.A. and the other biochemical units will be the principal subject of Professor Phillips' lectures. For the present, we have said enough about the structure and shape of molecules. These are the static models of a static world. But the world is not static, neither the restless world of the molecule nor that of man. We live in a world of change and change cannot occur without movement. Even in the absence of overall change, even at equilibrium, molecules are in ceaseless motion and the life we live, a dynamic affair, is the totality of this molecular motion.

CHAPTER TWO

Molecules in Disorder

Introduction

Few achievements of man have been more successful or more complete than his discovery of the laws which rule the chaos of atomic and molecular motion. These laws determine the direction of change and are quite general; they govern the universe. They are concerned with matters of everyday experience, with motion, change, heat and energy. Once understood they appear almost self-evident yet they are charged with deep philosophical meaning.

The chemist deals with millions of millions of particles in random chaotic motion, a world of apparent anarchy where the only laws are laws of disorder. True, each individual particle obeys the mechanical laws of motion but how are we to record the course of a number of particles far greater than the number of words in all the books of the world? Surely the answer to a question such as 'will this drop of water freeze?' cannot be as complicated as this. Certainly it is not; if we ask only for the averaged overall behaviour of molecules, the information required to answer the question is of a similar averaged kind. It is not with the mechanics of many separate molecules that we must concern ourselves, but with a statistical kind of mechanics of the whole chaotic motion.

There is another approach which ignores the existence of atoms and molecules altogether and is content to deal only with those bulk properties which can be directly measured; this is the great science of thermodynamics. Much has been written about thermodynamics and about its second law in particular; it has been quoted as the *sine qua non* of scientific enlightenment, rather as Shakespeare or Beethoven might be quoted to represent the arts. While few would deny a place to Beethoven and Shakespeare in a common culture of arts and sciences, the place of the second law of thermodynamics is less assured. Yet it is an idea of great beauty. It is true that a full understanding of all its implications and a facility in its application require some specialised training and a little mathematics, but a long technical training is necessary before one can interpret Beethoven fully. The essential point for most of us, is whether we can understand and appreciate the law well enough to derive some pleasure and satisfaction from it.

Like many of the most fundamental concepts, the spherical earth and the atomic theory for example, man's ideas about heat and energy developed painfully slowly and frequently progressed backwards. Carnot, a practical engineer and scientific genius, discovered the second law of thermodynamics but held a view of the nature of heat which was quite erroneous and inconsistent with the first law of that science. Rumford, with his cannon boring experiments at the Munich arsenal, Davy, Young and many other authorities stated clearly their belief that heat was a form of motion, but it was fifty years later in the middle of the nineteenth century before this was finally accepted and the science of thermodynamics was firmly founded by the two giants, Clausius and Thomson.

The final synthesis between the immaculate formality of classical thermodynamics and the great wealth of knowledge which we now possess about the atomic structure of matter came later, largely as a result of the work of Boltzmann in Germany, Maxwell in Cambridge, and of a reclusive Yale professor, Willard Gibbs. It is one of the most convincing syntheses in the whole of science. Now that the bridge has been made, however, it is no longer necessary to follow the long historical path through classical thermodynamics. Most of us, to-day, are more familiar with atoms than with the workings of heat engines, and we find the statistical approach to thermodynamics immediately acceptable, something we have known all along. And, after all, few people of this age would argue that the atomic theory is rash assumption.

The First Law

Change is not entirely a matter of substance, it is accompanied by exchanges of energy or heat. It is almost incredible that science should have progressed through Galileo and Newton into the middle of the nineteenth century without any clear understanding of the nature of heat or its relation to other forms of energy. The industrial revolution was, to a large extent, a result of the invention of the heat engine, a device for converting chemical energy into useful work, but the principles on which it worked were, at that time, obscure. These engines served man in two ways; they relieved him from the slavery of manual labour and they caused him to ponder on the nature of heat. In this way the science of thermodynamics was born.

Heat is another word for the random kinetic energy of molecules in motion. It is, therefore, itself a form of energy and when heat or other forms of energy such as potential, chemical or electrical energy are converted one to the other, they are converted according to rigorous laws of exchange. Just as wealth appears in many forms and can be exchanged from one form to another according to the laws of the realm, but not created from nothing, so it is with heat and energy; they may be exchanged according to the laws of thermodynamics but the total amount is always conserved. The analogy is not complete since, unlike the laws of the realm, the laws of thermodynamics have never been broken.

The substance of the first law of thermodynamics is given in the last paragraph; it is the law of conservation of energy. It recognises the equivalence of heat and other forms of energy and the impossibility of perpetual motion machines. When a change of any kind occurs in a system (a system is any bit of substance or part of the world in which we choose to be interested), three kinds of energy change may be involved. There will be an increase or decrease in the internal energy (U) of the system itself, heat (q) will be evolved or absorbed and work (w) will be done. The first law, the law of energy conservation, then tells us that

the increase in internal energy of the system	=	the heat absorbed by the system	−	the work done by the system

or, in symbols

$$\Delta U = q - w$$

where the Greek symbol delta (Δ) signifies, as always, the change in a quantity. It is an equation which is illustrated every time we pay our fuel bills.

Thermodynamics is not concerned with the motion of molecules other than its manifestation as heat, but the science of chemical change is very much concerned with this motion. How can we handle a problem as complex as the random motion of many molecules, each of which requires at least six numbers to describe it? We are not, fortunately, concerned with the motions of individual molecules as Maxwell's demon was supposed to be. If statistical, averaging methods are accurate enough to allow an insurance company dealing with a million customers to make a predictable annual profit, it is not surprising that predictions about the million, million, million, million, molecules in a flask of air can be made with a precision normally called certainty. Their average speeds, the distribution of these speeds, their number, size, their collisions with each other and with the walls of the flask are known nearly as accurately as the size of the vessel itself. There are few laws more precise than those of perfect molecular chaos.

Entropy and the Second Law

Why do things change in the direction they do? A lump of sugar spontaneously dissolves in a cup of tea but it would be quite an event if a cup of tea suddenly undissolved its lump of sugar and popped it back into the sugar bowl. But why not? Why does this sort of happening seem so ridiculous? Why do things fall to the ground but never rise up against gravity, why do gases expand, liquids mix and fuels burn but never the converse happens? In brief, what are the laws which govern the direction of a spontaneous change?

Let us make a hypothesis which at first sight seems quite promising; that all spontaneous change goes in such a way that the energy of the system decreases. But we are already in trouble; the total energy of the universe must be conserved, so that had we chosen to be interested in another part of the universe we should have found that the energy increased spontaneously. Try a modified hypothesis; it is the potential energy of the system which always decreases. This is better; it immediately explains most mechanical changes, why things roll down-hill, why springs unwind. But there are many changes where potential energy is not involved, and, worse still, there are many spontaneous changes where no energy exchange is involved at all; the simple mixing of two liquids or gases for instance. It is obviously not by considering energy that we shall find our answer; some completely new concept is required and this is the concept of randomness, chaos, disorder or mixed-upness.

The idea of disorder is familiar enough throughout life, from the building bricks of the nursery to the shuffling of a pack of cards. We know that things always seem to get mixed up and disordered spontaneously and rarely return to an ordered arrangement of their own accord. This is the principle which we have been seeking, and which determines the direction of all change. To make it useful we must be able to measure it, so we should begin by asking what we mean by disorder. Take any game of pure chance and ask how the player decides what is probable and what is improbable; the probable event is one which can happen in many ways. The same is true of order and disorder; an ordered arrangement is one conforming to some prechosen requirements, and all other arrangements we call disordered. Since there are far more arrangements in this latter category, disorder is the natural condition of things. The greater the number of ways in which a state can arise the more disordered is that state—and the more probably will it occur. This applies equally to arrangements of position, of motion and of molecules between discrete quantum energy levels.

Let us call the possible number of arrangements which can give rise to a particular state P; this is a measure of the probability of the state. A quantity which is more often used as a measure of the randomness of molecular motion is called entropy; it is given the symbol S and is related to the probability P by the equation of Boltzmann:

$$\text{Entropy} = S = k \log P$$

where k is a constant having the dimensions of energy.

When a substance is heated, the molecular motion becomes more violent and chaotic and we shall, therefore, expect the entropy to increase. This expectation is correct and the entropy increase is related to the heat absorbed in a very simple way. If a substance absorbs heat at an absolute temperature T (absolute temperature = temperature in degrees centigrade + 273), and no other changes occur, the increase in entropy of that substance is equal to the heat absorbed divided by the temperature or

$$\triangle S = q/T.$$

There are, therefore, two simple ways of calculating this quality called entropy, which is a measure of molecular disorder and the key to an understanding of the direction of change.

The second law of thermodynamics is one of the most comprehensive generalisations in the whole of science. Its basic theme is that disorder is all the time increasing and a precise general statement of the law is that *every system which is left to itself will, on the average, change towards a position of maximum probability.* The words 'on the average' take account of the fact that it is a statistical law, and fluctuations from the average are possible in principle although, in practice, the number of particles in the smallest bit of substance which we handle is so great that these fluctuations are insignificant.

From the relation between the number of ways of arrangement, or probability, of a state and its entropy, it is clear that if one increases so must the other and the second law can be stated in another way: *In any spontaneous change the total entropy increases.* The word 'total' is important here; the entropy of one part of the system may increase or decrease as long as the total entropy of everything involved, the whole universe if necessary, increases. Clausius recognised this in his succinct statement of the two laws of thermodynamics in the form 'The energy of the universe is constant, the entropy of the universe tends to a maximum'.

Another form of the second law, one of the earliest and still perhaps the most common, is the rather pragmatical statement that *heat cannot flow spontaneously from a colder to a warmer body.* This must be true if our earlier statements are correct, since otherwise the entropy increase, q/T_{hot}, of the hot body would be less than the entropy decrease, q/T_{cold}, of the cold body, giving a net total decrease of entropy which is impossible in a spontaneous change. These are all equivalent statements of the second law of thermodynamics; a strange principle, unlike most other scientific laws which deal with the conservation of something or the equality of one quantity to another. It applies equally well to a barrel rolling downhill and to the complex chemistry of a biological change.

Why is it that entropy and disorder always increase with time? Science can answer 'how' but not 'why?' and can only say, in reply to this question, that the entropy increases with time because time goes forwards. This answer is not as trivial as might at first appear; it provides a means of giving a direction to time. If, in some unfamiliar surroundings, we suddenly became concerned as to whether time was going forwards, backwards or just standing still we could quickly reassure ourselves by observing whether, in a spontaneous process, the entropy increased, decreased or remained the same. This is well illustrated by a moving film; only if the events depicted are accompanied by changes of entropy is it possible to tell whether the film is being run forwards or backwards.

The inevitable increase of disorder implies an irreversible trend in the universe itself, a trend towards total chaos and a heat death of timeless equilibrium. In so far as the universe is really a closed system this must follow, but science has learned to view long extrapolations of this kind with caution. It is a long step from the disorder of molecular motion to the chaos of the cosmos.

Equilibrium

Disorder and entropy cannot increase for ever, they must eventually reach the maximum possible value. According to the second law no further spontaneous changes are then possible; the system becomes static and permanent. This is the situation which we know as equilibrium.

A familiar spontaneous change is the melting of ice and an equally familiar one is the freezing of water. A fraction of a degree change in temperature, from just below 0°C to just above, is sufficient to change the face of a huge lake from solid to liquid and, at one temperature only, which we call the freezing point, it is possible to have both liquid water and ice in equilibrium. These are examples of nature's determination to increase the total entropy at all costs. The position of equilibrium can be calculated as the condition where the total entropy, of the system and its surroundings, is a maximum. But it is not always very practical to take into account the whole universe in these calculations; one must confess that it is not as simple a rule as that of minimum potential energy which applies to falling weights and similar mechanical changes. Cannot we find some characteristic of the system itself, the bit of the universe which interests us, which will tell us the direction of change and the position of equilibrium?

The property we want is called the *free energy,* the most important concept of thermodynamics after energy and entropy. The decrease in free energy is just the maximum work which can be got out of a change when it is carried out in the most efficient, reversible way. The free energy change $\triangle A$ is $\triangle u - T\triangle S$ and, for changes at constant temperature and volume, equilibrium is the condition where free energy is a minimum. There is a slightly different free energy change, given the symbol $\triangle G$, which is now commonly used because G is a minimum at equilibrium for a system at constant temperature and constant *pressure.* The two free energies are related by the equation $G = A + PV$.

The system strives to decrease its free energy and the struggle is one of lowering the energy u or H and raising the entropy S. The relative importance of each, and the best balance, will depend on temperature; at low temperatures the energy is important, at high temperatures the entropy takes over. That is why things break up at high temperatures, why solids melt, liquids evaporate, and the roast beef becomes tender. Equilibrium is a state of balance between the conflicting requirements for minimum internal energy and maximum entropy of a system. The distribution of molecules is rather like the distribution of people in a double decker bus; they may increase their energy as well as their freedom by going upstairs or they may remain in a position of lower energy, where the choice of seats or number of possible ways of arrangement is small. The distribution which is found on the average is neither that of lowest energy (all people on lower deck) or greatest randomness (equal numbers on upper and lower decks) but a compromise between the two, determined by the number of seats of each deck and the energy of the passengers.

From a molecular viewpoint equilibrium itself is a dynamic

affair. To the individual molecule it matters little whether the system is in equilibrium or not, it sees only a restless conflict of forward and backward change, a battle which rages as furiously at equilibrium as under other conditions, but one which has then reached complete stalemate.

Molecules at Work

A chemical reaction is usually carried out for one of two purposes: the manufacture of a useful chemical substance, or the production of energy. Judged by the sheer bulk of the material converted, the second purpose is by far the more important; most of the energy used by man is derived from chemical reactions, whether it be to drive his engines or his own body. The greatest industry of all is that which provides the chemical fuel used by man as food and, from the fossil fuels, petroleum and coal, man in the Western world enjoys energy equivalent to a labour force of thirty men to run his home and has at his command the energy of more than one thousand men to drive him over the ground.

Any spontaneous change can be made to do useful work, and the maximum work attainable is equal to the free energy available from the change. A few devices operate so efficiently that this theoretical maximum is very nearly achieved; the electrical cell and fuel cells for example. But these are expensive and the usual procedure is to convert the chemical energy into heat, by burning a fuel, and then to convert this heat into work as efficiently as possible. And here we find ourselves firmly ruled by the second law of thermodynamics in its original form, the form which was developed to answer this very problem of the efficiency of conversion of heat into useful work.

Consider this tantalising prospect: the sea is warm and extensive, it contains a vast amount of heat energy. Let us convert some of this into work and our energy problems are solved for evermore; man will attain a standard of living undreamed of to-day. As a start let us run a large ship's engine by taking heat from the sea and dropping the used up blocks of ice over the stern. But we must have a spontaneous process to work our engine and the second law tells us that heat can only flow from a warmer to a colder body. Application of the laws of thermodynamics quickly tells us that the efficiency with which the random disordered energy called heat can be converted into useful ordered energy or work is

$$\frac{T_h - T_c}{T_h}$$

when T_h and T_c are the temperatures (degrees absolute) of the hot and cold bodies. The struggle for efficiency therefore becomes, in practice, a struggle for a larger difference in temperature.

The engine of the body works at 37°C, and we soon feel miserable if it deviates even a few degrees from this. It therefore cannot afford the inefficiency which would result if it began by converting its chemical energy into heat, but carries out an intricate series of reactions to convert free energy directly into mechanical energy. It is remarkably efficient; of the 3,800 calories liberated by combustion of 1 gm. of glucose sugar, the body uses two thirds for doing work or providing stored energy. This is about twice the efficiency of a modern steam generating plant using a similar combustion reaction as its source of energy. The body is unsurpassed in the science of applied thermodynamics.

CHAPTER THREE

Patterns of Change

There are two questions we may ask about a chemical reaction; which way does it go and how fast? It is all very well to know that thermodynamics allows a chemical change to occur in a certain direction but this information is not very useful if, in practice, the reaction occurs so slowly that no product is formed in hundreds of years.

To understand the rates of chemical change we must first understand the mechanisms, we must know just how the chemical substances rearrange themselves to form the new products. How, for example, does petrol burn to form carbon dioxide and water? From an analysis we can check that, as we should expect, the total numbers of atoms are the same before and after the reaction and that the equation, for the burning of octane for example, is as follows:

$$2\ C_8H_{18} + 25\ O_2 = 16\ CO_2 + 18\ H_2O$$

Do twenty-seven molecules come together in a collision to bring about this change? Not in a million years; this reaction, like nearly all chemical changes, occurs in steps, each step involving one, two or, very rarely, three molecules at a time. It is the same in the building up of big polymer molecules and nature has devised the most intricate series of stepwise reactions to construct the complex molecules of biological chemistry and to utilise their energy.

Molecules may react alone or in pairs, in what are called unimolecular or bimolecular reactions. If molecules react alone in a simple one-step reaction, halving the number of molecules in the vessel will halve the number reacting in unit time, i.e. it will halve the rate of reaction. If the molecules react in pairs, halving the number of molecules in the vessel will quarter the rate of reaction (if the number of dancers on a ballroom floor is doubled, each dancer collides with other dancers twice as often and since there are twice as many dancers the rate of collision is four times as great as before).

If we heat a mixture of hydrogen and iodine some of it changes into hydrogen iodide and this goes mainly by simple bimolecular collisions as follows:

$$H_2 + I_2 \rightarrow 2HI$$

This seems obvious but, to illustrate how careful we have to be, consider the very similar reaction of hydrogen reacting with chlorine to form hydrogen chloride. It might be expected to go in a similar way

$$H_2 + Cl_2 \rightarrow 2HCl$$

In fact, it goes very slowly or explodes violently. The explosion is possible because a great amount of energy is released in the reaction but how does it occur? Like many chemical changes there are more substances involved than are shown in the equation, intermediate substances which provide the steps which make reaction possible. In this case, the intermediate substances are the atoms of H and Cl, in other reactions they may be bits of molecules, free radicals like methyl, CH_3 or hydroxyl OH, or they may be ions like H^+ or OH^-.

The reaction between H_2 and Cl_2 has another special characteristic—it is a chain reaction—and the pattern of the change is as follows:

$$Cl_2 \rightarrow 2Cl$$
$$Cl + H_2 \rightarrow H + HCl$$
$$H + Cl_2 \rightarrow Cl + HCl$$

The first step is difficult and needs to be started by energy from outside but once we have a few chlorine atoms in the vessel the pattern of change represented by these equations will continue until all the hydrogen or chlorine atoms have reacted. In fact, the chain sequence is not quite 100% efficient since the atoms can recombine or be lost at the walls so that the mixture does not explode unless we make chlorine atoms at least fast enough to make good these losses. The burning of petrol, which is also an explosive reaction, goes in a similar way, but the sequence of steps is much more complex, and involves dozens of different reactions and many different kinds of free radical intermediates. In some of these steps two radicals are formed from one, e.g. in the reaction of an oxygen atom with a hydrogen molecule

$$O + H_2 = OH + H$$

so that the total number of reactive centres grows during the reaction, leading to a branched chain reaction and, under the right conditions, a violent explosion.

We derive the energy for our cars and our power stations in this way. Furthermore, our bodies derive their energy from similar sources and by similar overall chemical reactions, by the combustion reaction of sugars, fats and similar fuels with oxygen. But there is an obvious difference. Whereas the temperature of a hydrogen-oxygen or a petrol-oxygen flame is about 2000°C, the temperature of our bodies is 37°C and we feel very uncomfortable if we exceed this by one or two degrees. Clearly the body does things differently. How does it manage to burn a fuel at body temperature when no such reaction occurs under ordinary laboratory conditions?

Obviously, although the overall reaction is the same, the mechanism or pattern of change must be different. The key to this pattern is the use of catalysts, and those particular biological catalysts called enzymes. To understand why we need a catalyst to make a reaction go at 37°C, we must first understand why it does not go at this temperature without a catalyst present.

Rates of Change

In a mixture of hydrogen and oxygen gases, near atmospheric pressure, each molecule is colliding with other molecules 10^{10} times a second. If every collision resulted in reaction we should not have to wait very long. But most molecules repel each other as their electrons approach and it is only after this initial repulsion has been overcome that the molecules are able to get close enough to react. There is, therefore, an energy barrier to most reactions called the activation energy. Only pairs of molecules which between them possess an energy greater than this activation energy will be able to react.

At a temperature of T° Kelvin, the fraction of molecules which have an energy greater than E is exp(–E/RT) and this may be a very small proportion. For example, the activation energy for the decomposition of ethane

$$C_2H_6 \quad = \quad C_2H_4 + H_2$$

is 300 kJ, and the fraction of molecules having this energy at room temperature (300°K) is 10^{-50}. The effect this large activation energy has on the time for half reaction is shown in Table 3.1.

TABLE 3.1

Thermal Decomposition of Ethane

Temperature (°C)	Time for Half Reaction
20	10^{30} years
300	30,000 years
500	2 weeks
600	5 minutes
700	10 seconds
1000	3 milliseconds
1500	1 microsecond

The activation energy is not the same as the energy of reaction and most exothermic reactions have an activation energy. If this were not so hydrogen and oxygen would explode immediately they were mixed and all our fuels would burn without being ignited. Nearly all the organic substances on the shelves of our laboratory as well as living things are unstable and react with the oxygen of the air, but fortunately the activation energy for the reaction is large enough to prevent the reaction going uncomfortably quickly. The relation between the activation energies of the forward and back reaction and the overall energy of reaction is shown in Figure 3.1 for the reaction A → B + energy.

The point of highest energy in the reaction path is called the activated state or the transition state and determines both the activation energy and the rate of reaction. For a simple gas reaction it is often satisfactory to regard this transition state as two molecules

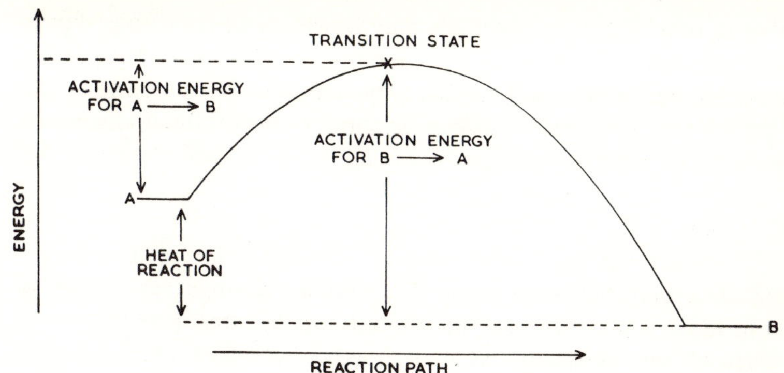

Figure 3-1. Relation between the activation energies of the forward and back reaction and the overall energy of reaction.

in collision having at that time energy greater than the activation energy and so to calculate the rate of reaction from the equation: number of molecules reacting = number of collisions x exp(–E/RT)

For more complex molecules, a more sophisticated treatment is necessary and one which has proved very valuable is to treat the transition state as a real chemical species, and to calculate its equilibrium concentration in the reaction mixture by the methods of thermodynamics and statistics which are used for ordinary chemical equilibrium. It is then only necessary to know the rate at which the transition complex crosses the top of the barrier to calculate the rate of reaction. It is found that the rate of crossing the barrier is a constant for all reactions and is given by the expression kT/h, where k and h are the constants of Boltzmann and Planck. It equals about 10^{13} sec^{-1} at normal temperature.

There are several things which we can do in order to increase the rate of a reaction. The first, as we have already seen, is to increase the temperature and this is the most common procedure in the laboratory, as well as in the kitchen. But there are obvious limitations to this method, especially in biological systems which have to be maintained very close to 37°C. Furthermore, it often happens that the temperature needed to give a convenient rate of reaction is so high that other reactions of decomposition occur first, reactions with still higher activation energy but ones which are more favourable in other respects. The only way in which we can bring about a large change in temperature in biological systems without killing them, is by cooling and in this way it is possible to decrease the rate of reaction very considerably and preserve living specimens for long periods of time.

Catalysts

There are two important ways in which the rate of biochemical reactions may be greatly increased without changing the temperature; these involve the use of catalysts or the use of light.

There will generally be many paths by which the molecules represented by A can change to the products B and the path with the lowest activation energy will be followed most frequently. If, by some means, we can provide a new path with a still lower activation energy this will increase the rate of reaction, just as we should probably find more people crossing from one valley to another on foot if we opened a new pass for them a few thousand feet lower than the older ones. A catalyst provides such a low energy path and thereby increases the rate of reaction (Figure 3.2).

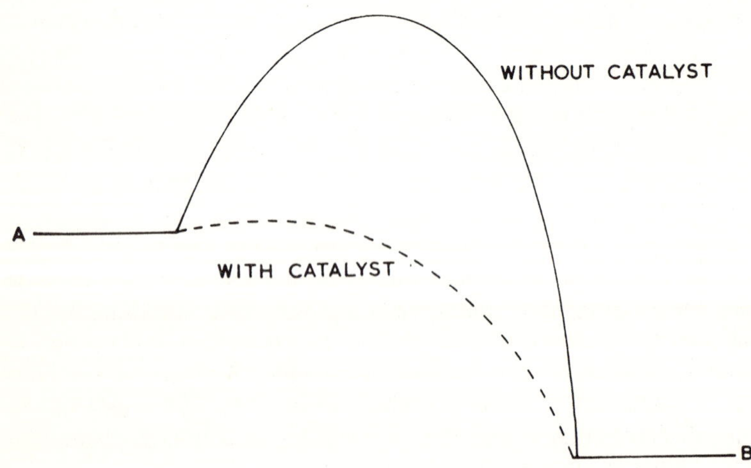

Figure 3-2. Effect of a catalyst on the energy barrier to reaction.

Catalysts may be simple molecules which react to form intermediates and are then regenerated or they may be surfaces to which the molecules are bound in some way which lowers the energy barrier to reaction. Metal surfaces are very important in this connection since the molecules bound to these surfaces may be split into atoms or other fragments. Thus a platinum surface will bring about combination of oxygen and hydrogen at room temperature. The biological catalysts called enzymes are complex protein structures to which the substrate reacting molecules can be bound by hydrogen bonds in a very specific way and they bring about reactions of a quite astounding complexity and specificity.

Catalysts increase the rate of the forward and back reactions in exactly the same ratio and do not alter the position of equilibrium. The best they can do is to lower the activation energy of the exothermic reaction to zero, they cannot themselves supply energy and therefore the activation energy of the backwards endothermic reaction must be at least as great as the heat of the overall reaction. If we wish to increase the rate of an endothermic reaction further, we must supply energy and nature has to do this without increasing the temperature appreciably. This can be achieved by using light to activate the reaction — by carrying out the reaction photochemically.

Photochemistry

When a molecule absorbs a quantum of light it receives a very great amount of energy for a short time, but in a quite different way from when it is activated by heat. A lamp used in a photochemical experiment is quite different from a Bunsen burner in its action.

The absorption of a quantum of light usually causes a single electron to jump from one orbital to another of higher energy. If the energy difference between the orbitals is E, then the frequency, υ, of the absorbed light quantum is E divided by Planck's constant, h, and the energy of the quantum is $h\upsilon$. The orbital defines the spatial distribution of the electron, when it is within the molecule, and, when an electron changes its orbital, the distribution of electrical charge within the molecule is also changed. This occurs instantaneously, immediately the quantum is absorbed, and any changes of position of the atoms within the molecule occur later, though they may be complete in 10^{-10} seconds or less. In this 'electronically excited state', the equilibrium position of the atoms in the molecule, as well as the electron distribution itself, may be very different from that in the normal ground state and we must regard excited molecules as new substances, which may have quite different physical and chemical properties from those of the original unexcited molecule.

Photochemistry is a study of the properties of molecules in electronically excited states. Since each molecule has only one ground state but has several excited states, it is clear that we are concerned with a field of study which is as great as the whole of normal ground state chemistry. It is important to realize that an electronically excited molecule is a quite different species from a 'hot' molecule which has obtained its energy by collisions with other molecules. Although a molecule which has absorbed a quantum of light may have a high total energy, it is 'cold' in the sense that, after a very short time, it is in thermal equilibrium with its surroundings in all respects except for the excess energy of its excited electron.

Orbitals of Electrons

A molecule like methane, CH_4, which has all its outer electrons in orbitals forming two-electron covalent single bonds, does not absorb visible or near-ultra-violet light. The electrons are strongly bound in the molecule and to excite them requires a quantum of light in the far-ultra-violet region. In photobiology we are concerned with coloured substances, or molecules which absorb at energies not much greater than those of the visible quanta, and therefore the molecules of interest always have at least one electron which can be excited to a higher orbital fairly easily, i.e. by means of a low-energy long-wavelength light quantum.

There are a number of different kinds of orbital which result in visible absorption—for example, metal atoms and ions having only partly filled d and f shells are usually coloured. But nearly all molecules of interest in photobiology owe their long wavelength absorption to electrons in one or other of two kinds of orbital, which are known as 'π orbitals' or 'n (non-bonding) orbitals'.

π Orbitals are those which form double and triple bonds in organic molecules. For example, in ethylene, $CH_2{=}CH_2$, two atomic p orbitals, one in each of the carbon atoms, overlap to form the π orbital between the carbon atoms. If there is a conjugated series of double bonds, as in butadiene, $CH_2{=}CH{-}CH{=}CH_2$, each carbon atom has one p orbital associated with it which is able to contribute to a π orbital of the whole molecule. As the conjugated chain becomes longer, so the electrons in the π orbitals become more delocalized, they are essentially spread over the whole conjugated system. As the electrons become more delocalized the energy between orbitals decreases and so an electron jump to an upper orbital can occur with quanta of lower energy. Eventually, with molecules such as vitamin A, which have a conjugated chain of ten carbon atoms (five 'double' bonds), absorption occurs in the blue region and the substance is therefore coloured yellow. Carotene, which is double the length of vitamin A, is more highly coloured still. Finally, the conjugated chains of carbon atoms can be in a ring without changing the situation very greatly. Benzene, naphthalene and anthracene absorb in the ultra-violet region but tetracene (with four condensed benzene rings) is yellow and pentacene is red. Of particular interest in photobiology are the porphin ring systems, such as chlorophyll, which have eighteen conjugated atoms in a single ring. Electron groups which occur as a result of excitation of an electron from a π orbital to a higher π orbital (called π^*) are described as 'π-π^* transitions'.

There is one other type of orbital in which the electron is relatively loosely bound and can be excited by low energy quanta, though usually in the ultra-violet region. These are the non-bonding electrons (n) which occur as pairs in atoms like oxygen and nitrogen. The non-bonding electrons on the oxygen atom in a carbonyl group or on the nitrogen atom in pyridine can be excited into the π^* orbital by absorption of a quantum in the near ultra-violet region and the transition is therefore called an n-π^* transition. In small molecules, which do not have long conjugated π-electron systems, the n-π^* absorption will usually be at longer wavelengths than the π-π^* absorption but it is a 'forbidden' transition and therefore usually rather a weak absorption. All transitions between singlet and triplet states are 'forbidden'; that is they are very slow compared with transitions between two singlet states or between two triplet states.

Singlet and Triplet States

There is one other characteristic of any electronic state which is of fundamental importance in photobiology. This is its 'multiplicity' which, in the case of most organic molecules, means the question of whether the state is a singlet or a triplet. Each electron is spinning and if we have two electrons in the same orbital they have no option, according to the strange rules of quantum mechanics, but to spin in opposite directions. If they are in different orbitals, as they will be in an excited state which has been formed by excitation of one electron from a doubly filled into an empty orbital, then the two electrons can spin either in the same direction (a triplet state) or in opposite directions (a singlet state). The ground state is nearly always a singlet state—this must be so if every orbital is either occupied by two electrons or is empty (see Figure 3.3). When the electrostatic interaction of the electrons with each other is taken into account it turns out that the triplet state lies at somewhat lower energy than the singlet of the same configuration, i.e. T_1 lies below S_1 in energy and the excited state of lowest energy in most organic molecules will therefore be a triplet.

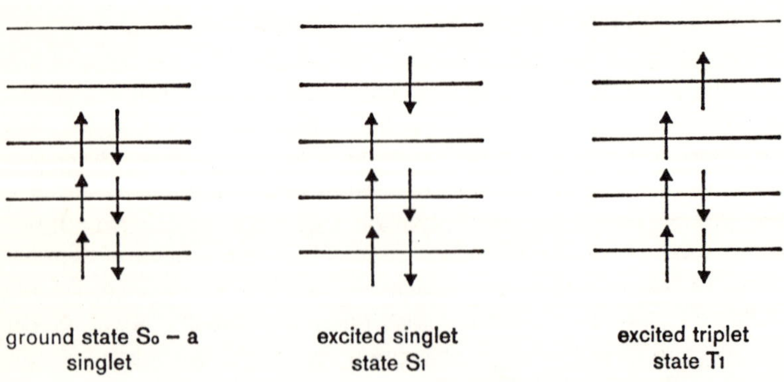

Figure 3-3. Electron orbitals and spin in three states.

We may now consider the various processes which occur when an organic molecule absorbs a quantum of light in terms of the diagram (Figure 3.4). This figure is the basic starting point for almost every discussion about photochemistry or photobiology—it might well be placed permanently on the wall of a photochemistry laboratory along with the periodic table. On the left are the singlet states S_0 (ground state) S_1 and S_2 and on the right are the triplet

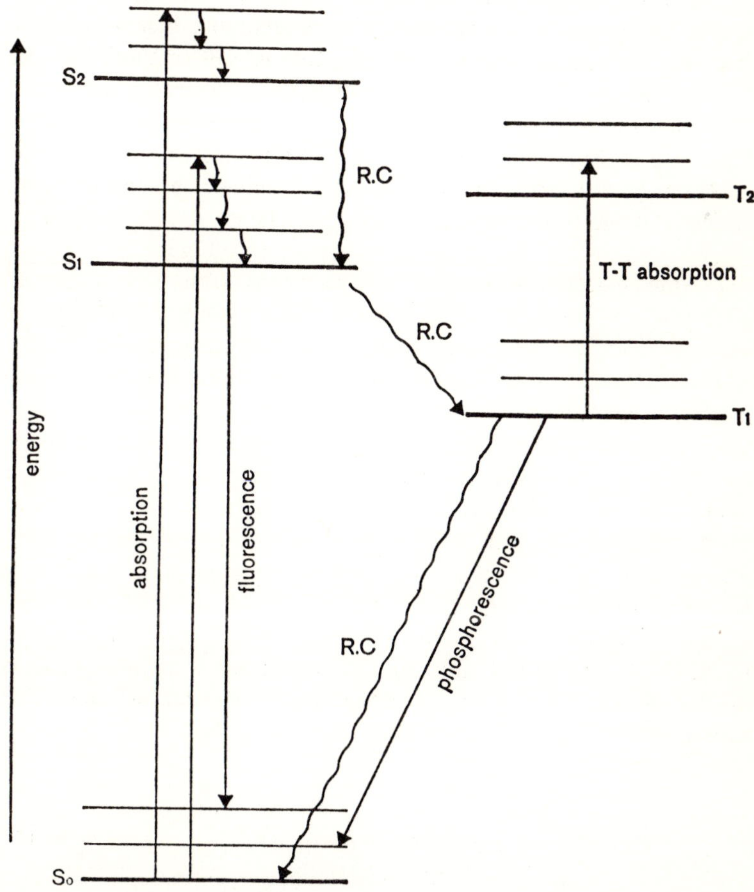

Figure 3-4. Physical processes following absorption of light by a molecule.

states T_1 and T_2. Absorption of light can only occur with reasonable intensity when the transition occurs to upper singlet states (because the ground state is also a singlet). The absorption corresponding to the transition from S_0 to T_1 is very weak and will normally only be seen in very long paths.

Any excess vibrational energy will be rapidly lost. There are now a number of possibilities for radiationless conversion and these are shown as wavy lines marked R.C. in Figure 3.4. Alternatively the molecule may lose energy by emission of a photon as fluorescence or phosphorescence. Reaction may occur at any point in this scheme and in some cases the intermediate states may be directly observed, e.g. the triplet state by T-T absorption.

The chemical changes which may occur in electronically excited states are at least as diverse as those which occur in the ground state, but most of the primary processes of importance fall into one or other of the following categories.

One Molecule Reactions

1. *Dissociation*

The molecule breaks into two fragments, which may be ions or radicals. Particularly important are loss of a proton or loss of a hydrogen atom. In solution there is a 'cage' effect which prevents the two fragments of a molecule from separating and it is often found that dissociation processes which occur readily in gases have a low yield in solution. This is particularly the case when both fragments of the molecule are big, but when one of them is a hydrogen atom the cage effect is less important and the small hydrogen atom can diffuse away before re-combination occurs. Fission of C–H, O–H, and N–H bonds is therefore rather a common process in the photochemistry of solutions and biological systems. For example:

HO–⟨◯⟩–OH ——→ OH ⟨◯⟩ O. + H
ultra-violet light

Hydroquinone Semiquinone

$$—CH_2—\langle\bigcirc\rangle—HO \xrightarrow{\text{ultra-violet light}} CH_2—\langle\bigcirc\rangle O. + H$$

Tyrosine

This latter dissociation is observed not only in phenols and in tyrosine itself but is one of the effects observed when *proteins* such as egg albumin are irradiated with ultra-violet light. The free radicals have a characteristic absorption spectrum which can be observed immediately after irradiation ceases, for about one thousandth of a second. The reaction does not reverse itself exactly but various secondary reactions of the hydrogen atom and radicals lead to complex changes and inactivation of the protein.

2. *Cis-trans isomerization*

The simplest type of isomerization which occurs on irradiation

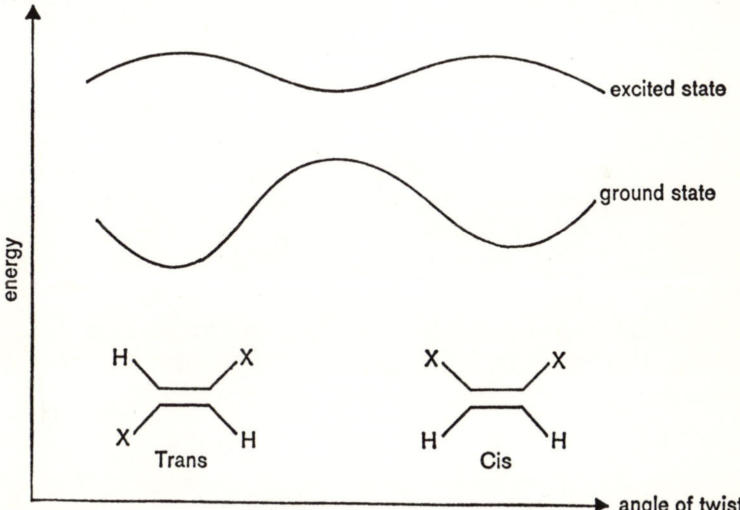

Figure 3-5. Energy/angle relationship in a cis-trans isomerization process in ground and excited states.

is of cis-trans type. The cis-trans equilibrium in the excited state will generally be different from that in the ground state and one may have, for example, a situation such as that shown in Figure 3.5. In this case, the most stable molecule in the excited state is one where the molecule is neither cis nor trans but is bent through 90°. From this equilibrium position it can return to either the cis or trans configuration of the ground state so that irradiation of such a molecule, which is mainly trans under normal conditions, will convert much of it to the cis form.

Photo-isomerization of the cis-trans type is important in the process of *vision*. The visual pigments are aldehydes of vitamin A, the colour of which has already been mentioned and arises from a π-π^* transition in the long conjugated chain. There are many possible cis-trans configurations in these molecules, since each double bond may occur in either cis or trans form. Two isomers of one of the visual pigments—retinal—are shown below

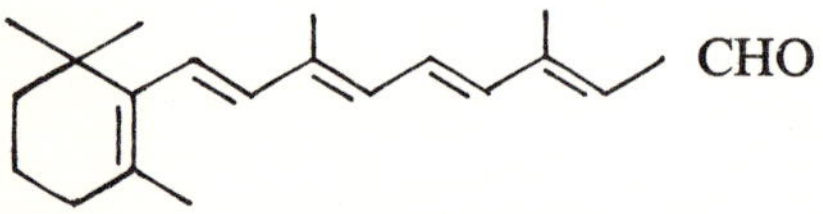

CHO all-trans

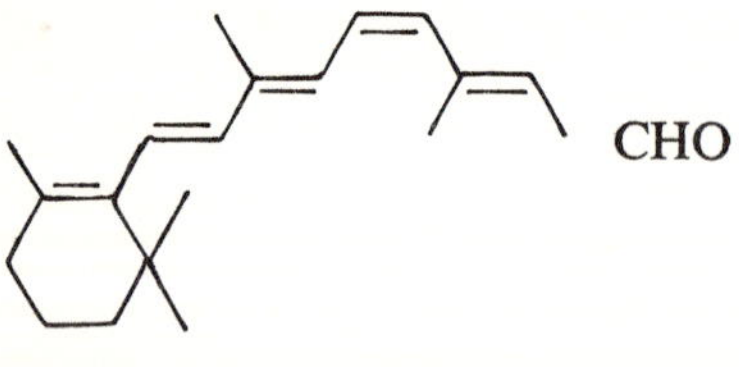

CHO 11 cis

This apparently trivial geometrical change in structure has a quite profound effect on the colour or wavelength of the transition. We have already referred to the fact that the energy of the first electronic excitation of molecules of this kind is lowered (i.e. moves to longer wavelengths) as the extent of conjugation increases. The conjugation is greatest in the linear all-trans form and introduction of cis configurations decreases the conjugation and shifts the absorption to shorter wavelengths. The effect of light on the visual pigment of the eye may be to convert the 11 cis to the all-trans configuration and this is then the precursor of the physiological effects which ensue. The retinal, in the 11 cis form, was attached to a complex protein, rhodopsin. In the all-trans form, it will clearly not be able to fit into the protein in the same way as before and structural changes in the protein enzyme will occur which eventually make it possible for the enzyme to catalyse the visual process. This is an attractive idea since one light quantum produces one active enzyme site which would then be able to effect the catalysis of many molecules.

Reactivity of the Excited State with Other Molecules

Since the electron distribution in the excited state has been changed we shall not be surprised to find a changed chemical reactivity and different equilibrium constants with respect to such common reactions as electron transfer ('redox reactions'), and proton transfer ('acid-base reactions'). In both of these we regard the change simply as the establishment of the new equilibrium whilst the molecule is in its electronically excited state and a return to the original equilibrium after return to the ground state. The extent to which the equilibria are established and to which other reactions will occur will depend on the relative rates of the physical and chemical changes involved.

1. *Acid-base equilibria (proton transfer reactions)*

It is found quite generally that there is a profound change in acidity constants (pK) upon electronic excitation, often corresponding to six units of pH or even more. If the material BH is a stronger acid in the excited state than in the ground state, the following processes may occur:

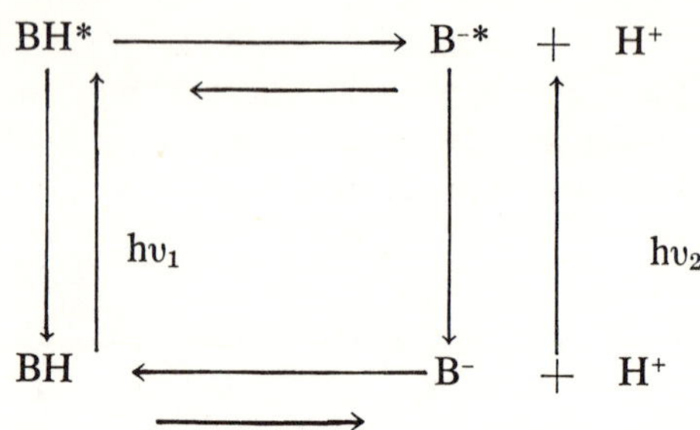

The establishment of the new equilibrium in the excited state may occur within the lifetime of the excited state. In the example above, this would result in the striking observation, first discovered by Förster, that absorption of light by BH, in acid solution, results in fluorescence of B^- although this is present in negligible concentration in ground state equilibria at the acid strength employed. Many examples of changed proton affinity, resulting from changed electron distribution in the excited state, have subsequently been found.

2. *Electron transfer reactions*

Electron transfer is the principal means by which biological systems transport energy and convey messages. An electronically excited molecule may have a quite different electron affinity from that of the same molecule in its ground state and, again, the photochemical change is best understood in terms of the establishment of a new equilibrium in the excited state. For example, quinones and dye molecules such as methylene blue have a greater electron affinity in the excited state (in this case the triplet state T is the state of importance) than in the ground state and will establish a quite different equilibrium with, for example, the ferrous-ferric ion couple:

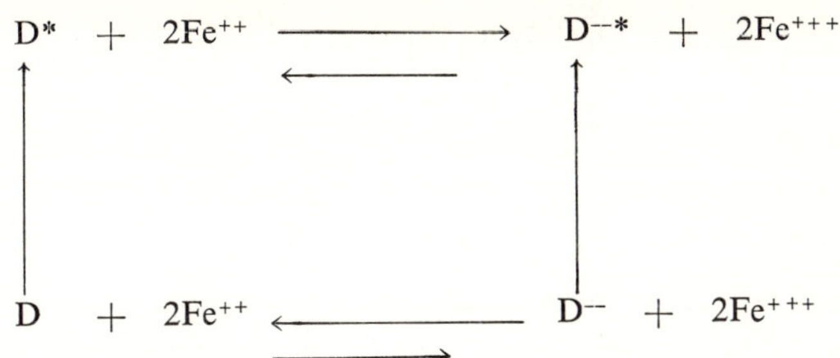

The excited dye (D*) abstracts electrons from ferrous ion and is converted to its colourless 'leuco' form (usually in two stages via a semiquinone). Upon de-activation to the ground state, the reverse process occurs as the leuco dye reacts with ferric iron and is re-oxidized to its original form. Such processes are common in carbonyl derivatives; one particular case of photobiological interest which has been thoroughly studied is the photo-reduction of flavins. This, and other photo-reductions of quinone type molecules, occur exclusively via the triplet state of the excited quinone.

3. *Hydrogen atom transfer*

Transfer of a hydrogen atom is very similar to transfer of an electron, the difference being a matter of whether the transfer is accompanied by a simultaneous transfer of a proton. In non-polar media, the electron and proton are transferred simultaneously and again the reaction occurs via the triplet state in such cases as carbonyl compounds. One of the best studied reactions of this kind is that of ketones and quinones in organic solvents, RH:

$$>C=O + \text{light} \rightarrow \overset{n\text{-}\pi^*}{>C=O}\ (\text{singlet}) \rightarrow >C=O\ (n\text{-}\pi^*\ \text{triplet})$$

RH

$$R\cdot + >\dot{C}-OH$$

The radicals R and >C–OH subsequently react by a series of recombination and exchange reactions which depend on the nature of the solvent and the carbonyl compound.

4. *Other reactions of excited states*

One of the more common reactions of an excited state is dimerization with a similar molecule in its ground state. Anthracene, which in its excited state reacts with a normal anthracene molecule to form dianthracene, was one of the earliest examples known of this type of reaction. The dimerization of thymine occurs readily in rigid media at low temperatures and is probably of importance in the photochemistry of DNA.

```
          O                                     O              O
          ||                                    ||             ||
   H      C      CH3                     H      C  CH3    CH3  C      H
    \    / \    /        ultra-violet     \    / \  |      |  / \    /
      N       C         ------------->      N      C  —  C       N
      |       ||        <-------------      |      |     |       |
      C       C             light           C      C  —  C       C
    //  \    /  \                         //  \   /  \  /  \    /  \\
   O      N      H                       O     N      H H    N       O
          |                                    |                |
          H                                    H                H

       Thymine                                    Thymine dimer
```

Hydrolysis and solvolysis are other examples of reactions whose rate may be greatly modified in the excited state. Cytosine is hydrolysed in aqueous solutions in the excited state and, like the dimerization of thymine, the process may be reversed in the ground state. Such events occurring in the nucleic acids themselves may be of fundamental importance in photobiologically induced mutations.

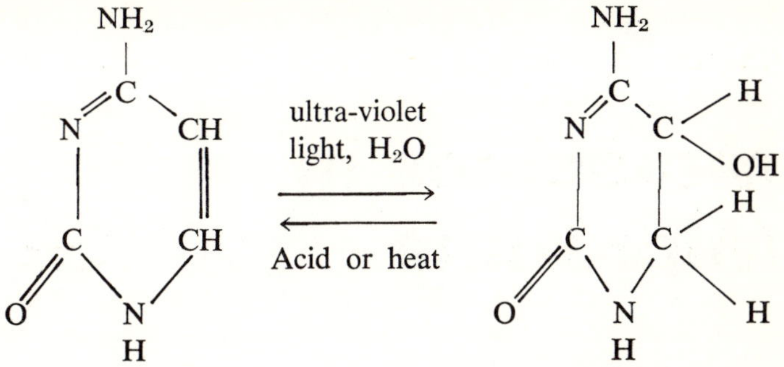
NH2
N
CH
C
CH
O
N
H
ultra-violet
light, H2O
Acid or heat
NH2
C
H
N
C
OH
H
C
C
O
N
H
H

Cytosine Photoproduct

The patterns of change brought about by light are above all responsible not only for the maintenance of all life on our planet but very probably also for its origins. This we shall pursue further in the fifth lecture.

CHAPTER FOUR

Molecules in Microtime

In Figure 4.1, I have set out, on a logarithmic scale, the range of times with which man is in some way or other concerned. It extends from the shortest time to which we could give any meaning, the time for light to travel across a fundamental particle, to the age of the universe itself. Chemical changes occur over the whole of this range of times down to 10^{-15} s. Shorter times than this are in the realm of physics since the uncertainty principle tells us that if we carry out measurements in 10^{-15} s, the uncertainty in the measurement would be 400 kJ/mole which is as great as the energy of the chemical bonds themselves. In nuclear physics uncertainties of this magnitude would be negligible compared with the total energies associated with the nucleus so that meaningful and precise measurements can be carried out in much shorter times.

At the other end of the scale, the first stages in the formation of the universe may also have been largely physical, with no chemical molecules present, though this is by no means certain. We do know, however, that chemistry on the earth has been going on since its formation 4600 million years ago. In these terms, the times of change which we are able to study in the laboratory cover a very small range. If a reaction lasts more than a few days, we become rather impatient and even in the stress of a practical examination a student would be hard pressed to carry out a titration in less than a few seconds. But special methods are now available for extending this range of studies of chemical change and there is great interest in the new fields of chemistry which are revealed in both directions of time. In this lecture I shall discuss reactions in very short times—split second chemistry—and in the next lecture I shall be concerned with chemical reactions over periods of billions of years.

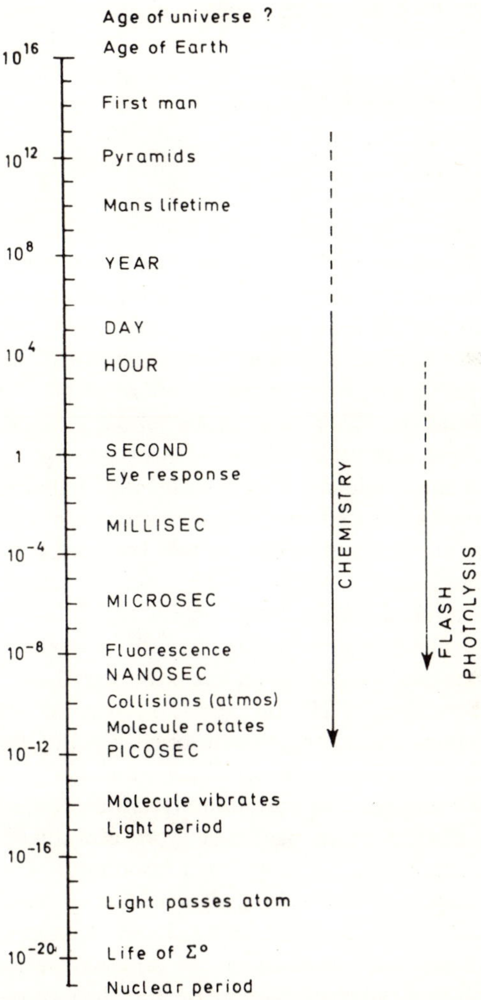

Figure 4-1. A time scale, in seconds, showing the range of chemistry and of flash photolysis.

Fast Reaction Studies

Two problems arise in the study of the reaction which is taking place in a few milliseconds or less. First, we must measure some physical property over a very short time range and this presents few problems nowadays since most physical quantities can be recorded electronically in very short periods of time, although of course one will lose sensitivity as one increases the rate of measurement. The second problem is a more serious and restrictive one. If the reaction is complete in one thousandth of a second then it must be started very quickly because if we take several seconds to start the reaction then it will be nearly over before we have finished starting it (a statement which will be understood by any Irish scholars present!). This second problem which we have to face therefore, is the problem of initiating a chemical reaction in a very short period of time.

Quite regardless of whether the reaction is fast or not, there are very few ways of initiating chemical change. All the variety of chemical reactions which we bring about in our chemical laboratories involve the use of one of three processes of initiating the chemical change. We can start the chemical reaction by mixing, by heating, or by some form of radiation, or electrical energy. Each of these methods of initiation has been used for the study of fast reactions and the first mixing was historically the first method to be applied to fast reactions.

Hartridge and Roughton in the 1920s introduced the *flow technique* for the study of fast reactions between two components in solution (Figure 4.2). The two solutions were fed into a mixing chamber and then flowed together through a tube, along which a change in colour or absorption spectrum was observed. Knowing the rate of flow and the concentration at various points in the tube, measured by the colour change, they were able to determine the velocity of reactions taking place in a hundredth of a second or more. Refinements of this technique, particularly the introduction of the stopped flow technique by Gibson, Chance and others (Figure 4.3), has made it possible not only to use much smaller quantities of liquid but also to increase the time resolution below 1 millisecond.

The technique is valuable for the study of reactions in solution

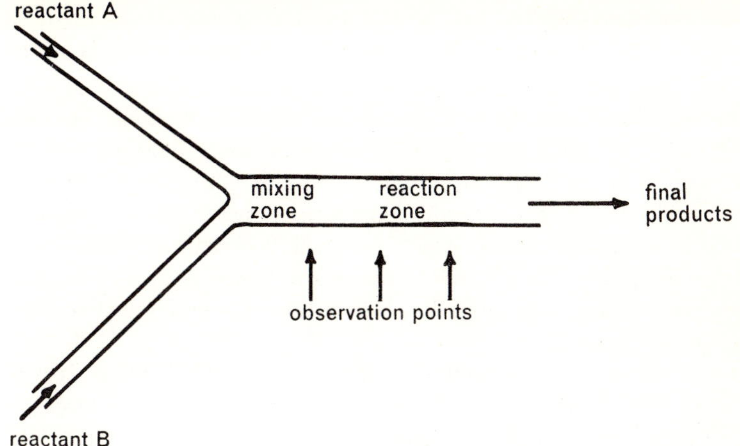

Figure 4-2. Stationary state flow technique for the study of a fast reaction between substances A and B.

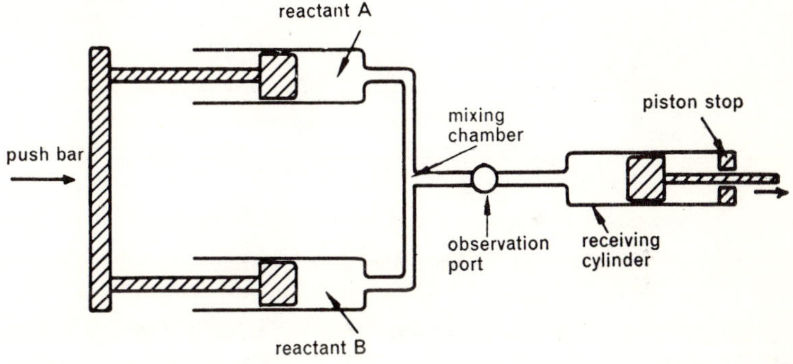

Figure 4-3. The stopped flow method.

in this time range and particularly for the study of many reactions of biological importance.

Although it is a flow method in which a stationary state is set up, so that no fast recording is necessary, it is nevertheless a direct method and in principle the transient intermediates which are formed in the reaction can be observed directly. I should point out at this stage that all the methods which I am describing are direct methods in which the reaction intermediates can, in principle, be observed just like any other chemical compound of longer life. There is a whole variety of methods for the study of fast reactions which are indirect or competitive methods. In these methods, the chemical reaction competes with some other fast process such as diffusion, or another chemical reaction, and by measuring the success of the competition, one obtains a ratio of two rate constants, from which one can be obtained if the other is known. Whilst very useful for many purposes these indirect methods usually give no information other than the relative time or rate of the reaction.

Two rapid *heating* methods have been devised which are of great utility. The first of these, which is applicable to gases, is the *shock tube technique* (Figure 4.4). Here a gas at high pressure is separated by a membrane from a gas at low pressure, and when the membrane is burst a shock wave travels through the gas and raises its temperature. The shock front is very sharp and the

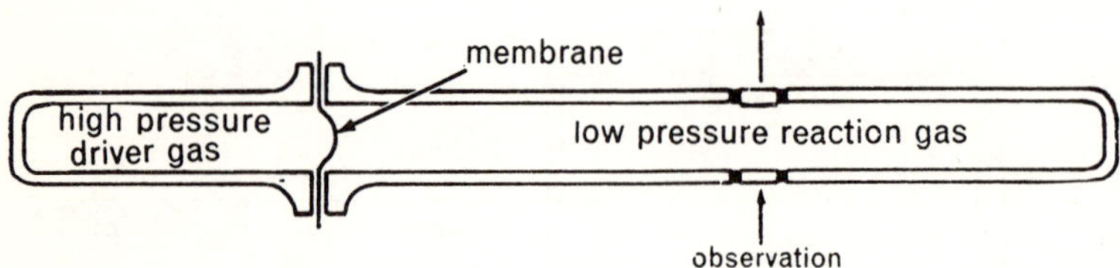

Figure 4-4. The shock tube method.

temperature rise occurs during the time of only a few collisions. If one observes some parameter, such as the absorption spectrum, at a fixed point in the tube, the temperature jump is complete in one or two microseconds and the ensuing chemical reaction or relaxation can be followed immediately afterwards, so that reaction times of a few microseconds can be studied.

For solutions, Eigen in Gottingen, introduced the *temperature jump method*. Here the solution is heated by an electrical current which is passed rapidly through the liquid by discharging a condenser. The effect is purely thermal since no discharge occurs but only a heating caused by the degradation of the kinetic energy of the ions moving in the electrical field. The apparatus (Figure 4.5) consists of a small cell with two electrodes and the solution must be a good conductor so that the method is confined to strong ionic solutions. A temperature jump of about ten degrees temperature rise produces a sufficient shift of equilibrium, the relaxation to the new temperature can then be followed. Again the method is applicable to reaction times of a few microseconds or greater.

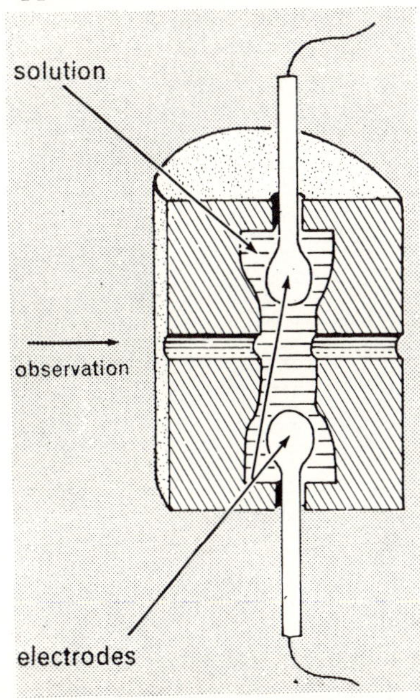

Figure 4-5. The temperature jump apparatus.

The third method of initiation of chemical reactions which I mentioned was by *radiation,* that is to say, by light, X rays, γ-rays or pulses of electrons. This is the method which interests me the most, so that I shall spend the rest of my time talking about it. I shall, however, be talking mainly about the use of light rather than ionizing radiations, because most of the work and nearly all of my own has been photochemical. Pulses of electrons of a few microseconds duration have also been used to initiate chemical change and the subsequent reactions have been followed by absorption spectroscopy. So far the results which have been obtained are very similar to those obtained using light, but this technique is of particular value to chemists who are interested in the primary processes of radiation chemistry.

Flash Photolysis

The flash photolysis technique is one in which the chemical change is brought about by a brief intense flash of light, and this is now very widely used for the study of all kinds of reactions as well as for the direct study of spectroscopic and other properties of transient intermediates. The technique is very simple in principle and for most cases it is very simple in practice as well. The material under study, which can be gas, liquid or solid, is contained in a reaction vessel, usually between 10 and 100 centimetres long (Figure 4.6). Alongside the reaction vessel are placed one or more electronic flash lamps which consist of a long quartz tube with an electrode at either end, filled with an inert gas at a pressure of a few centimetres. Across these electrodes is connected a large condenser, the energy of which, when charged is $\frac{1}{2}CV^2$, where C is

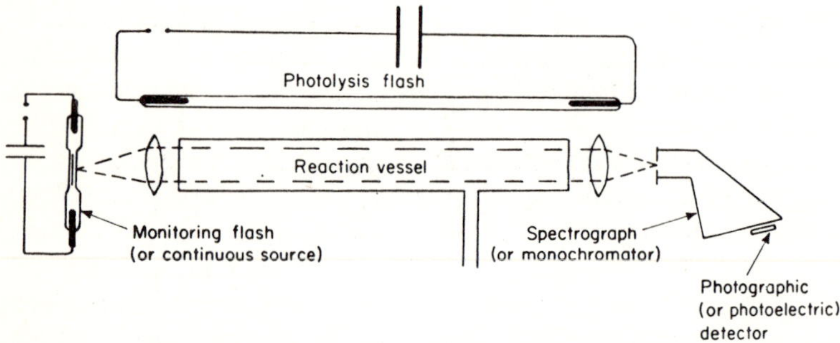

Figure 4-6. The flash photolysis apparatus.

its capacity and V the voltage to which it is charged. This energy may be anything from about 1 joule to half a million joules, although a typical energy for most purposes is about 1000 joules. Many of the experiments which I shall describe were carried out with very small condensers, and indeed the tendency over the last few years has been to use smaller and smaller flashes.

When the condenser is charged, the flash lamp is fired by applying a small high voltage pulse to the lamp or, if high voltages are used, a spark gap is introduced into the circuit and the lamp is fired by shorting this gap with a high voltage spark. The lamp then fires and photochemical change occurs in the reaction mixture, and our next problem is to study this change as a function of time after the flash. Almost exclusively, absorption spectroscopy in the visible and ultra-violet regions is used for this purpose. In principle, other physical parameters could be measured but it turns out that, because of the high sensitivity of the photographic plate and the photoelectric cell, no other method is quite as powerful for rapid recording as electronic absorption spectroscopy.

The spectrum can be studied in two ways. First, we can use a second electronic flash at one end of the tube and a spectrograph with a photographic plate at the other. The slit of the spectrograph is left open since scattered light from the main flash is no trouble, and the second flash is timed to go off at some predetermined interval after the first one. By repeating this procedure one can obtain a whole series of spectra at various times after the flash and I will show examples of this type of record in a moment. The second method uses a continuous source of light in place of the flash and a monochromator and photoelectric cell in place of the spectroscope. In this way one records only one wavelength but this wavelength is recorded at all times after the flash. The first method of flash spectroscopy is most useful for spectroscopic studies and for preliminary work, and the second, photoelectric method, is much more useful for kinetic studies once the spectra have been identified.

The rate of reaction, or the time which can be resolved, depends of course on a number of factors but is determined in the last resort by the duration of the flash because one cannot begin kinetic observations until the photolysis is nearly completed. The duration of the flash itself depends on its energy and the smaller energy pulse one can use, the faster the reaction which can be studied. Since it is found that for most photochemical reactions sufficient change is brought about by flashes of between 50 and 5000 joules, this is the normal working range and flash durations are usually between 1 and 100 microseconds.

The Free Radical Cl O

I will first illustrate the use of the flash photolysis method by describing two studies of very simple gaseous reactions, one carried out by using the flash spectroscopic method and the other by the photoelectric method. The first reaction happens also to be the first which was studied by flash photolysis.[3] We found that, on flashing a mixture of chlorine and oxygen gases, a new transient spectrum appeared just after the flash. This was rather a surprise because there is no photochemical reaction between chlorine and oxygen which can be observed in the usual way and long illumination of this mixture with an ordinary light source has no effect. The transient spectrum which we observed is shown in Figure 4.7. It is a nice single progression of bands and is fairly clearly the spectrum of a diatomic molecule. It is no very great feat of deduction to identify this spectrum therefore with the radical Cl O. Since the spectrum was previously unknown its interpretation was of interest and immediately led to an accurate value for the dissociation energy of the Cl O radical which is very useful in the subsequent interpretation of its kinetics. The next step is to record this spectrum as a function of time and Figure 4.7 is a typical record of this kind. The spectrum disappears in about 5 milliseconds and the kinetics of its disappearance are studied by photometering the plate in one of the regions of continuous absorption.

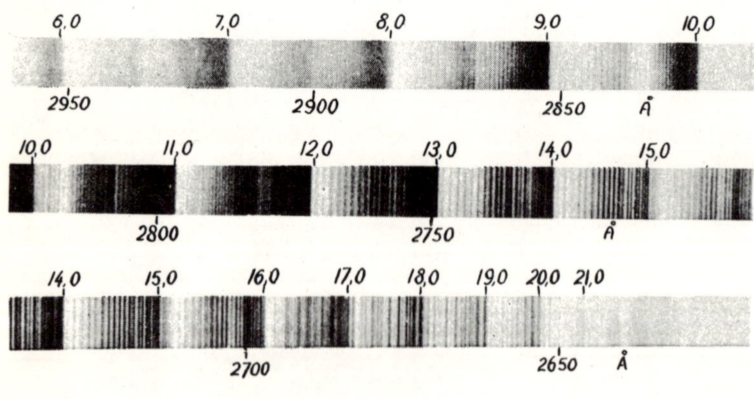

Figure 4-7. Absorption spectrum of the Cl O radical.

This reaction is found to be bimolecular and the whole scheme of reactions is as follows:

$$Cl_2 + h\upsilon \rightarrow 2Cl$$

$$2Cl + O_2 \xrightarrow{k_1} 2ClO$$

$$2ClO \xrightarrow{k_2} Cl_2 + O_2$$

The value obtained for the rate constant k_2 is $8 \cdot 6 \times 10^7$ 1. mole^{-1} s^{-1}.

Atom Recombination

Another investigation which is typical of the use of this method is the study of the simplest of all chemical reactions, although as we shall see, it is not quite so simple as might at first appear. It is the recombination of two atoms and of iodine atoms in particular. Iodine happens to be particularly convenient because of the region of absorption of I_2. The procedure is quite a simple one; iodine vapour at a pressure of about a tenth of a millimetre is mixed with some inert gas at a high pressure, is dissociated by a flash so that about 20% of the molecules are dissociated into atoms, and then one watches the reappearance of the I_2 spectrum as a function of time, recording this on an oscilloscope.

Now it has been known for a long time that in theory two atoms cannot combine except in the presence of a third body or chaperon molecule which is necessary to remove at least part of the energy. The first thing which it was interesting to do was to see whether this theoretical requirement really was borne out in practice, that is to say, to check that the recombination was indeed termolecular. Figure 4.8 shows oscilloscope traces of the recombination of iodine atoms in the presence of argon at various pressures and, if the reaction is third order, then a plot of the reciprocal of iodine atom concentration against time should be linear and the slope of this plot should be proportional to the argon pressure. That this is the case is shown in the figure where the results from the oscilloscope traces are plotted in this way.

As the accuracy of the determinations was improved, it was found that there were two complicating factors which caused some

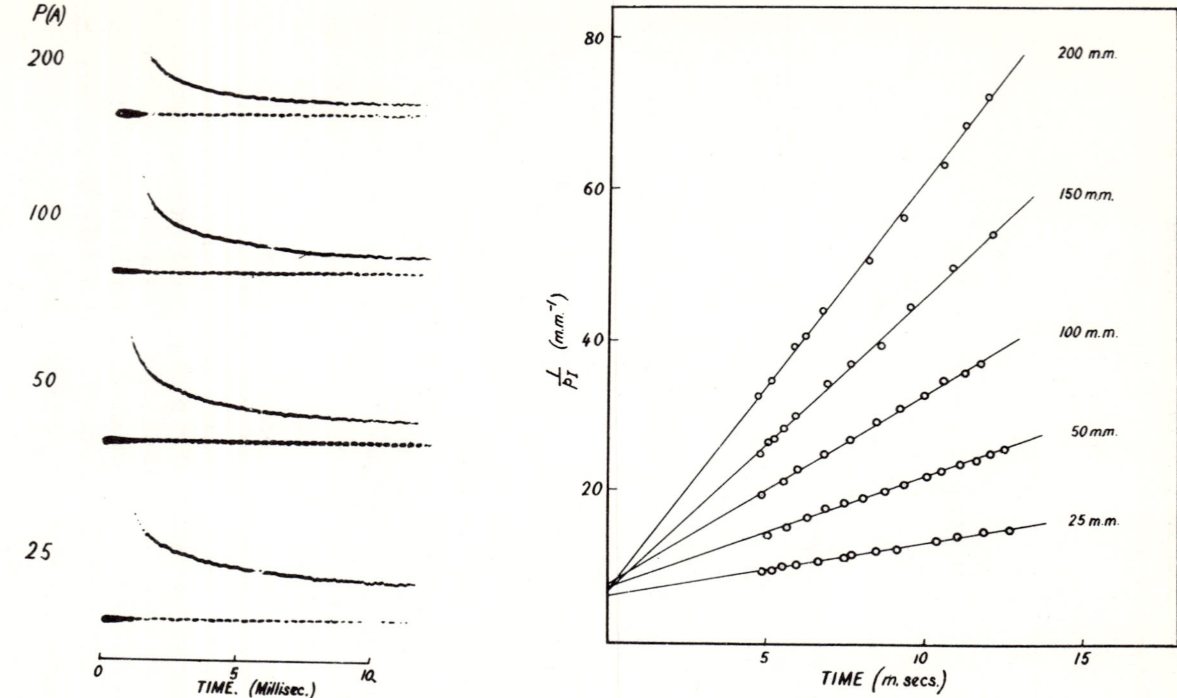

Figure 4-8. Oscilloscope traces of I_2 *absorption after flash photolysis of iodine in excess argon, and second order plots from the data. P(A) is argon pressure in mm Hg and P(I) is iodine atom partial pressure.*

deviation from linearity in these plots. The first one was a temperature effect which is common to all fast reaction studies. The effect is very complicated and the only satisfactory procedure is to eliminate it completely by working always with a very large excess of inert gas. The second deviation from linearity was a real one, caused by the very great efficiency of I_2 molecules as chaperons. It had been assumed that because the pressure of argon was about ten thousand times greater than the pressure of I_2, any effect of the latter could be ignored. It turned out, however, that I_2 was a thousand times more efficient as a chaperon than argon and so was contributing as much as 10% to the total third body effect.

Subsequent work has been concerned with the measurement of rate constants of recombination in a number of gases and as a function of temperature. These measurements have been carried out in four different laboratories and the agreement is now very satisfactory. The two most striking things about the results which have been obtained are firstly that the temperature coefficients of recombination are negative and second, that there is an enormous variation in the efficiency of different chaperon molecules. The temperature effect is the opposite of what is found in nearly all other chemical reactions. The way that the rate constant varies from one chaperon to another is shown in Table 4.1 where it will be seen that the rate constant varies over a factor of a thousand in going from helium to I_2. What is the explanation of these observations?

If we write the recombination in two steps as follows:

$$I + I \leftrightarrow I_2^*$$

$$I_2^* + M \rightarrow I_2 + M$$

then there is no reason either why the temperature coefficient should be negative or why there should be very much difference between different third bodies because the collision diameters do not vary by a factor of more than 2 or 3, and we can hardly explain the results in term of energy transfer probabilities or the number of available degrees of freedom when we find that I_2 is so much more efficient than, say, benzene. If, however, we write the sequence of reactions as follows:

$$I + M \rightleftharpoons IM$$

$$IM + I \rightarrow I_2 + M$$

we can immediately understand both these observations. We can see why the rate of reaction should decrease at higher temperatures, because the equilibrium will be shifted to the left as IM dissociates. Secondly, different chaperon molecules will be expected to have very different energies of complexing with the iodine atom and so a large variation in their efficiency is not unexpected. It is possible to calculate, from the measurements, the heats of formation of the complex IM, which are very nearly equal to E_a given in table 4.1.

TABLE 4.1

M	k_{27} (1^2 mole^{-2} s^{-1}) x 10^{-9}	E_a(kcal)
He	1.5	0.4
A	3.0	1.3
H_2	5.7	1.22
O_2	6.8	1.5
CO_2	13.4	1.75
C_4H_{10}	36	1.65
C_6H_6	80	1.7
CH_3I	160	2.55
$C_6H_5CH_3$	194	2.7
C_2H_5I	262	2.4
$C_6H_3(CH_3)_3$	405	4.1
I_2	1600	4.4

Table 4-1 Termolecular recombination constants (k_{27}) at 27°C and temperature coefficients expressed as negative activation energies (E) for the recombination of iodine atoms in the presence of various chaperon molecules M.

These studies of Cl O and iodine atom recombination are representative of most of the work using flash photolysis which has followed, although there has naturally been a trend towards the study of more complex molecules.

In addition to the increased complexity of the molecules investigated, the flash photolysis technique has been increasingly applied to solutions, solids and even to biological systems so that these applications are now more extensive than those in the gas phase. I have been particularly interested in the transient species which appear upon excitation of larger organic molecules, many of them of interest in organic mechanisms and in biological processes. The first aromatic free radical, triphenyl methyl, was discovered, quite unexpectedly, by Gomberg in 1900. Many similar resonance-stabilised free radicals, which are stable enough to exist in observable concentrations at equilibrium, were subsequently reported but the direct spectroscopic observation of reactive, short-lived, aromatic free radicals was only achieved relatively recently.

In 1955, we detected a series of aromatic free radicals in the gas phase by flash photolysis of aromatic vapours. For example, spectra in the region of 3000 Å were attributed to benzyl (from toluene), (See Figure 4.9) anilino (from aniline) and phenoxyl (from phenol), (Table 4.2). These radicals are isoelectronic with each other, and form a type of seven π-electron system which, in aromatic free radical chemistry, is of comparable importance with the six π-electron benzene ring in normal molecules.

The spectra of benzyl, anilino and phenoxyl in solution are

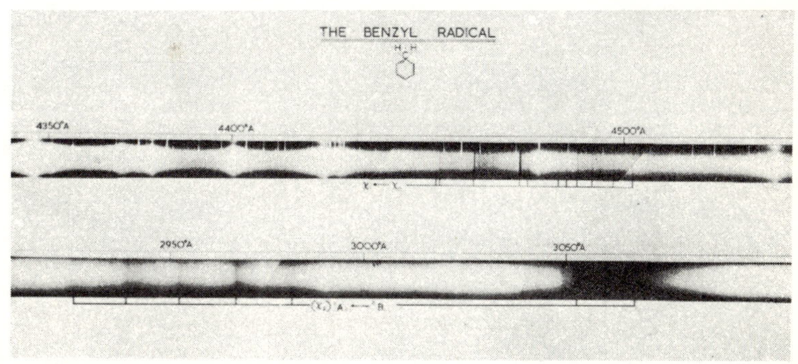

Figure 4-9. Absorption spectrum of the benzyl radical in the vapour phase after flash photolysis of toluene. Path length = 8 metres.

diffuse, but quite characteristic as will be seen from the phenoxyl radical spectra, in various media, shown in Figure 4.10. An

BENZYL, ANILINO AND PHENOXYL

CH_3 → CH_2•

4530 A

3180 A

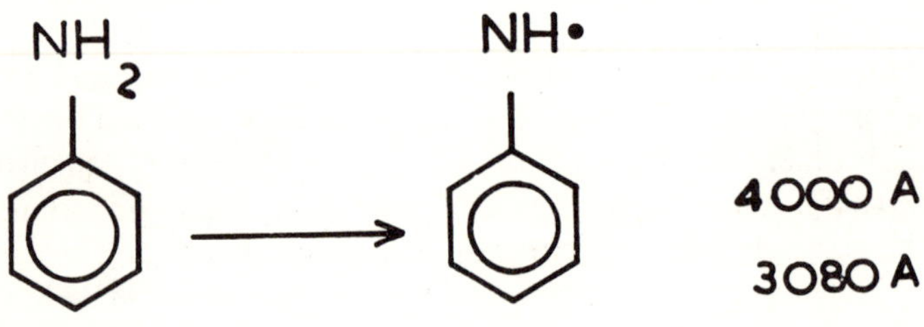

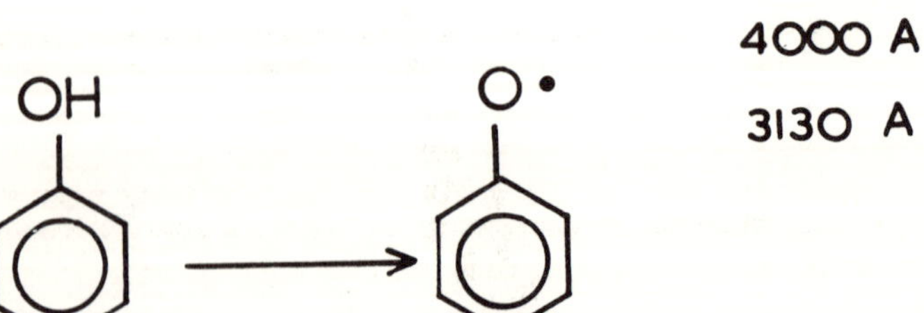

Table 4-2. Photochemical reactions leading to radicals and their absorption maxima.

interesting complication arises in the anilino radical, where two quite different spectra are observed depending on solvent or, in

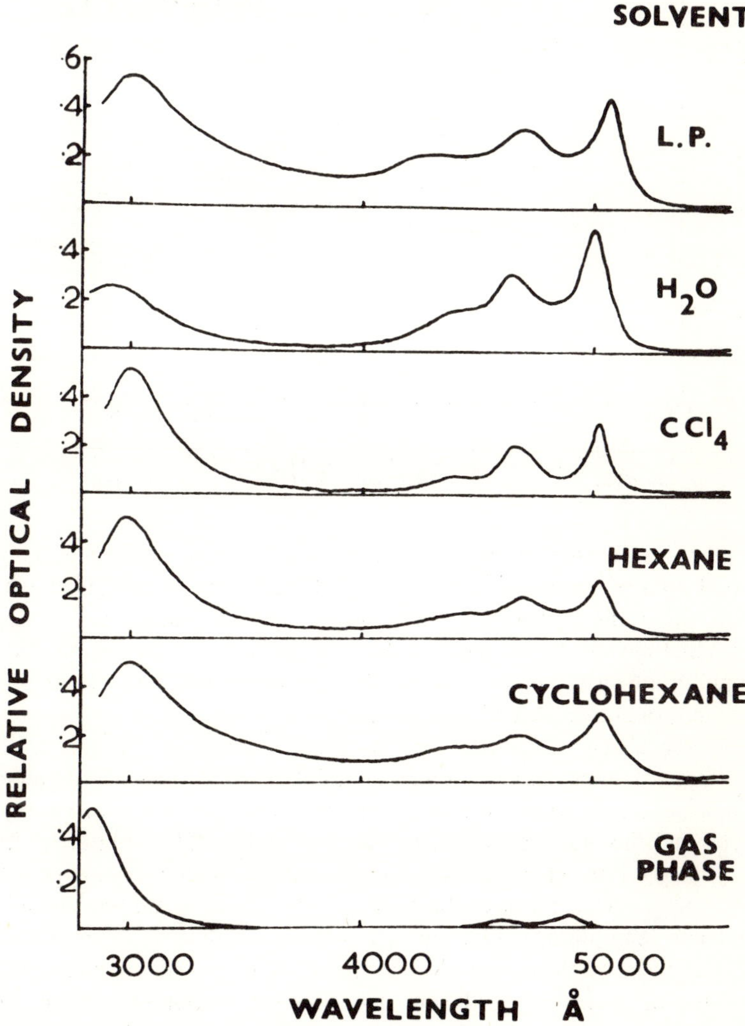

Figure 4-10. Absorption spectrum of the phenoxyl radical in the gas phase and in various solvents after flash photolysis of phenol.

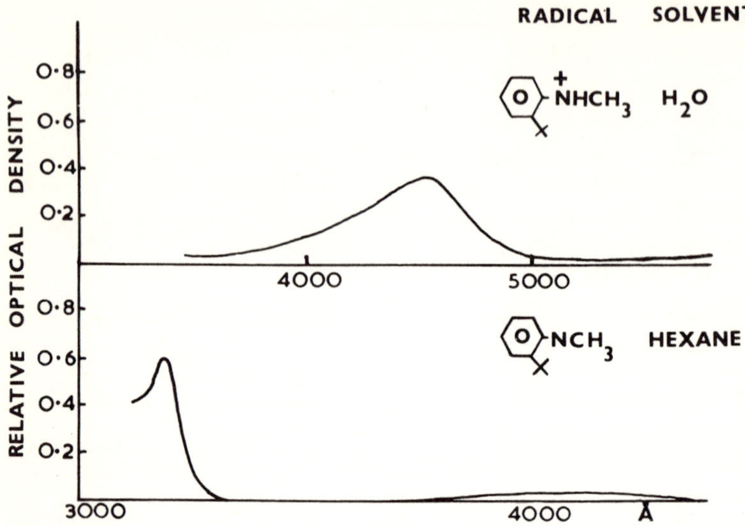

Figure 4-11. Absorption spectrum of a substituted anilino radical showing acidic form in water and deprotonated form in hexane solution.

aqueous solvents, on pH (Figure 4.11). The two spectra correspond to the protonated and unprotonated forms, the radical ion and the radical, and the equilibrium can be established within the lifetime of the radical making it possible to determine the equilibrium constant of this acid-base equilibrium. The pH of the anilino radical is found to be 7.0.

These seven π-electron radicals are the prototype of many of the most important, resonance stabilised radicals of organic chemistry. For example, benzyl is the prototype of the Gomberg type radicals such as triphenyl methyl whilst OH substitution in the β position gives ketyl radicals, anilino is the prototype of Würster radical cations whilst phenoxyls, on substitution by OH, become semiquinones. The spectra and physico-chemical properties of all these radicals are closely similar and provide a large and interesting field of study, of considerable importance both in chemistry and biology. For example, the flavins have been shown to behave similarly and to yield semiquinone radicals on flash photolysis and the principal transients observed on flash photolysis of proteins such as ovalbumin are phenoxyl-type radicals derived from tyrosine groups.

Flash photolysis studies carried out more recently on aromatic vapours at high resolution have succeeded in detecting many other aromatic free radicals of interest, which are not of the seven π-electron type. The most important of these is phenyl, obtained from benzene and halogeno benzenes and substituted phenyls derived from disubstituted benzene derivatives.

The assignment of spectra in all these cases is made on the basis of studies of a series of related substituted compounds, and the method is both convenient and reliable in the aromatic series, since so many compounds are available with the possibility of cross checks in most cases. The data on cyclopentadienyl radical formation, Figure 4.12 and Table 4.3, illustrate the method. Although little or no assistance in the identification is possible from the spectra, apart from general similarity and positions of electronic transitions in related radicals, none of the assignments which have been made so far has subsequently had to be revised; a situation which is not always found in the assignment of spectra, even those of much simpler compounds.

The field of aromatic free radical spectroscopy is relatively new and what has been done already is little more than an indication

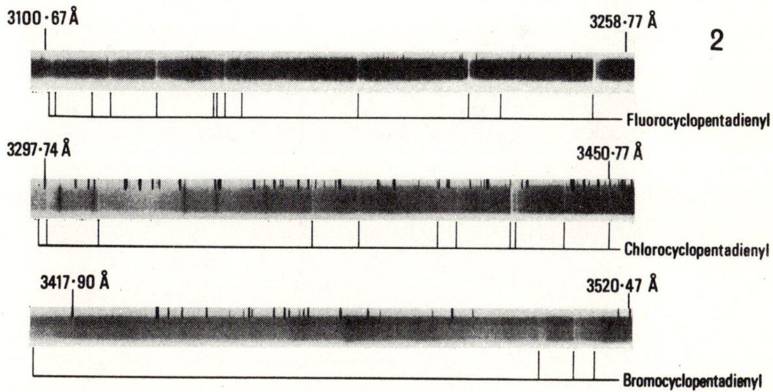

Figure 4-12. Absorption spectra of three halogenated cyclopentadienyl radicals.

of what will be done by flash photolysis techniques in this field in the future. The hundred radicals so far assigned are prototypes of thousands of others which may be observed whenever there is a reason for wanting to study them. Many of the spectra show fine

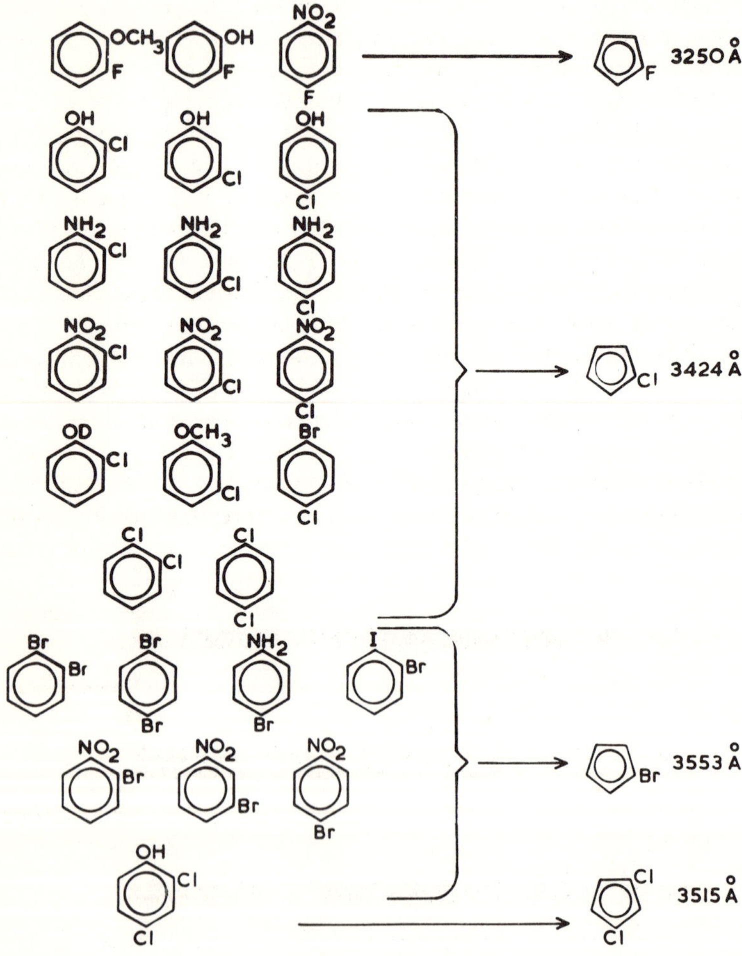

Table 4-3. Observed fission processes of benzene derivatives to form cyclopentadienyl radicals.

detail which, with the help of computer programmes, should eventually yield information about the radical structures. The chemical problems are almost untouched, and represent a much more extensive series of problems than mere identification. Extinction coefficients are known in only a few cases, usually in solution. Finally, there are the intriguing photochemical problems of the primary processes in the excited state by which these often remarkable transformations take place.

The Triplet State

In Chapter 3, I explained how the lowest excited state of most molecules is a triplet which under certain conditions (rigid, solvent or solid) shows its presence by the emission of phosphorescence.

In 1952 we decided to attempt the observation of triplet absorption spectra of organic molecules by flash photolysis in ordinary fluid solvents at normal temperatures. The experiments were successful; it transpired that triplet state lifetimes under these conditions were of the order of a millisecond, ideal for studies by flash photolysis, provided oxygen was excluded. The flash photolysis records of triplet states of aromatic molecules in solution are shown in Figure 4.13.

Any discussion of mechanism in organic photochemistry immediately involves the triplet state and questions about this state are most directly answered by means of flash photolysis. It is now known that many of the most important photochemical reactions in solution, such as those of ketones and quinones, proceed almost exclusively via the triplet state and the properties of this state therefore become of prime importance. Its relatively long lifetime, compared with the time of a flash experiment, has made it possible to study the triplet state almost as readily as the ground state, and in many systems its physical and physico-chemical properties and its chemical reactions are now as well characterised as those of the ground state.

Electron and hydrogen atom transfer, particularly from solvent to triplet state of ketones, aldehydes and quinones have been the subject of very extensive investigations in a number of laboratories. My interest in this type of reaction first arose in a rather practical way when I was consulted about a technical problem known as

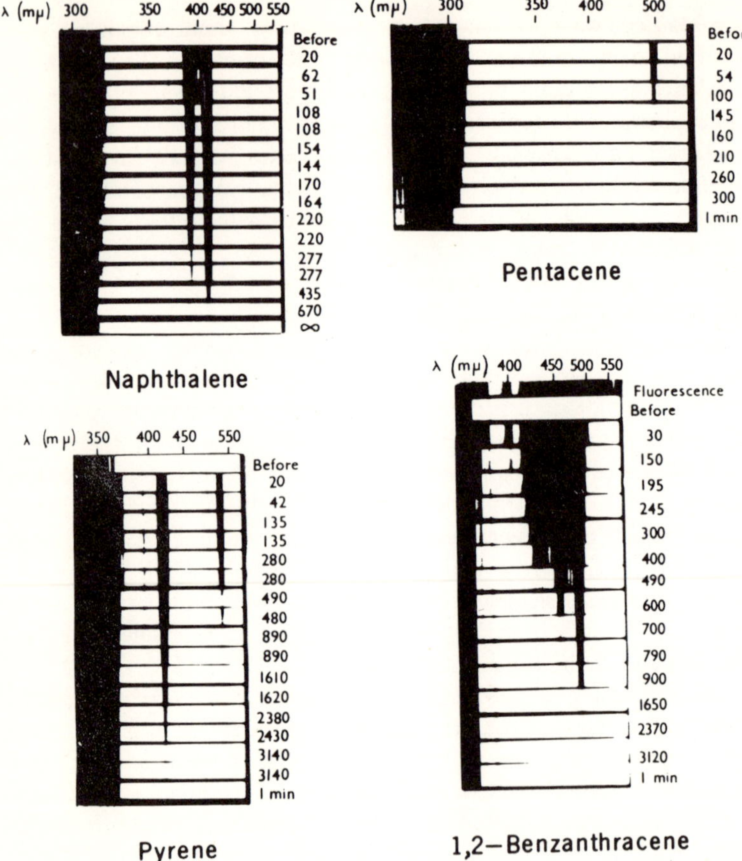

Figure 4-13. Sequences of spectra after flash photolysis of four aromatic hydrocarbons in solution showing formation and decay of their triplet states. Delays in microseconds.

phototendering, in which a dyed fabric, such as cellulose, becomes degraded under illumination in sunlight. The mechanism of this presented few problems; it was merely abstraction, by the excited dye molecule, of a hydrogen atom from the cellulose followed by addition of oxygen to the resulting radicals followed by degradation of the cellulose chain. What was surprising was that dyes, such as the anthraquinones, fell into two classes, one very active and one almost completely unreactive, the difference between the two classes being caused by the apparently small effects of substitution. After a long and interesting series of investigations, mostly carried out on benzophenone derivatives which show exactly the same phenomena, the matter became quite clear. It is always the lowest triplet state which reacts and the electronic structure of this state is, therefore, the prime consideration. Depending on substituents and solvent, this lowest triplet state may be n-π, with electrophilic oxygen and therefore reactive or π-π, with considerable charge transfer character in the opposite sense to that of the n-π state, (CT) and therefore unreactive.

These studies, both of proton and hydrogen atom transfer illustrate how the excited electronic state must be treated as a new species, with its own structure, electron distribution and chemical reactivity and how flash photolysis techniques make it possible to study these characteristics of the excited triplet almost as readily as those of the ground state. Since each molecule has only one ground state, but several excited states, it is clear that this field of investigation is, in a real sense, a bigger subject than the whole of conventional chemistry.

Nanosecond Laser Photolysis

Ordinary flash photolysis is limited to the study of intermediates with lifetimes of microseconds or more and, in spite of much effort, it has not proved possible to make gas discharge flash lamps with adequate energy having shorter flash times than this. There seemed to be little hope of further progress to shorter times until, a few years ago, the laser, and in particular the pulsed laser, was discovered. As a result of this the time resolution of flash photolysis techniques has been improved by a factor of a thousand in the last two years and will very probably soon be extended by another similar factor.

Of the various types, solid-state lasers (ruby, Nd) have so far found the most extensive use in this field. The laser cavity consists of the lasing material, such as a ruby rod, with an excitation lamp fixed parallel to it. At one end of the cavity there is a totally reflecting mirror and at the opposite end, a partially transmitting mirror. The excitation lamp 'pumps' the lasing material through the state responsible for absorption into an excited state from which light emission is possible, achieving, in that state, a population inversion over the terminal state of the emission. It is this condition—the population inversion—which is necessary for laser action to occur. Much of the energy from this upper state may be lost through ordinary fluorescence, depending on the lifetime of the state. However, some of the photons move along a path between the mirrors giving rise to oscillations which increase in energy through stimulation of further emission as they pass through the excited material. On each trip, some of the energy passes out of the cavity through the partially transparent mirror. The unique characteristics of this radiation are its monochromaticity, coherence and polarization—all of great value to the photochemist.

The solid-state laser, when operated as described, produces stimulated radiation of irregular intensity over a period which may extend to hundreds of microseconds and it does not provide the time resolution desired. But by applying a Q-switching technique to the laser (ruby for this discussion), it is possible to produce a highly reproducible flash of a few nanoseconds (10^{-9} s) which has enough energy to meet the excitation requirements of the method. To effect Q-switching, the path between the two mirrors in the cavity is optically blocked with a cell of bleachable dye, Pockels cell or other device. This prevents oscillations within the cavity and allows relevant energy states to be populated much in excess of the lasing threshold. The switch is then opened temporarily allowing the stimulated emission to build up rapidly and with high amplification due to the excess excitation. The result is a giant pulse of several joules which may be confined to under 20 ns duration.

The primary emission of the ruby laser is at 694 nm and so lies too far towards the red end of the spectrum to be absorbed by most organic systems. However, by passing the giant pulse through certain types of crystal (ammonium dihydrogen phosphate or potassium dihydrogen phosphate for ruby) in the proper orientation, up to 20 per cent of the emerging radiation will be of the first harmonic (347 nm), which is much more useful as an exciting source. This process is called ‘frequency doubling’. It is not possible to use the laser pulse directly in place of the conventional flash, because of the need for monitoring over a broad spectrum, but it can be used indirectly.

The monitoring method developed by Porter and Topp follows the conventional method of building up such pictures from a series of spectra, each taken at a pre-chosen delay after the excitation flash. We used a delay unit based on the speed of light, thereby eliminating electronics completely from the system except for the laser itself (Fig. 4.14). The U.V. pulse is sent through a beam splitter and a portion of the light is directed to the reaction vessel. Through focusing, the light may be made to excite only a small volume of solution and so a high radiation density is achieved.

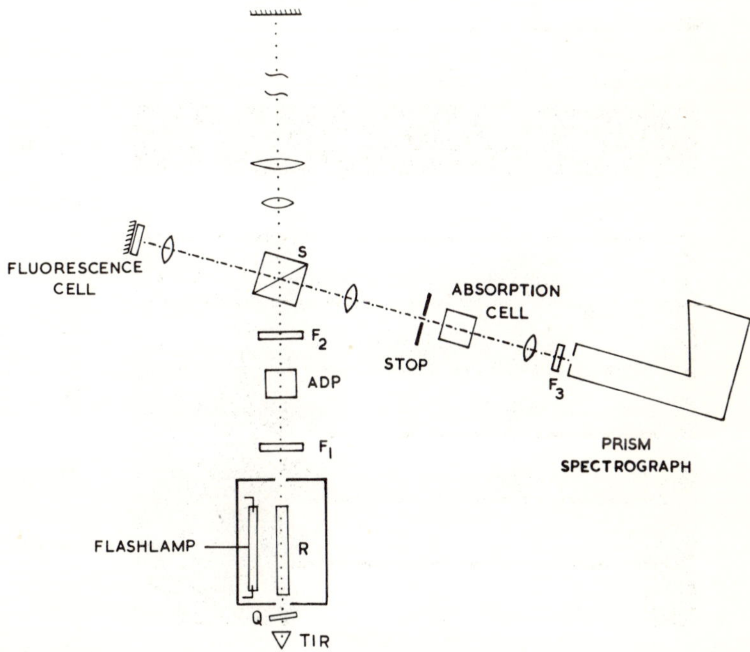

Figure 4-14. Nanosecond flash spectrographic apparatus.

F_1, F_2, F_3 = filters;	*S = beam splitter;*
ADP = ADPh crystal;	*R = ruby;*
Q = passive Q-switch;	*TIR = quartz t.i.r. prism.*

The rest of the light travels out from the beam splitter to a mirror and it is the distance to the mirror, which the beam must traverse twice, that determines the delay. On returning from the mirror, the beam enters the beam splitter again and a portion of the light is reflected to a scintillator solution whose fluorescence provides a spectral continuum in the wavelength region to the red, i.e. to longer wavelengths, of the exciting wavelength. The fluorescence lifetime of the scintillator is also very short, assuring that the time shape of the monitoring pulse will be essentially that of the laser pulse. This induced 'flash' now travels through the beam splitter, the reaction vessel and into the spectrograph, recording the spectrum of the transient at the pre-chosen delay.

One of the principal interests in the nanosecond time region was

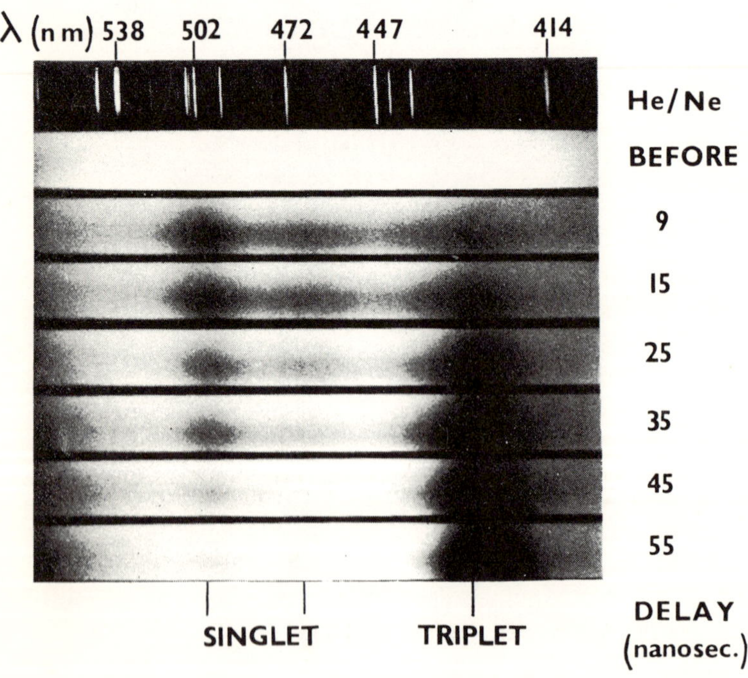

Figure 4-15. Time delay spectra of triphenylene transients in benzene.

the possibility of observing absorption from excited singlet states of molecules, which were known to have lifetimes in the nanosecond region, in the same way as the much longer lived triplet states were observed by conventional flash methods. This was soon achieved for a number of aromatic hydrocarbons and one example, that of absorption by the excited singlet state of triphenylene, which decays into the triplet state, is clearly shown in Figure 4.15.

A new development in laser techniques now makes it possible in principle to extend flash photolysis into the picosecond region. Under certain conditions, known as mode locking, it is found that the Q-switched laser pulse is in fact a train of much shorter pulses, each only a few picoseconds (10^{-12} s) in duration. By selection of just one of these followed by amplification in a second laser it should then be possible to provide enough energy in a few picoseconds to produce detectable concentrations of transients.

Already lasers have extended a thousandfold our ability to study fast chemical reactions. The nanosecond region is accessible and the picosecond region is now in sight. These are very short times. If we look at them on our scale we see that there are as many nanoseconds in a second as there are seconds in a man's lifetime. If we carried out one experiment every nanosecond, the results of a few seconds' work would fill all the books in the world.

It is interesting to note that the time of events is closely related to their complexity. The elementary particle processes are all that appear at the fastest times, chemical primary events follow, then the more complex processes of molecular biology, then the whole biological system of man himself. This is followed by the still greater organisation of many men into societies and civilisations and finally evolution of the universe itself.

Nowhere is this association between complexity and time more apparent than in biology. Man is a lumbering, slow moving, giant. A midge beats its wings once every few thousandths of a second. Life quickens to a fantastic pace in the world of bacteria and similar organisms seen through the microscope. The biochemical processes controlling this activity must be faster still and an enzyme can carry out its highly organised catalysis of these processes in a few microseconds. The primary processes of photosynthesis begin in the nanosecond time range.

Two of the great growing points of knowledge at the present time, cosmology and the elementary particles, appear at the extreme ends of our time scale. The third, molecular biology, occurs principally in the range where the new techniques for the study of fast chemical reactions are likely to play an important role in its advance. When the natural history of man is fully written, the period between one nanosecond and one second will occupy many volumes.

CHAPTER FIVE

Molecules in Evolution

We have considered chemical changes which occur in millionths of a second, others which occupy minutes or days, and now, finally, we shall look at chemical change over several billion years; not any single change but the trend of chemistry over evolutionary time.

We have already noticed one such trend, that which is dictated by the second law of thermodynamics. But there is another trend which seems diametrically opposed to this; the creation and evolution of life. Whether we take the story from the book of Genesis or from that of Darwin, this is a long story of increasing order and decreasing entropy, from the chaos of the first day to the supreme organization of man himself. Even beyond this the trend continues as man progresses unsteadily but discernibly, towards a unity of communities and of nations.

What is life and how did it begin? These are questions which the first philosopher must have asked himself. I shall not spend much of our time on the first question which has many answers, none of them completely satisfactory. The fact is that, without much difficulty we can all agree about a classification of most of the objects around us as living or dead. It is true that there is a misty region which includes such objects as viruses, where we may be unsure in our classification just as there are misty regions of classification of certain living things between the animal and vegetable kingdoms. But nobody, except the most pedantic, worries very much about whether we call euglena plant or an animal, it is recognised to have some characteristics of both. There is no reason whatever why we should not take the same attitude with the classification between living and non-living, and recognise that it is perfectly all right to have objects which are intermediate. Indeed, the whole matter of life and death is to a large extent a matter of degree; amongst things which are quite clearly living, some are more "alive" than others.

The question about the origin of life is now seen, not as a question about how something alive appeared suddenly from something dead, but how the characteristics which we all recognise as life gradually evolved to their present state of wondrous complexity.

First, let us look at the time scale in which we have to work. Various lines of evidence agree in dating the origin of the earth at 4600 million years ago and fossil evidence in the oldest rocks indicates that recognisable forms of life have been present on the earth for half this period, about two and a half thousand million years. These are big numbers; to imagine them better let us divide our time scale by a million so that the origin of the earth would have occurred 5000 years ago, about the time of the first Pharaoh; life then appears five hundred years before Christ, but man appears only last year, the first signs of civilisation, such as the cave drawings, two weeks ago, and the first great civilisations thirty six hours ago. With such examples it is possible to imagine almost anything in the enormous period of time before life began.

The evolution of complex living species, and of man, from the most primitive forms of life, is, of course, the substance of the "Origin of Species" of Darwin, and whether or not one wishes to include a little Lamarkian inheritance of acquired characteristics into Darwin's pure 'survival of the fittest' explanation of natural selection, we have here an explanation of the evolution of life from the most primitive forms which provides, on a grand scale, a satisfactory account of what we see around us. Those who still find this theory difficult to accept do not usually question the order of appearance of the species, or the occurrence of processes of natural selection, but are concerned with the improbability that such a complex and wonderful thing as man could arise out of a series of accidents. I shall return to this point later. It is one of the two great problems which we have to face in trying to understand the origin of life. The other is how even the simplest form of life could have arisen in the first place.

The problem is that of the chicken and the egg. To say that one or other of these came first is a nonsense. Yet, as far as we know, all living things arise to-day from other living things. We can see how they might change over the course of time, but how can they appear from nothing. The problem is even more difficult. We may try to say that we must trace Darwinian evolution back to simpler and simpler things until they can hardly be called living at all. But it is our experience that such things would be rapidly destroyed, being organic in nature and, worse still, it is also our experience that all organic matter on our earth also arose from living things. How then could living things arise from organic matter?

It is because of these difficulties that the various theories of spontaneous generation found favour. Since there was nothing, i.e. no organic matter, to start with and since life certainly started, it must have started from nothing. That is spontaneous generation. And we see obvious signs of it around us if we look casually but not too carefully . . . maggots from meat for example. Aristotle clearly stated his belief. "Such are the facts" he says "everything comes into being not only from the mating of animals but from the decay of earth." He believed in an active principle which was necessary to produce a living thing from dead matter . . . an idea which has been used again and again throughout history as a cloak for ignorance. It was as late as 1828 when Wöhler first dispelled the idea of the necessity for any vital force in chemistry when he synthesised urea from the inorganic compound ammonium cyanate.

Belief in spontaneous generation continued for centuries after Aristotle. By the early 13th century, people believed that geese originated from fir trees, known as goose trees, which had been in contact with the ocean and the belief was still in existence only 250 years ago. It was not only ignorant people who formulated such ideas, even trained observers like Paracelsus, the 16th century physician, described a series of observations on the spontaneous generation of mice, frogs, eels and turtles from decaying matter and van Helmont, the famous seventeenth century Belgian physician, who did some excellent experiments in plant physiology, actually gave a recipe for the producing of mice. One needs a dirty shirt which is left in contact with wheat kernels for twenty-one days. It is important that the shirt be dirty because the "active principle" is human sweat.

Shortly after these reports, the principles of modern scientific investigation and the controlled experiment were developed. At the same time as Galileo was looking at the heavens through his new telescope, Francesco Redi in Florence, carried out "Experiments on the generation of insects" and showed that flies, as well as meat, were necessary for the generation of fly larvae or maggots. He stated his conclusions thus "Although content to be corrected by anyone wiser than myself, if I should make erroneous statements, I shall express my belief that the earth, after having brought forth the first plants and animals at the beginning, by order of the Supreme and Omnipotent Creator, has never since produced any kinds of plants or animals, either perfect or imperfect". In other words, he believed in spontaneous generation of a special kind, a once and for all happening by intervention of a creator.

But the controversy was not over. Soon after Redi's experiments, Anton Leeuwenhoek perfected the simple microscope and discovered a new world teeming with micro-organisms. It seemed hard to believe that sexual reproduction could account for the exceedingly rapid appearance of these tiny organisms and controlled experiments were more difficult to perform. At first, experiments by Needham in London seemed to prove that spontaneous generation of these organisms did occur, but more careful experiments by the Abbé, Lazzaro Spallanzani, showed that Needham's experiments were inadequate. Finally, really conclusive experiments were performed by Louis Pasteur who, using his simple but very elegant swan-necked flash technique, showed that the active principle by which bacteria were generated in nutrient media was nothing more than similar bacteria borne in the dust particles in the air all around us.

Thus the idea of spontaneous generation was dispelled and its passing left man once more with the original problem which had given rise to the idea. How did life arise in the first place?

Modern Theories of the Origin of Life

In 1924, Oparin in Russia, and independently in 1927, J. B. S. Haldane in England, made a suggestion which immediately pointed the way to a satisfactory theory of life's origin. They pointed out that conditions on our planet were not always as they are to-day; that, although spontaneous generation may never occur again because the present conditions are unsuitable, conditions might have been eminently suitable for it in the past. In one form or another, these ideas are the basis of nearly all serious work on the origin of life which is in progress at the present time, although, at the time, they received very little attention.

There is much evidence that our atmosphere was not always the same as it is to-day. It would be very surprising if it were, because to-day the processes of photosynthesis carried out by green plants which cover much of the surface of the earth, re-cycle all the carbon dioxide in our atmosphere every 400 years and all the oxygen every 2000 years. But, in the first place, our atmosphere was formed as a part of the creation of the earth and one can easily understand that it must have undergone many changes since that time as the chemical equilibria between the earth's mantle and the atmosphere became established and, above all, as the sun set up a series of photochemical changes on an immense scale in the gases above the earth.

According to the book of Genesis, God's first words were "Let there be light" and without that light of the first day the rest could not have happened. Not only is the light of the sun the source from which all life derives its sustenance at the present time, it is also most probably the source which created the organic molecules from which life arose in the first place.

There is some evidence that at an early stage before life began, the atmosphere of the earth contained no oxygen but consisted of methane, ammonia and water. It certainly had to be one or the other extreme because a mixture of these gases with oxygen would, at the first stroke of lightning, be brought to equilibrium in a spectacular way. The composition of the earliest rocks indicates that, when they were laid down, the atmosphere was reducing and planets in an earlier stage of development than ours, e.g. Jupiter, to-day have an atmosphere containing much methane and ammonia.

It is obvious that life, as we know it to-day, could not have existed in this atmosphere of marsh-gas and smelling salts, not because of the need for oxygen to breathe, indeed anaerobic forms of life which do not require oxygen are well known, but because of the absence of the protecting blanket of oxygen to shield such living things from the lethal rays of the sun. The sun's radiation contains a good deal of energy in the ultra-violet region. Oxygen does not filter out any but the very shortest wavelengths below 220 nm, but it is responsible for setting up a screen for us in another way. When very short ultra-violet light falls on oxygen gas, a photochemical change occurs which converts this oxygen into ozone. This is our salvation because, although there is only a very thin layer of ozone 15 miles up in our atmosphere, so thin that it corresponds to a thickness of a few millimetres of atmospheric pressure, nevertheless, this is enough to filter out the short wavelength, ultra-violet light below 300 nm which would be lethal to simple living organisms and subsequently lethal to man, for it would destroy all the vegetation on which our existence ultimately depends.

In passing, it is interesting to note how delicate is the thread by which our life continues on this planet, and how carefully we should think before we interfere with our environment. A few years ago, a popular rocket experiment was to fire cans of sodium into the atmosphere and watch them glow; the experiment was of some interest in understanding the composition of the upper atmosphere. Sodium reacts violently with chlorine and there were also proposals to fire chlorine gas up there as well. Now every photochemist knows that chlorine upsets the photochemical equilibrium between oxygen and ozone so that most of the ozone is destroyed. It is conceivable that the hole, 100 square miles in extent, produced by one ton of chlorine could have some startling local effects on our planet!

To return to our primitive atmosphere of methane, ammonia and water, these gases do not absorb light from the sun as we receive it to-day but they do absorb ultra-violet light below about 200 nm very strongly and, in the absence of oxygen and the consequent protective ozone layer, there would be lots of these short wavelengths falling in on the atmosphere. Furthermore, it is easy to show by experiments in the laboratory that photochemical reactions occur when these gases absorb ultra-violet light.

We are now ready to take the final and crucial step in our speculations about the creation. The principal elements of which the proteins and nucleic acids are composed are carbon, hydrogen, oxygen and nitrogen. These are also the four elements of which are composed the gases methane, ammonia and water. The other elements which occur in proteins and nucleic acids are phosphorus and sulphur and two gases which we might expect to find in the primitive atmosphere in addition to those already mentioned would be phosphine and hydrogen sulphide. It took the genius of a Haldane to put these facts together in the first place and to draw the obvious deduction; the first complex organic molecules necessary for the genesis of simple living organisms were formed by photochemical action of the light of the sun on the gases of the primitive atmosphere. These gases, although themselves inorganic, contained the elements necessary for the synthesis of the basic organic structures like amino acids.

Even at this stage, there are many unanswered questions and possible modifications in our theory. There was some tendency to favour, in the original theories, electric discharges of lightning, or even thermal reactions, in place of the light of the sun but, whilst it is certainly possible that the chemical synthesis could also have occurred in these alternative ways, the ultra-violet light of the sun is such a potent and universal source that it would probably be of predominant importance.

We may then speculate as to what happened subsequently. Still perhaps under the influence of the sun's radiation, the organic molecules, now built up to a significant concentration in the oceans or accumulated to very high concentrations in local pools, became a primeval soup in which other more complex reactions could occur. Agglomerations, at first random, but quite early having membranes formed by the concentration of surface active molecules at the surface of the water, would be formed in great variety.

Whilst these events were occurring, photochemical reactions would be changing, gradually, the composition of the atmosphere. The decomposition of water and ammonia would yield, among other things, oxygen, nitrogen and hydrogen and the latter gas, being much lighter than the others, would be held less firmly by the gravitational attraction of the earth. So, by slow photochemical change, our atmosphere would lose hydrogen and accumulate its present constituents nitrogen and oxygen. At this point, the synthesis of organic compounds from the atmosphere would have to cease but by then, a large enough quantity of them could have been accumulated to ensure that life got off to a good start. The new oxygen-containing atmosphere would, by this time, be an advantage in offering protection from the sun and also making possible the development of respiratory processes as a means of energy conversion.

There is still a long way to go but, if we could be confident that we are on the right lines up to this point, the gap between a dead world and one full of living things becomes one that we can see across, if not actually bridge. The next step is to test as much of our hypothesis as possible in the laboratory.

In 1953, Miller, a young colleague of Urey's in Chicago, performed the following simple experiment. He took a mixture of methane and ammonia gases in a bottle containing water. He passed an electrical discharge through the gases and boiled the water so that its vapour fell through the electric spark like rainfall through a thunderstorm in the primitive atmosphere. After several hours, he opened up the tube and analysed the products. He found a complex mixture of substances, some of them composed of quite large molecules. By modern, but standard, methods of analysis he showed that amongst these substances were several amino acids, those very building blocks which above all were necessary as a first stage in the chemical synthesis of life.

This was a very important experiment and it is encouraging to think that it is one which could, at that time, have been performed in most well equipped school laboratories and required no more mental equipment to understand its significance than is possessed by any bright schoolboy. It is a fallacy that, because in certain well established fields, like nuclear physics, many experiments cannot be performed without extremely complex apparatus or theoretical background, all future scientific progress will depend on such equipment. I have no doubt whatever that there are many other experiments of the greatest significance which could be performed to-day with the simplest equipment and not too much background. But they will all of them require a mind of imagination and originality.

Very rapidly, however, after the key discovery, a field of research tends to become highly sophisticated and it would be unwise for a schoolboy now to try to improve on Miller's first experiment because its consequences are being examined all over the world. Other forms of energy, particularly ultra-violet irradiation, have been shown to be effective and almost every important type of organic compound including sugars, porphyrins and the components of nucleic acids, has been shown to be capable of synthesis in this way, given the right conditions. Furthermore, Fox and others have shown that, in addition to molecules, aggregates of molecules appear in these mixtures, visible under the microscope and looking for all the world like simple cells, some of them apparently undergoing division, though nobody yet suggests that the next thing to be seen will be little green men. This is now an entirely reputable and serious branch of scientific study and one in which very rapid progress is being made in bridging the gap from both directions. As more is learned about the structure of the viruses and ever simpler forms of life, if indeed our definition of life extends even now to such simple bodies as crystalline viruses, the smaller the gap which has to be bridged by those trying to synthesise such bodies. When one reads a newspaper report that somebody has created life spontaneously in a test tube it is a reflection of the ignorance or at best, the over-enthusiasm of the reporter. An informed person would know that the most that can have happened is that one more piece of the jig-saw puzzle of life's origins has been put into place, that already the puzzle has been completed to the extent that the outlines of the whole picture are discernible but that countless pieces remain to be fitted into almost every part of it.

Sequence of Molecular Evolution

At the present time, the most likely sequence of events leading to the appearance of living things seems to be the following.

1. The synthesis of amino acids, simple sugars, fatty acids, heterocyclic bases like adenine, porphyrins and other relatively simple molecules by ultra-violet photolysis of the molecules in the reducing atmosphere, i.e. methane, water, and ammonia.
2. The solution of these molecules in the seas and consequent protection from destruction by further photolysis by ultra-violet light, though photochemical reactions with light of longer wavelengths may have continued.

3. The synthesis of polymeric molecules based on these substances by dehydration-condensation reactions, for example, peptides from amino acids:

$$\underset{\displaystyle R_1}{H_2N\cdot\underset{|}{CH}}\cdot\overset{\displaystyle O}{\overset{\|}{C}}\cdot\boxed{OH + H}\cdot\overset{\displaystyle H}{\overset{|}{N}}\cdot\underset{\displaystyle R_2}{\underset{|}{CH}}\cdot CO_2H \rightarrow \underbrace{H_2N\cdot\underset{\displaystyle R_1}{\underset{|}{CH}}\cdot\overset{\displaystyle O}{\overset{\|}{C}}\cdot NH\cdot\underset{\displaystyle R_2}{\underset{|}{CH}}\cdot CO_2H}_{\text{dipeptide}} \rightarrow \text{Polymer}$$

(amino; carboxyl)

and similarly polysaccharides from sugars, and lipids from alcohols and fatty acids, also by elimination of water. The nucleic acids also are formed by condensation reactions; for example, elimination of water from a molecule of adenine and a ribose sugar forms the nucleoside adenosine, further elimination of a second water between this and phosphoric acid (present in solution) forms the nucleotide adenylic acid, and elimination of water between molecules of the nucleotide forms the polynucleotide.

These dehydration reactions might occur by

(a) Heating the components (dry) to a temperature of 150°C at which temperature Fox has shown that protein-like polymers are formed from amino acids. This does not seem to be a likely process on the earth's surface.

(b) Dehydration by molecules with groups such as $-N=C=N-$ (e.g. carbodiimide $HN=C=NH$ (a tautomer of cyanamide NH_2CN) formed from HCN and NH_3) which reacts to remove water and form urea NH_2CONH_2. This reaction has been shown to occur by Calvin.

(c) Photochemical dehydrogenation sensitised by heterocyclic bases. For example, the very important dehydrogenation of orthophosphate to give pyrophosphate

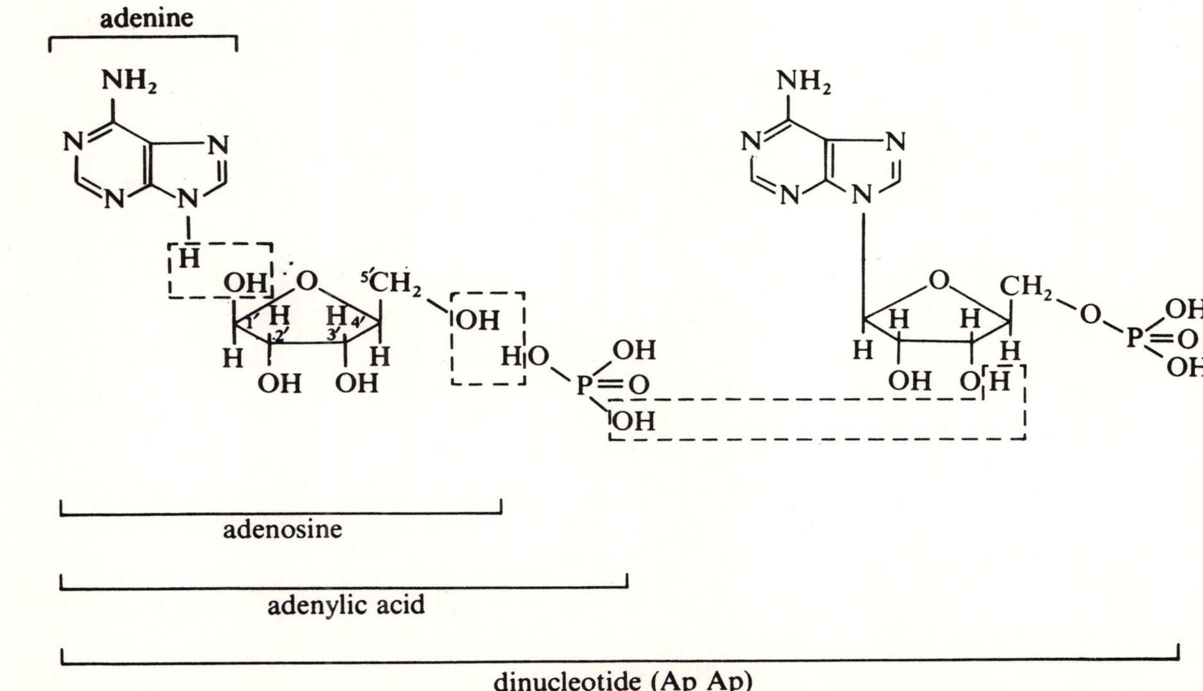

Figure 5-1. Dehydration condensation reactions in the formation of ribonucleic acid. DNA lacks the OH group on the 2' carbon of the ribose.

$$\mathrm{HO{-}\overset{\overset{\large O}{\|}}{\underset{\underset{\large OH}{|}}{P}}{-}OH + HO{-}\overset{\overset{\large O}{\|}}{\underset{\underset{\large OH}{|}}{P}}{-}OH \rightarrow HO{-}\overset{\overset{\large O}{\|}}{\underset{\underset{\large OH}{|}}{P}}{-}O{-}\overset{\overset{\large O}{\|}}{\underset{\underset{\large OH}{|}}{P}}{-}OH + H_2O}$$

is brought about by irradiation of mixtures of phosphoric acid and calcium ions with pyridine as the photosensitiser.

4. Autocatalysis of polymer formation leads to replication of specific sequences, especially in the polynucleotides, in a similar way to that which occurs to-day, but with general dehydrating agents rather than specific enzyme catalysts which are not yet available.
5. Coupling of the polypeptide and polynucleotide synthesis by a reflexive autocatalytic sequence of several steps, probably only involving a triplet of amino acids in the first place. The general concept of reflexive catalysis is, that the products of one reaction catalyse a second reaction, the product of which catalyses a third and so on until the product of the last reaction in the sequence catalyses the first reaction again (Figure 5.2).
6. Natural selection of certain sequences results from the success of the more efficient reflexive catalytic sequences over the less efficient, and these sequences, now present in high concentration are themselves built into more complex molecular sequences by further reflexive catalytic mechanisms. When two stereoisomers were possible an unstable situation would arise until one predominated.
7. These molecules, being surface active, become concentrated at an interface, probably a water-air surface, and the layers fold over so as to form membranes which isolate the contents within a primitive "cell" or coacervation droplet. Such droplets have been observed in the products of laboratory irradiation of primitive atmosphere mixtures.

So far, the free energy to drive these processes has been in the monomer molecules, such as sugars, synthesised by the sun from the primeval atmosphere. These processes became more sophisticated with the development of energy transfer

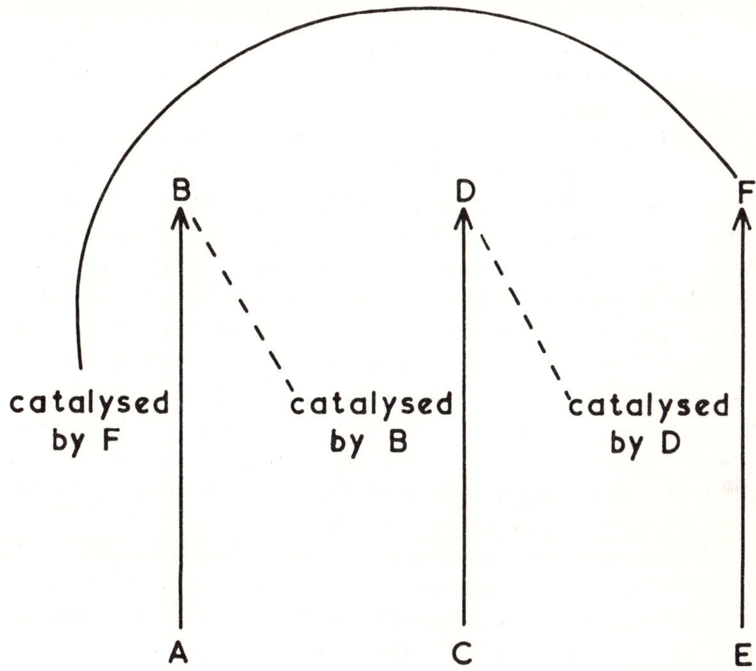

Figure 5-2. Reflexive catalysis—a cyclic sequence of catalysed reactions.

and storage molecules like A.T.P. (adenosine triphosphate). The cells may continue to feed on this form of energy but later they must develop a mechanism of utilising the energy of the sun themselves directly.

8. The cells developed photosynthetic mechanisms leading eventually to the use of chlorophyll and similar molecules as absorbers of sunlight in the visible region.
9. The reducing atmosphere gradually changed to an oxidising one by
   (a) photochemical reactions in the atmosphere and loss of hydrogen, and
   (b) the evolution of oxygen by photosynthesis in the biosphere.

10. The final energy producing mechanism could now be developed, the process used by animals to-day, which involves respiration and oxidation of the organic molecules synthesised by the plants.

From now on "life" is well established and further evolutionary developments by natural selection of the objects best able to survive and multiply in their environment links up with Darwinian theories of evolution of the biological species. The cell coacervates became more and more complex, not only in chemical composition but also in physical shape so as to be able to absorb their food more efficiently. Reproduction was by cell division, in the first place probably not unlike the splitting of a soap bubble into two smaller bubbles. A method of causing chemical energy to move a fibre formed of long chain molecules was developed which provided vibrating hairs or cilia. This and other forms of mobility led to a better chance of finding food, even though the mobility was in the first place probably entirely random.

All this is, of course, highly speculative and some of it will doubtless be modified by further study. But this does not mean that a more complete understanding of the origins of life is beyond our grasp. Experiments in the laboratory can show the feasibility or otherwise of the reactions which are proposed and "molecular archaeology" can be greatly extended to tell us more about the sequence of chemical events on our developing planet. Above all, we have much to learn about the nature of living things to-day, the structures of enzymes, which Professor Phillips will talk about.

Many of the problems about the origin of life no longer present any difficulty. The apparent contradiction to the second law of thermodynamics disappears when we take into account the whole system including the sun, with its increase in entropy far greater than the local entropy decrease during evolution. The difficulty that a complex system like man could never have arisen by the chance assembly of millions of molecules does not arise. Certainly this could never happen, the probability is impossibly small if all arrangements are given equal weight. But the dice are loaded, the synthesis proceeds in simple steps each of high probability and the natural selection process eliminates those steps which are in the "wrong" direction for survival in the environment. If the success of this process up to the present time seems incredible, we must remember that it is the result of two billion years of experiment.

Is, then, the life of man merely a complex example of chemical change, subject to the same laws and predictable in the same way? Certainly, in the last resort, economics, psychology and the social sciences are subject to the same basic laws as chemistry. As to predictability, however, the situation is quite different; the number of similar bodies is relatively small and the statistical laws have become correspondingly less exact. Some progress can be made when numbers are large and the behaviour to be predicted is of a simple kind. But, whereas the removal of one molecule from a million will have no observable effect, the removal of one man may change the course of history.

FURTHER READING

Angrist, S. W. and Hopler, S. W. *Order and Chaos* (New York: Basic Books, 1967).

Bernal, J. D. *The Origin of Life* (London: Weidenfeld and Nicolson, 1967).

Bragg, Sir W. L. and Porter, G. (Eds.) *The Royal Institution Library of Science: Physical Sciences* (*10 vols.*) (London: Applied Science Publishers, 1970).

Caldin, E. F. *Fast Reactions in Solution* (Oxford: Blackwell, 1964).

Calvin, M. *Chemical Evolution* (Oxford: Clarendon Press, 1969).

Clayton, R. K. *Light and Living Matter* Vol. 1: The Physical Part, 1970. Vol. 2: The Biological Part, 1971. (New York and London: McGraw-Hill).

Cundall, R. B. and Gilbert, A. *Photochemistry* (London: Nelson, 1971).

Frost, A. A. and Pearson, R. G. *Kinetics and Mechanism* (London: John Wiley, 1961).

Horspool, W. M. and Sutherland, R. G. "Organic Photochemistry, Parts 1 and 2" *Educ. in Chem.*, Vol. 4 (1967), p. 71–75, and Vol. 5 (1968), p. 253–256.

"Light" (Special Issue) *Sc. Amer.*, Sept. 1968 (Vol. 219, No. 3).

Mark, H. F. et al *Giant Molecules* (("Life" Science Library) Time-Life International, 1967).

Pauling, L. and Hayward, R. *The Architecture of Molecules* (San Francisco and London: W. H. Freeman, 1964).

Porter, G. "Flash photolysis and some of its applications" in *Les Prix Nobel en 1967* (Stockholm: Norstedt & Söner, 1969).

Young, J. Z. *An Introduction to the Study of Man* (Oxford: Clarendon Press, 1971).

Structures, Activities and Evolution of Biologically Important Macromolecules

by

D. C. Phillips

CHAPTER ONE

Nucleic Acids and Heredity

In these talks we shall be considering the chemical and three-dimensional structures of the kinds of molecules from which we, and other living creatures, are made, and we shall be trying to see how far our detailed knowledge of these molecules can take us towards understanding the larger and more complex structures and properties of living cells and organisms. Compared with the situation even ten or twenty years ago, we now know a great deal in general outline about the molecular processes underlying life and the properties of the living systems. We also have much detailed knowledge about the important families of molecules that are involved and the ways in which they interact. Most of the details and some of the most important general principles, however, remain to be discovered and I hope that this brief survey of the current state will not obscure the fact or contrive to make the prospect of the next twenty years seem less exciting than the recent past has been.

Although many relatively small molecules play essential parts in the chemistry of life our main concern will be with two different kinds of macromolecules, the nucleic acids which act as the store of biological information that has to be passed on from one generation to the next and the proteins which are the means of translating that information into action.

Proteins are really very familiar to us, whether we know it or not, since it is mainly proteins that we see when we look at one another—especially if we are patriotic enough to be wearing wool! Our hair and skin are mainly protein and when we explore our bodies in a little more detail we find that the molecules transported in our blood—such as the red pigment, haemoglobin—and the principal constituents of the cells from which we are made, are also protein molecules of various kinds. We are constructed mainly of protein, even bone has a protein base overlaid with a mineral, and virtually every chemical reaction that takes place in our bodies is affected by a protein catalyst or enzyme. Clearly proteins are of prime importance and we shall be principally concerned with examining their properties. It happens, however, that they are not the molecules which act as the fundamental store of biological information and in order to understand the molecular mechanisms that underlie heredity we must first consider the nucleic acids.

Our knowledge of nucleic acids dates back almost exactly one hundred years to the work of Friedrich Miescher, a Swiss physician. As a young man of 22 in 1868 he extracted a substance from pus cells which he called nuclein. Over the years that followed—helped by the ready availability of pus cells from the bandages of wounded soldiers—Miescher devised a way of isolating cell nuclei and characterised the nuclein derived from them as an acid substance of high molecular weight containing a large proportion of phosphorus. As early as 1884 Hertwig, who demonstrated that the fertilisation of a sea urchin egg involved the fusion of nuclei from the egg and a sperm, recognised from his own and Miescher's work that "nuclein is the substance responsible not only for fertilisation but also for the transmission of heredity characteristics," but this view was not generally accepted until the 1940's, so strong was the belief in the prime importance of proteins. Soon afterwards, thanks to the detailed chemical studies of Todd and many others, nuclein—now known as deoxyribonucleic acid (DNA)—was fully defined chemically and studies of its mode of action were begun.

DNA is a long chain molecule made up from small molecules of several different kinds: adenine and guanine, nitrogen containing bases derived from purine; cytosine and thymine, nitrogen bases derived from pyrimidine; deoxyribose, a sugar molecule with a five-membered ring; and phosphoric acid. In DNA these molecules are linked together as shown in Figure 1-1 to form long chains in which sugar molecules carrying one of the four bases attached to their 1′-carbon atoms are linked together by phosphate groups attached to their 3′- and 5′-carbon atoms. From this structure we can see at once the possibility of encoding information in the arrangement of the bases. The language of the genes is written in a four letter alphabet, A, G, C, T, determined by the nature of the bases, adenine, guanine, cytosine and thymine.

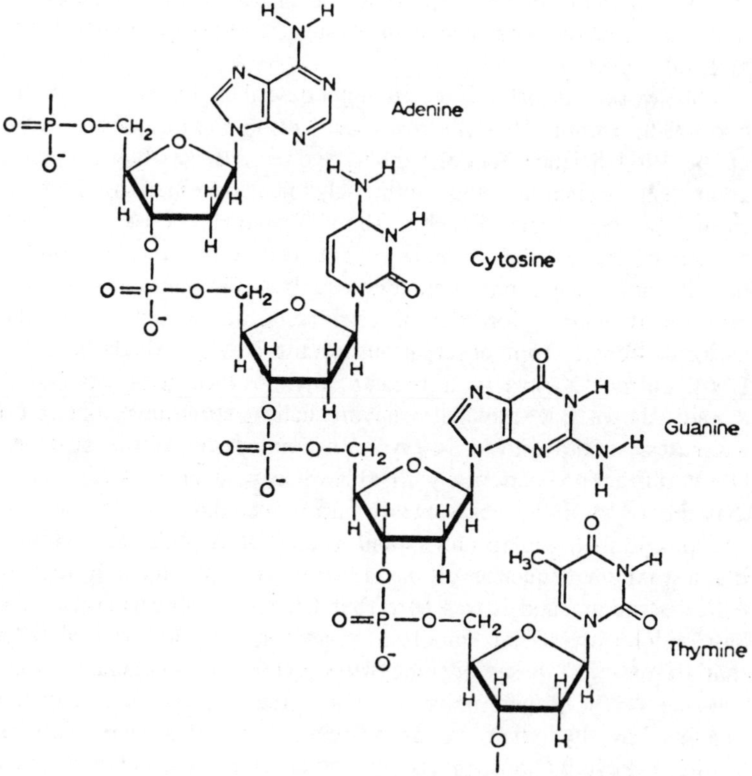

Adenine

Cytosine

Guanine

Thymine

Figure 1-1. The chemical structure of DNA.

Replication of the Genetic Message

The first requirement of the gene-molecule is that it should be easily copied so that daughter molecules, exactly the same as the original, can carry the genetic information to daughter cells in a growing organism and to the egg or sperm cells leading to the next generation. The chemical structure of DNA provides no indication of how this might be achieved and it is here that the importance of three-dimensional structural studies becomes apparent in a most dramatic way. The study of the structure of DNA by Crick, Watson, Wilkins, Franklin and their colleagues in the early 1950's provides the best possible example of how the structure of a biologically important macromolecule can itself suggest the way in which the molecule functions.

This work has been entertainingly described by Watson in his book "The Double Helix." (Professor Watson lectured on this work at the 1964 Science School here in Sydney.) It was based upon a great deal of chemical and physico-chemical information—perhaps especially the observation by Chargaff that in DNA from any particular species the amount of adenine equals the amount of thymine and the amount of guanine equals the amount of cytosine—but the evidence for the three-dimensional structure of the molecules came from observations of the way in which fibres of DNA diffract X-rays in a regular way. A theoretical study has already shown how molecules having helical structures, like spiral staircases, should diffract X-rays and comparison of the observed DNA diffraction patterns with this theory showed straightaway that the DNA molecules somehow are coiled into regular helices. The puzzle then was to understand how a DNA molecule embodying a random sequence of bases could possibly form a regular helical structure and it was here that Chargaff's observation of the base-ratios provided an important clue. Watson and Crick showed that the pairs of bases adenine with thymine (AT), guanine with cytosine (GC) formed units, in which the two bases are linked together by hydrogen bonds (discussed in Professor Porter's lectures), having almost precisely identical dimensions *(Figure 1-2)*. This led to their suggestion of the structure in which two complementary DNA chains are intertwined to form a double-helix *(Figure 1-3)* in which pairs of bases provide a regular core with the sugar phosphate chains coiled around them.

This structure was shown to be quite consistent with all of the available evidence and it immediately suggested a way in which DNA may be copied. The two chains are complementary in that whenever a particular base occurs in one chain its partner in the other chain is the base that makes up either an AT or a GC unit

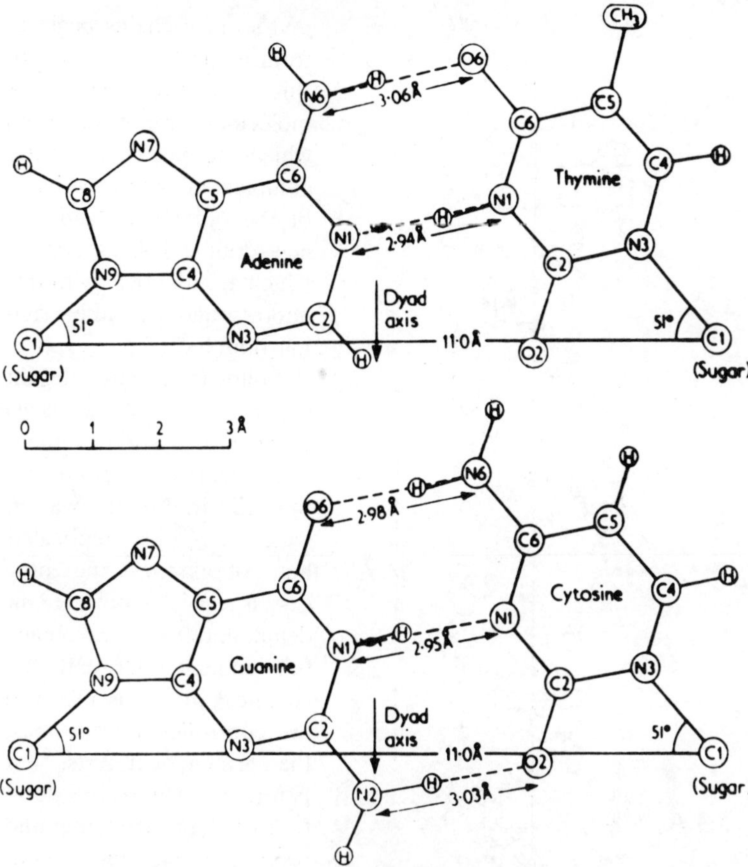

Figure 1-2. Hydrogen-bonding between adenine-thymine and guanine-cytosine bases in DNA. Note that the pairs have closely identical overall dimensions and stereochemical properties.

in the double helix. This suggests the scheme shown in Figure 1-4. It is supposed that the hydrogen bonds joining the bases in pairs are broken and the two chains then separate. If this occurs in a pool of the four possible nucleotides (the individual sugar, phosphate, base units) which have already been synthesized, one can imagine the bases of these nucleotides jostling to form hydrogen bonds with the bases of the two chains. When the correct Watson-Crick pairs are formed the sugar-phosphate molecules knit together

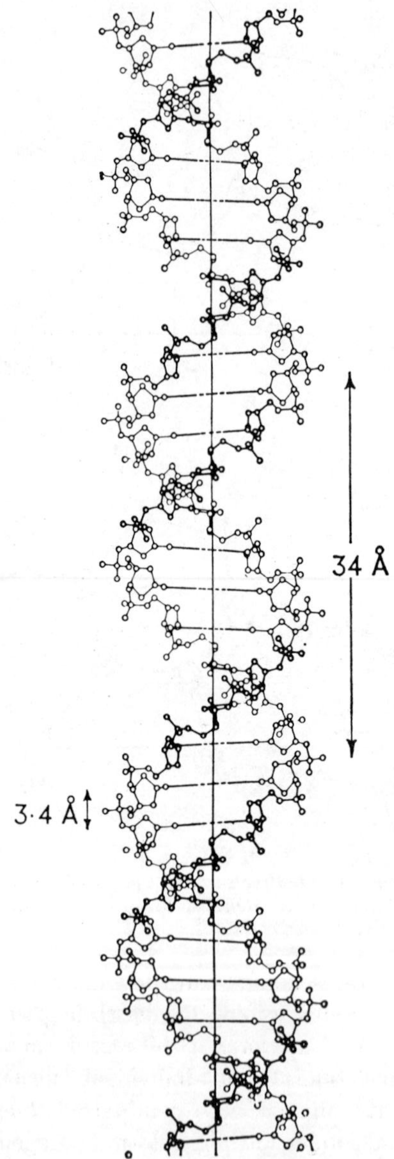

Figure 1-3. The double-helix structure of DNA.

and so new chains begin to form along the old ones. In this way two new DNA molecules are formed in which the bases are exactly complementary to the bases of the old chains and two new double helices emerge which are identical with one another and with the original double helix.

Following many elegant experiments by a large number of workers there now seems no doubt that this is in outline the way in which DNA is replicated but, of course, the process is very complicated in detail, and it is by no means fully understood. Clearly the chemical reactions involved are controlled by enzymes, the protein catalysts, but before we turn to consideration of their structures and properties we must consider the second requirement that must be satisfied by the genetic material, it must be capable of expression in the form of functional and structural molecules.

Expression of the Genetic message: Protein synthesis

The main structural and functional molecules in our

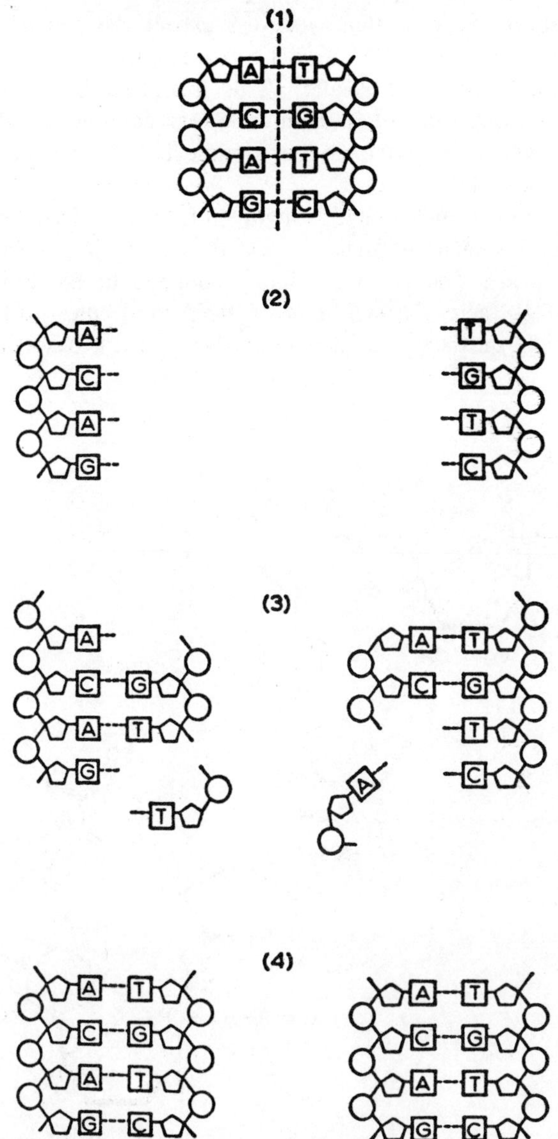

Figure 1-4. Schematic representation of the replication of DNA showing the initial separation of the two strands of a double helix and the subsequent growth of two identical daughter helices.

bodies are proteins so that we must expect the genetic message that is encoded in the DNA somehow to control the structures and properties of protein molecules. This is indeed how the genetic message is expressed but it requires a very complicated molecular apparatus which is understood in outline but not yet in anything like adequate detail.

A new type of nucleic acid is involved in this process, ribonucleic acid (RNA) shown in Figure 1-5, which plays a part in several different forms. The DNA is mainly confined to the cell nucleus and controls there the synthesis of RNA molecules carrying the same genetic message (encoded in almost identical bases, A, U, G,

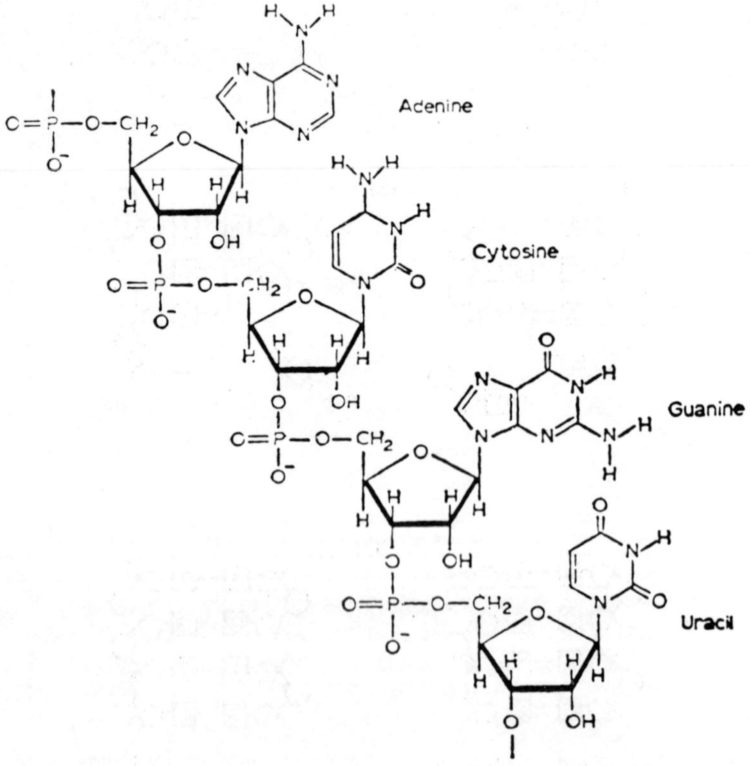

Figure 1-5. The chemical structure of RNA.

C) by an enzyme-controlled process similar to that involved in replication of the DNA itself. These messenger RNA molecules travel to the cytoplasm of the cell, where protein synthesis takes place, and they are associated there with small cellular organelles, the ribosomes, which are composed partly of protein sub-units and partly of RNA (often written rRNA for short). This complex between the ribosomes and messenger RNA (mRNA) forms the assembly line upon which the proteins are synthesized.

The key feature of this process is the way in which the genetic message now encoded in the base sequence of the mRNA, is translated into a protein structure. Since proteins, as we shall see, are composed from small molecules (the alpha-amino acids) of twenty different kinds this is essentially a question of translating a four letter code (corresponding to the four different bases of the nucleic acids) into a twenty letter code. Yet another type of RNA, transfer RNA, (tRNA) is used to solve the problem.

Clearly, since the two codes involve different numbers of "letters" in the two alphabets, there cannot be a one-to-one correspondence between them and more than one base in RNA must be involved in coding for each amino-acid in the linear sequence of amino-acid residues found in a protein. Before the definitive experiments had been carried out there was much discussion of the nature of the "Genetic code"—the relationship between the nucleotide sequences of nucleic acids and the amino-acid sequences of proteins—and many ingenious suggestions were made of overlapping and non-overlapping codes involving various numbers of nucleotides for each amino-acid. The code is now known to be of the simplest non-overlapping type with three nucleotides in sequence defining each amino acid residue in the corresponding protein chain.

A sequence of two nucleotides, each chosen from four different possibilities, clearly may have any one of $4 \times 4 = 16$ different structures. This is not quite enough to encode the twenty different kinds of amino-acid. Similarly a sequence of three nucleotides may have any one of $4 \times 4 \times 4 = 64$ structures and this is more than enough to provide "codons" for each type of amino-acid. Detailed studies by many workers have shown that this is the code, nevertheless, so that in general there are several different triplet codons for each amino-acid and two are available as punctuation marks, indications of the beginning and end of genetic messages that correspond to the initiation and termination of protein chains. The code is shown in Figure 1-6.

SECOND LETTER

FIRST LETTER	U		C		A		G		THIRD LETTER
U	UUU	Phe	UCU		UAU	Tyr	UGU	Cys	U
	UUC		UCC	Ser	UAC		UGC		C
	UUA	Leu	UCA		UAA	OCHRE	UGA	?	A
	UUG		UCG		UAG	AMBER	UGG	Tryp	G
C	CUU		CCU		CAU	His	CGU		U
	CUC	Leu	CCC	Pro	CAC		CGC	Arg	C
	CUA		CCA		CAA	GluN	CGA		A
	CUG		CCG		CAG		CGG		G
A	AUU		ACU		AAU	AspN	AGU	Ser	U
	AUC	Ileu	ACC	Thr	AAC		AGC		C
	AUA		ACA		AAA	Lys	AGA	Arg	A
	AUG	Met	ACG		AAG		AGG		G
G	GUU		GCU		GAU	Asp	GGU		U
	GUC	Val	GCC	Ala	GAC		GGC	Gly	C
	GUA		GCA		GAA	Glu	GGA		A
	GUG		GCG		GAG		GGG		G

Figure 1-6. The genetic code.

Transfer RNA (tRNA) molecules have a complex three dimensional structure that is not yet known in detail but the chemical structures of several different tRNA molecules are known and they all have the general forms shown in Figure 1-7. One loop in this structure has a nucleotide sequence that is complementary to the "codon" for the corresponding amino-acid. This amino-acid, appropriately activated for incorporation in the protein chain, is attached—again under the influence of an enzyme—to the end of

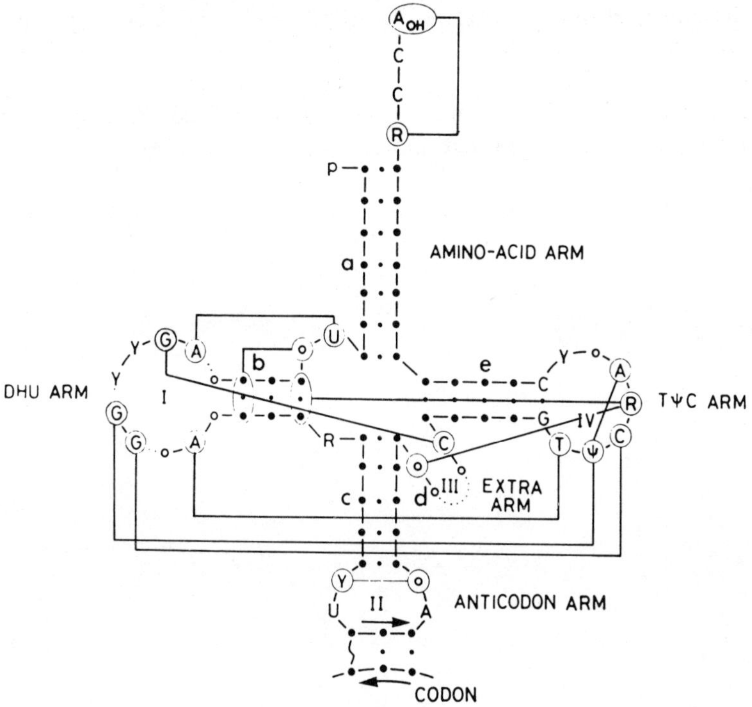

Figure 1-7. General structure of tRNA showing probable contacts in three dimensions.

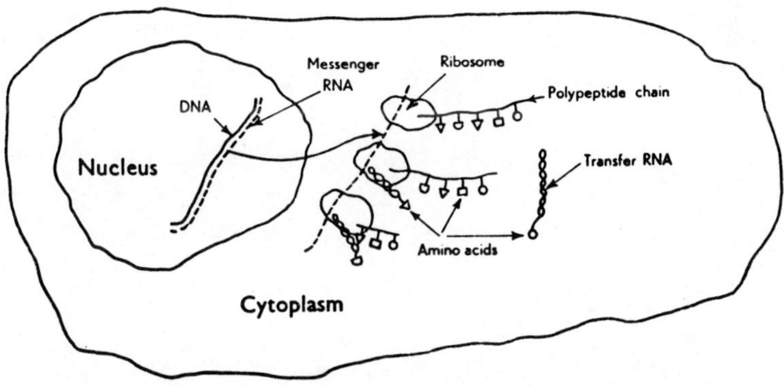

Figure 1-8. Schematic representation of the course of protein synthesis.

the polynucleotide chain of the tRNA molecule. During protein synthesis on the mRNA-ribosome complex *(Figure 1-8)* the tRNA molecules bind in turn to the corresponding codons on the mRNA and so deliver their amino-acids in a defined sequence for incorporation in the protein molecules.

CHAPTER TWO

The Structure of Proteins: Fibrous Proteins

Chemically proteins at first seem to be very simple and this perhaps explains why chemists have often found them too unexciting for study. They are polymers in which rather simple small molecules, the α-amino acids, are joined end to end in long chains. These amino acids have a common general structure *(Figure 2-1)* in

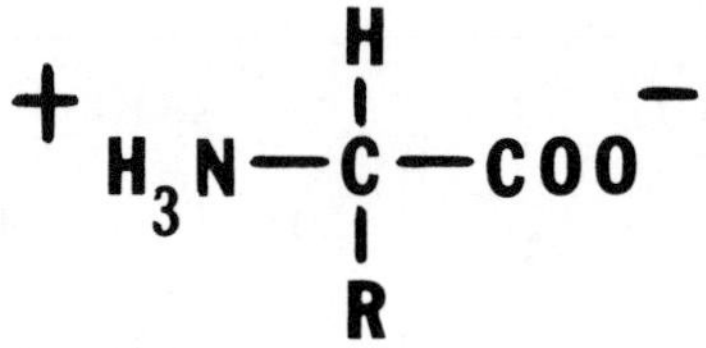

$$^{+}H_3N-\underset{\displaystyle R}{\overset{\displaystyle H}{\overset{|}{\underset{|}{C}}}}-COO^{-}$$

Figure 2-1. The structure of α-amino acids.

which the central α-carbon atom is covalently linked with a hydrogen atom (— H); a carboxyl group (— COO^-); an amino group (— NH_3^+); and one of the twenty different side chains that are set out in detail in Table 2-1. Clearly these side chains fall into a number of different chemical types and very much affect the chemical properties of the amino acids. They include aliphatic, aromatic, polar, acidic and basic groups and the amino-acid proline. Proline deserves special mention because in this molecule the side chain is linked directly to the nitrogen atom as well as to the α-carbon atom forming an imino-acid with a closed ring. As we shall see, the versatility of proteins and their very considerable chemical interest arises from the wide variety of ways in which these different side

TABLE 2-1

1. Glycine (GLY)

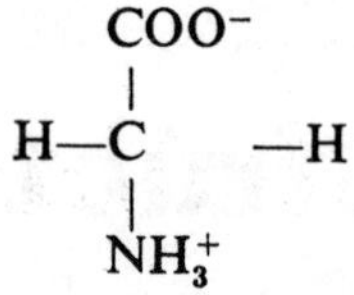

$$\begin{array}{ccc} & COO^- & \\ & | & \\ H— & C & —H \\ & | & \\ & NH_3^+ & \end{array}$$

2. L-alanine (ALA)

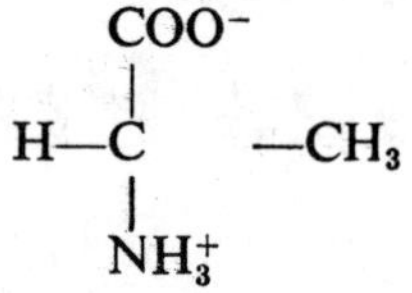

$$\begin{array}{ccc} & COO^- & \\ & | & \\ H— & C & —CH_3 \\ & | & \\ & NH_3^+ & \end{array}$$

3. L-valine (VAL)

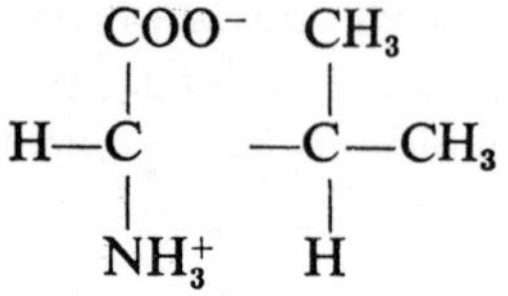

$$\begin{array}{ccccc} & COO^- & & CH_3 & \\ & | & & | & \\ H— & C & — & C & —CH_3 \\ & | & & | & \\ & NH_3^+ & & H & \end{array}$$

4. L-leucine (LEU)

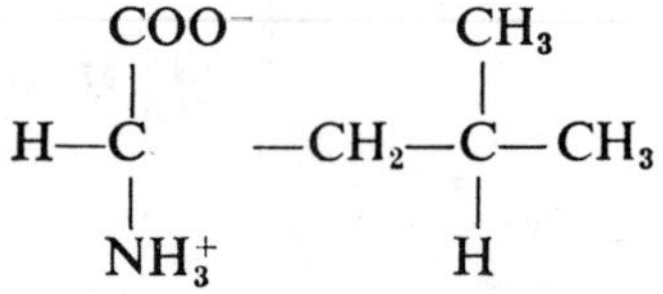

$$\begin{array}{ccccc} & COO^- & & CH_3 & \\ & | & & | & \\ H— & C & —CH_2— & C & —CH_3 \\ & | & & | & \\ & NH_3^+ & & H & \end{array}$$

5. L-isoleucine (ILE)

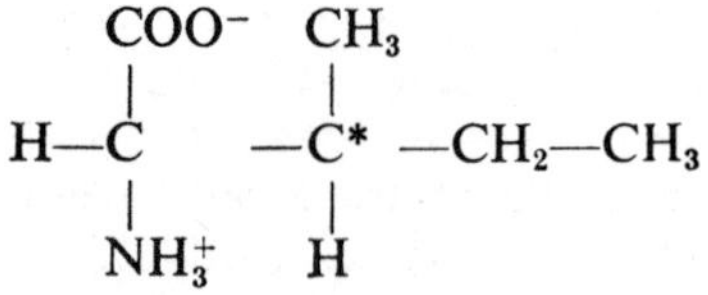

$$\begin{array}{ccccc} & COO^- & & CH_3 & \\ & | & & | & \\ H— & C & — & C^* & —CH_2—CH_3 \\ & | & & | & \\ & NH_3^+ & & H & \end{array}$$

6. L-serine (SER)

$$\begin{array}{ccc} & COO^- & \\ & | & \\ H— & C & —CH_2—OH \\ & | & \\ & NH_3^+ & \end{array}$$

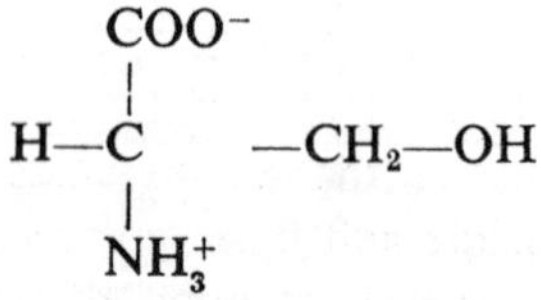

7. L-threonine (THR)

$$\begin{array}{ccccc} & COO^- & & CH_3 & \\ & | & & | & \\ H— & C & — & C^* & —OH \\ & | & & | & \\ & NH_3^+ & & H & \end{array}$$

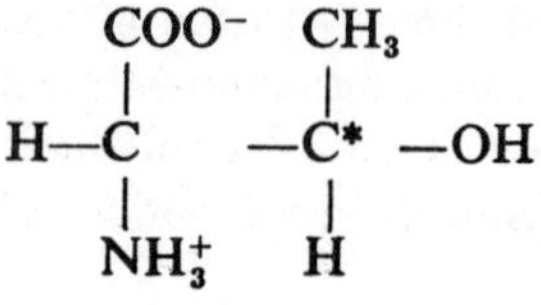

TABLE 2-1

Amino acid	Structure
8. L-cysteine (CYS H)	$H-\overset{COO^-}{\underset{NH_3^+}{C}}-CH_2-SH$
9. L-cystine (CYS)	$H-\overset{COO^-}{\underset{NH_3^+}{C}}-CH_2-S-S-CH_2-\overset{NH_3^+}{\underset{COO^-}{C}}-H$
10. L-methionine (MET)	$H-\overset{COO^-}{\underset{NH_3^+}{C}}-CH_2-CH_2-S-CH_3$
11. L-glutamic acid (GLU)	$H-\overset{COO^-}{\underset{NH_3^+}{C}}-CH_2-CH_2-C(=O)-OH$
12. L-glutamine (GLN)	$H-\overset{COO^-}{\underset{NH_3^+}{C}}-CH_2-CH_2-C(=O)-NH_2$
13. L-aspartic acid (ASP)	$H-\overset{COO^-}{\underset{NH_3^+}{C}}-CH_2-C(=O)-OH$
14. L-asparagine (ASN)	$H-\overset{COO^-}{\underset{NH_3^+}{C}}-CH_2-C(=O)-NH_2$
15. L-lysine (LYS)	$H-\overset{COO^-}{\underset{NH_3^+}{C}}-CH_2-CH_2-CH_2-CH_2-NH_2$
16. L-arginine (ARG)	$H-\overset{COO^-}{\underset{NH_3^+}{C}}-CH_2-CH_2-CH_2-\underset{H}{N}-C(=NH)-NH_2$
17. L-histidine (HIS)	$H-\overset{COO^-}{\underset{NH_3^+}{C}}-CH_2-C$ (ring: $C-NH-CH=N-CH=C$)

TABLE 2-1

Advances in Protein Crystallography

18. L-phenylalanine (PHE)

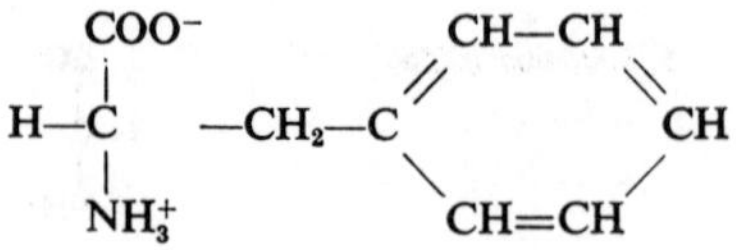

```
      COO⁻              CH—CH
      |                //    \\
   H—C    —CH₂—C            CH
      |                \     /
      NH₃⁺              CH=CH
```

19. L-tyrosine (TYR)

```
      COO⁻              CH—CH
      |                //    \\
   H—C    —CH₂—C            C—OH
      |                \     /
      NH₃⁺              CH=CH
```

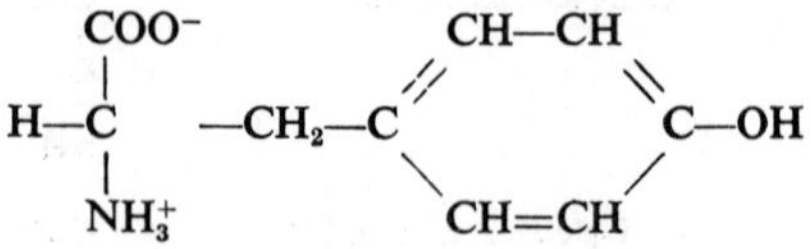

20. L-tryptophan (TRY or TRP)

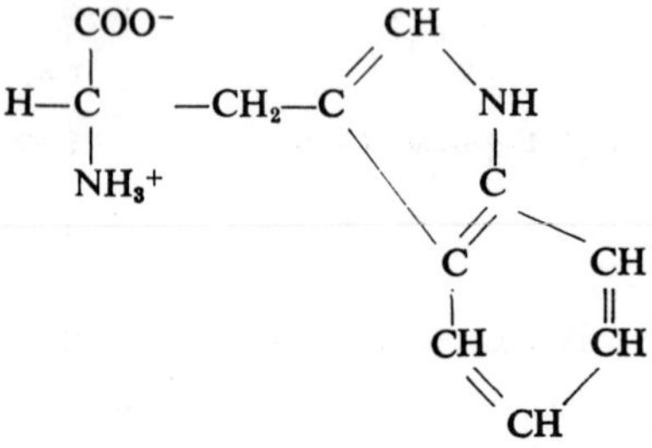

```
      COO⁻              CH
      |                //  \
   H—C    —CH₂—C          NH
      |                \    |
      NH₃⁺              \   C
                         \ // \
                          C     CH
                          |     ||
                          CH    CH
                           \\  /
                            CH
```

21. L-proline (PRO)

```
      COO⁻
      |
   H—C    —CH₂
      |        \\
      NH₂⁺       CH₂
        \       /
          CH₂
```

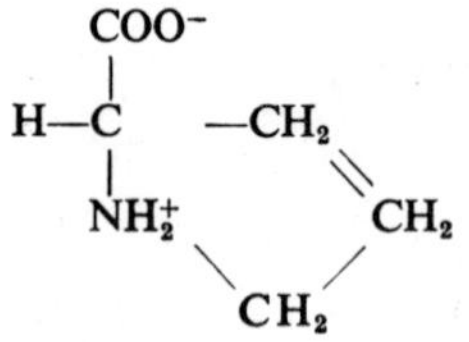

22. L-hydroxyproline (HYP)

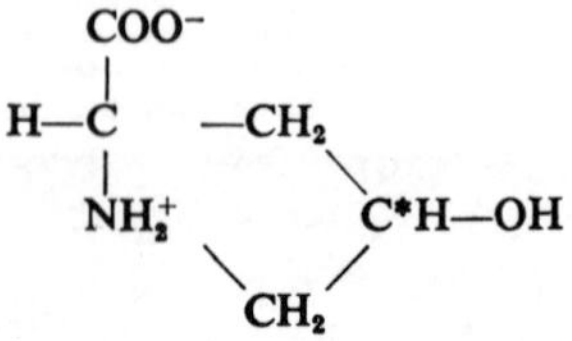

```
      COO⁻
      |
   H—C    —CH₂
      |        \
      NH₂⁺       C*H—OH
        \       /
          CH₂
```

chains can interact with one another and create special environments when they form part of a very large macromolecule.

The configuration of the other atoms around the α-carbon atom is shown in Figure 2-2 (a) which represents a view of the α-carbon atom along the direction of the H — Cα bond. Only L-amino acids are found in naturally occurring proteins and this means that when viewed along the H — Cα bond, the other groups of atoms appear in the order CO → R → N when read clockwise. In D-amino acids, which are found occasionally in nature but not incorporated in proteins, the arrangement of the atoms is the mirror image of this one. Threonine and isoleucine also have asymmetric centres at their β-carbon atoms: their absolute configurations are shown in Figures 2-2 (b) and (c).

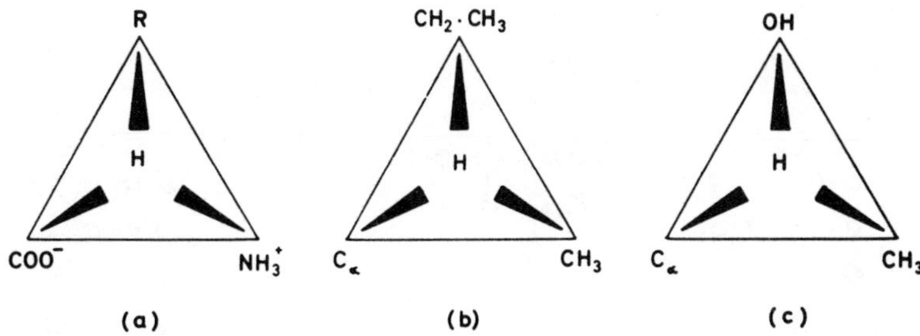

Figure 2-2. The absolute configuration of α-amino acids: (a) when viewed along the H-C_α bond the groups of atoms attached to C_α occur in the clockwise order CO.R.N; (b) the configuration about C_β of Ile; (c) the configuration about C_β of Thr.

The amino acids are joined in long chains by the condensation reaction shown in Figure 2-3 which leads to the formation of a peptide bond joining two amino-acids, which are then called amino-acid residues. The product shown here is simply a dipeptide glycyl-L-alanine or Gly-Ala for short, but true protein molecules consist of much longer polypeptides, perhaps comprising several hundred amino acid residues in a single chain with a free amino group at one end and a free carboxylic acid group at the other. In such a chain the amino acid residues occur in a quite definite order, the amino-acid sequence, as was first shown by Frederick Sanger in his studies of insulin, and as we have seen, we now know in some detail how this order is determined for each protein molecule by the chemical constitution of the nucleic acids from which genes are made. A typical amino sequence is shown in Figure 2-4 which illustrates also the fact that additional covalent connections can be made between the different parts of a polypeptide chain by the formation of disulphide bonds between cysteine residues.

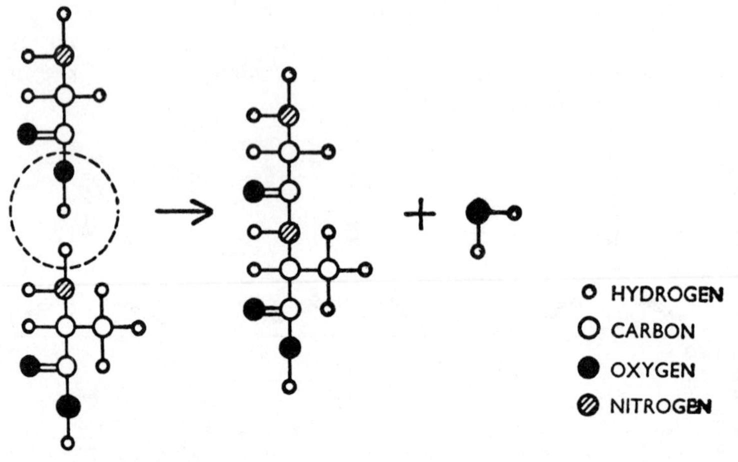

Figure 2-3. Peptide bond formation in Glycyl-L-alanine.

The properties of a protein molecule depend upon how the amino-acid residues are arranged in space. The 3-dimensional structures of these molecules have therefore been studied intensively and a number of their structures have been described in detail. From these studies some of the principles which determine the shapes of the molecules are beginning to emerge. It is clear that the versatility of proteins depends on the fact that polypeptide chains are relatively flexible and can adopt many different shapes or conformations. The conformation that is adopted by any particular molecule depends upon its amino-acid sequence, so that, in principle, it should be possible to predict the conformation of a folded polypeptide chain from its chemical formula. In practice, however, this is not yet possible but a great deal of work is in progress towards this end and some progress is being made. This began with the X-ray crystallographic studies mainly by Linus

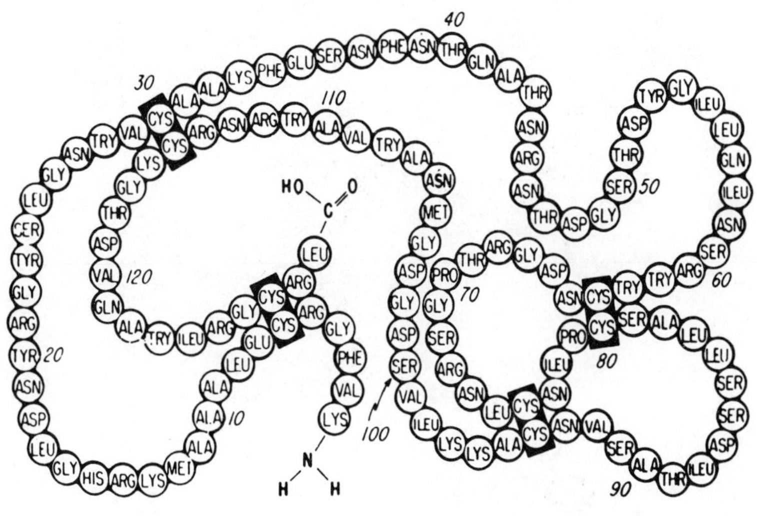

Figure 2-4. The amino-acid sequence of Hen egg-white lysozyme.

Pauling, Robert Corey and their colleagues at the California Institute of Technology, of the 3-dimensional structures of individual amino acids and simple peptides. Results typical of these analyses are shown in Figure 2-5. From a study of the molecular dimensions and interactions revealed in this way, Pauling and Corey were able to suggest that: a) the six atoms in the peptide group linking amino-acid residues tend to lie in a plane because the peptide bond has some double bond character; b) the bond

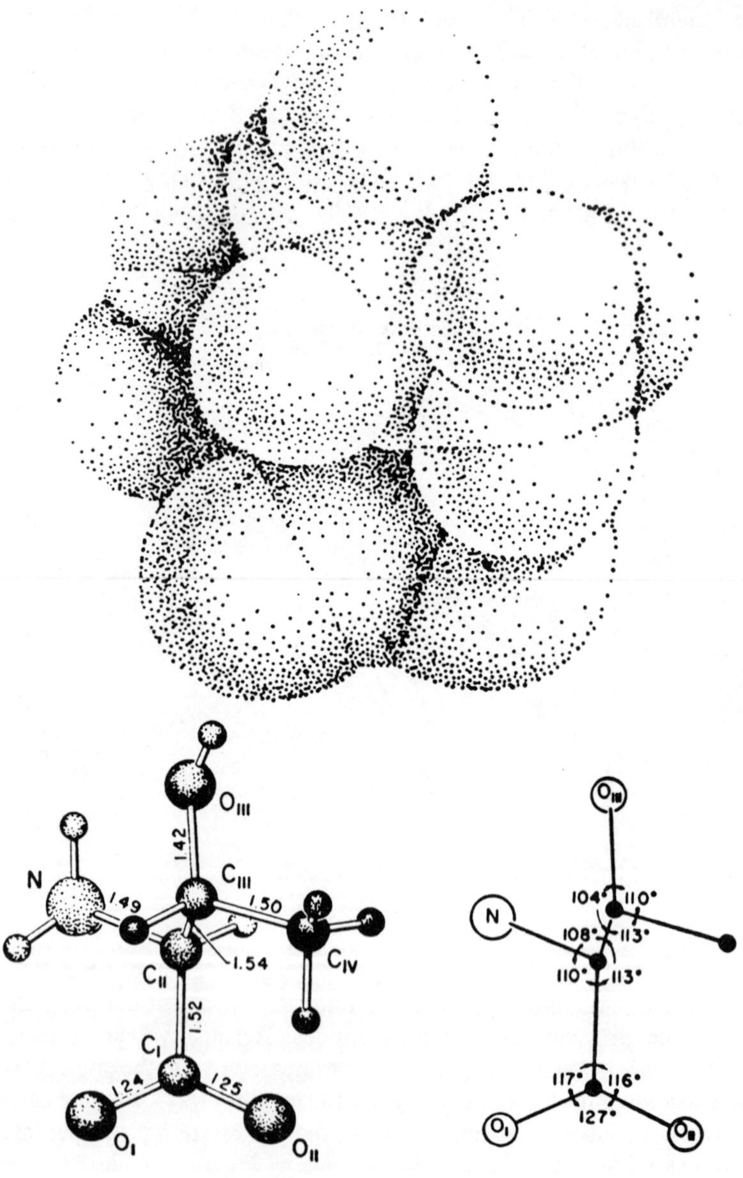

Figure 2-5. The structure of L-threonine.

links and angles in the peptide group have values close to those shown in Figure 2-6 and c) hydrogen bonds can be expected to link the main chain carbonyl oxygen atoms and amino groups of different peptide groups within the same molecule or in neighbouring molecules.

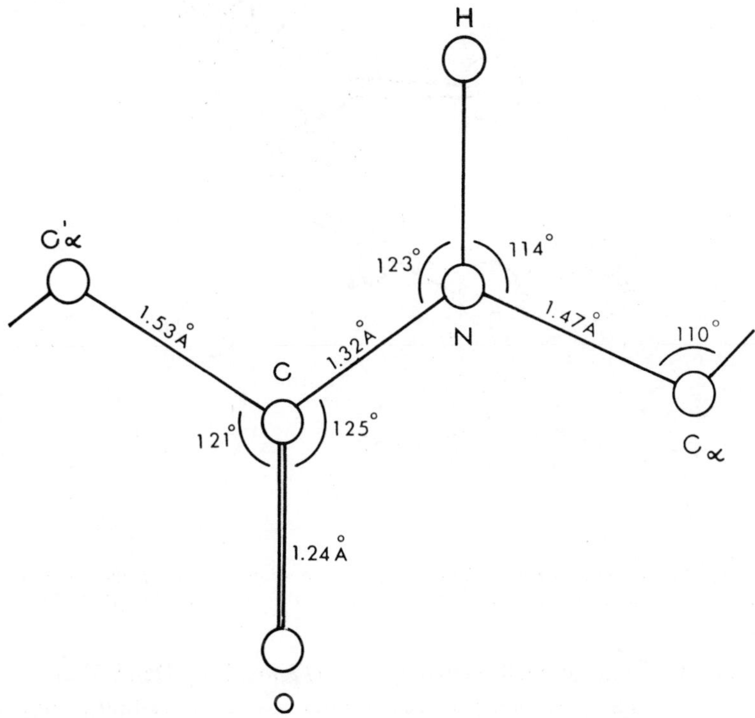

Figure 2-6. Bond lengths and angles in the planar peptide group.

Armed with their conclusions from studies of simple molecules, Pauling and Corey next considered the possible conformations of polypeptide chains that would satisfy their criteria of peptide group planarity and optimum hydrogen bonding. Given a planar peptide group it is clear that the flexibility of a polypeptide chain arises from the possibility of relatively free rotations about N — Cα and

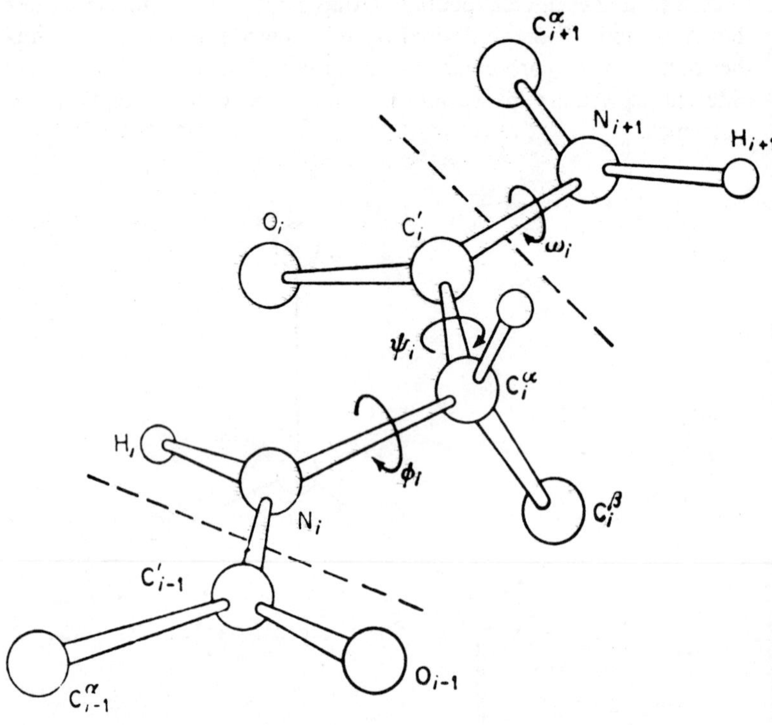

Figure 2-7. Diagram of two planar peptide units showing the variable dihedral angles ϕ and ψ which together define the conformation of a polypeptide chain.

Cα — C′ bonds at each α-carbon atom *(Figure 2-7)*. Detailed analysis shows that these are not free rotations, since at certain settings of the angles some atoms are brought impossibly close together but a wide enough range of angles is accessible to allow a variety of different conformations at each α-carbon atom. Ramachandran has shown how this information can be summarised neatly in a diagram *(Figure 2-8)*.

Pauling and Corey were particularly concerned however with repeating structures in which the conformation of which each carbon atom is the same and they examined all such structures in a search for those in which good hydrogen bonds are made. Two

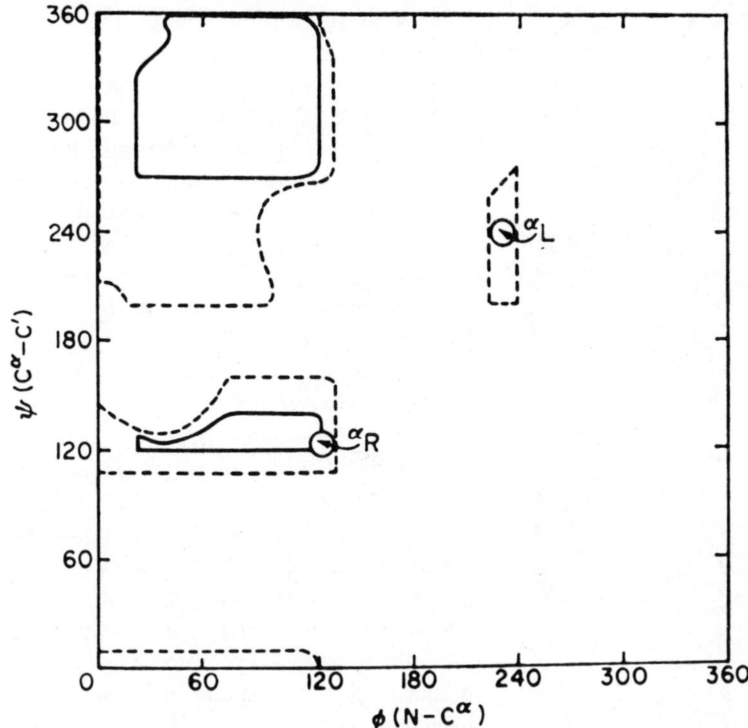

Figure 2-8. Diagram showing the accessible values of ϕ and ψ for an alanyl residue according to the calculations of Ramachandran and his colleagues who treated the atoms as hard-spheres.

general schemes of hydrogen bonding were considered: (a) one in which the bonds are between residues that are relatively close together in the same polypeptide chain; (b) those in which the hydrogen bonds are between residues in different polypeptide chains or widely separated in the same chain. Two standard conformations of fundamental importance emerged.

The α-Helix

With particular values of the torsion angles at each α-carbon atom a right-handed structure is formed in which hydrogen bonds of optimum length are made between the imino group of amino acid residues n and the carbonyl group of n-4, that is four residues nearer to the amino end of the chain. This is the famous right-handed α-helix *(Figure 2-9)* which careful analysis has shown to be favoured not only by the hydrogen bonding pattern but also by the advantageous close-packing of all the atoms in van der Waal's contact. The related left handed helix is less stable because of unfavourable close contacts between the carbonyl oxygen atoms and the first (β) carbon atoms in the side chains. Similar helices with hydrogen bonding between different pairs of residues have been considered but calculations show that the right-handed α-helix is much more stable than any of the others that are sterically possible and experiment has shown that this conformation is common in proteins of all classes whereas only minor examples of the others have been observed. In particular the right-handed α-helix is the principal conformation in the fibrous protein α-keratin of which skin and hair are largely composed.

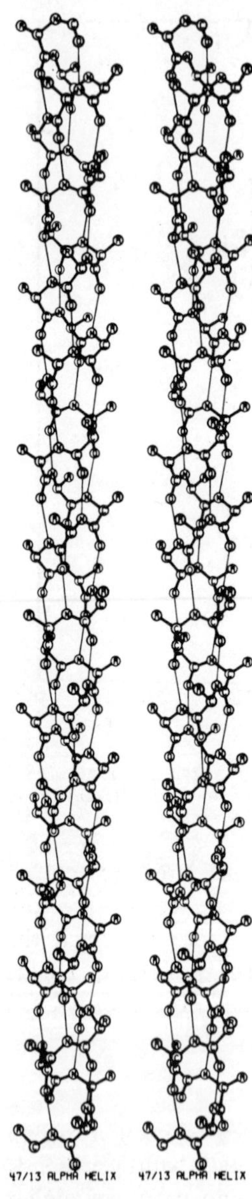

Figure 2-9. Stereodiagram of the α-helix.

Before considering Pauling and Corey's second standard conformation we should note in passing that the nature of the side chains also affects the stability of the α-helix. Clearly proline residues cannot be involved in α-helices since they have no hydrogen bond donor to contribute. More subtly some side chains, such as valine and isoleucine, which branch at the β-carbon atom, are rather too bulky to fit comfortably on the surface of an α-helix. Furthermore, side chains which include the polar hydroxyl

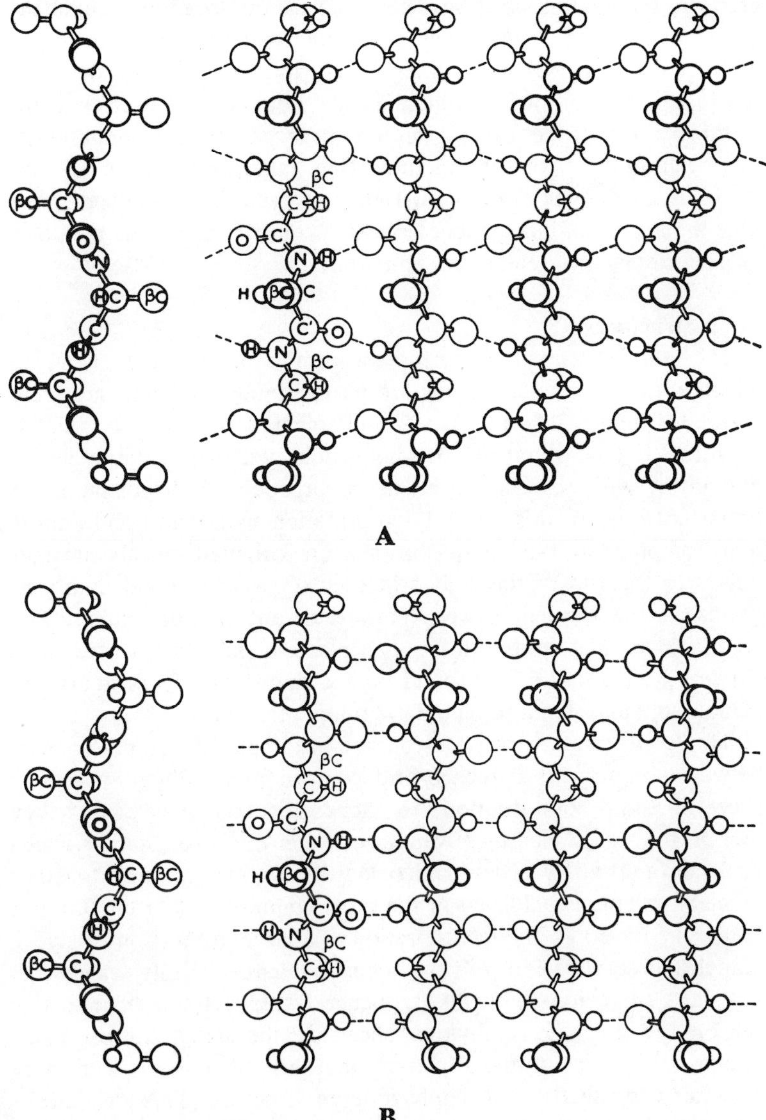

Figure 2-10. (a) The parallel and (b) the anti-parallel pleated sheet (β) structures.

group in the gamma position such as serine and threonine can adopt conformations in which they tend to weaken the main chain hydrogen bonds. On the other hand, the side chains of some residues, for example, alanine, leucine, glutamic acid pack comfortably around the α-helix and tend to shield the main chain hydrogen bonds from interactions that might weaken or disrupt them. Here we see for the first time, therefore, an indication of how the actual amino sequence might favour or inhibit the adoption of a particular molecular conformation.

(b) **β-Structures**

The fully extended conformation of the polypeptide chain, in which adjacent peptide groups are parallel and co-planar, is barely allowed *(Figure 2-8)* but not found, whereas the closely related conformation in which the peptide groups are tilted alternately up and down with respect to the main direction of the chain is an important one. In this pleated conformation hydrogen bond donors and acceptors of the peptide groups are oriented closely at right angles to the run of the chains in such a way that good hydrogen bonds can be formed between parallel or anti-parallel chains. The resultant parallel and anti-parallel pleated sheet or β structures *(Figure 2-10 a and b)* are found in silk among fibrous proteins and as components of many globular protein structures.

This conformation provides another example of the relationship between amino-acid sequence and conformation. When successive sheets in the β conformation are packed together in layers, as they are in silk, it is clearly advantageous for the side chains (which protrude alternately above and below the sheets) to pack together efficiently. This is achieved in the most common silk from *Bombyx Mori* (the moth) by the incorporation of a large proportion of glycyl, alanyl and seryl residues in the repeating sequence (gly-ser-gly-ala-gly-ala)$_n$. In regions with this sequence the glycyl side chains stick out on one side of an individual sheet and the alanyl and seryl side chains stick out on the other so that adjacent sheets can pack together very neatly in a highly ordered structure. This structure is disrupted by the inclusion of any more bulky side chains which lead therefore to the presence of some relatively disordered regions in the mainly well ordered and inextensible fibres of silk.

Collagen

Although its essential structure was not worked out in quite the same way by systematic conformational analysis it is appropriate here to mention collagen, the main constituent of tendons and other tissues in which high tensile strength and rigidity are required. The basic structure of this fibrous protein is a triple helix in which three polypeptide chains, each in a rather extended helical conformation involving no intra chain hydrogen bonds, are coiled together in a three-stranded rope-like structure with hydrogen bonds between residues in the different strands. This structure, which was worked out by Rich and Crick with contributions from several other workers, is closely related to that adopted by polyproline and polyglycine and it is strongly promoted by amino-acid sequences of the type $(gly\text{-}x\text{-}pro)_n$ in which x represents any residue type.

Chemical analysis of collagen shows that about one third of the amino-acid residues are glycine and about a quarter are proline on the slightly modified hydroxyproline; again there is a clear connection between the amino-acid constitution of the protein and the conformation which it adopts.

CHAPTER THREE

The Structure of Proteins: Globular Proteins and Enzymes

The conformations of globular proteins are inherently more complicated than the basic conformations of the fibrous proteins, as might be expected from their much less repetitive amino-acid sequences. Nevertheless α-helices and β-pleated sheets are found in the structures of these molecules and this helps to simplify their description.

Globins

The first globular protein structure to be worked out in detail was that of myoglobin from the muscle of the sperm-whale. Myoglobin is closely related to haemoglobin, the substance which transports oxygen from the lungs to the tissues and is responsible for the red colour of blood. The function of myoglobin is to store oxygen in the muscle tissues until it is required for use. In his classic studies, Kendrew chose to work on sperm-whale myoglobin because diving mammals of this sort need a large local store of oxygen and their muscles are therefore rich in myoglobin which can be extracted relatively easily.

In addition to a polypeptide chain of 153 amino-acid residues the myoglobin molecule includes a planar group of atoms with an iron atom at the centre which is called the haem group. Many globular protein molecules include *prosthetic* groups such as this which are usually very closely concerned with the activity of the molecule. It is no surprise, therefore, that the oxygen molecule stored by a myoglobin molecule binds to the haem-iron atom. The protein part of the molecule provides an environment in which this binding takes place reversibly in a way that is well adapted to the physiological needs of muscle tissue and quite different from the interaction between oxygen and free haem in solution.

Kendrew's crystallographic structure analysis showed that most of the polypeptide chain in myoglobin is in a α-helical conformation having very closely the characteristics predicted by Pauling and Corey. In order to create a nest for the haem group, however, the molecule comprises eight helical regions connected by lengths of chain in less regular conformations and these helices are packed together in a way that seems, at first sight at least, to be rather irregular *(Figure 3-1)*. The whole molecule illustrates well the

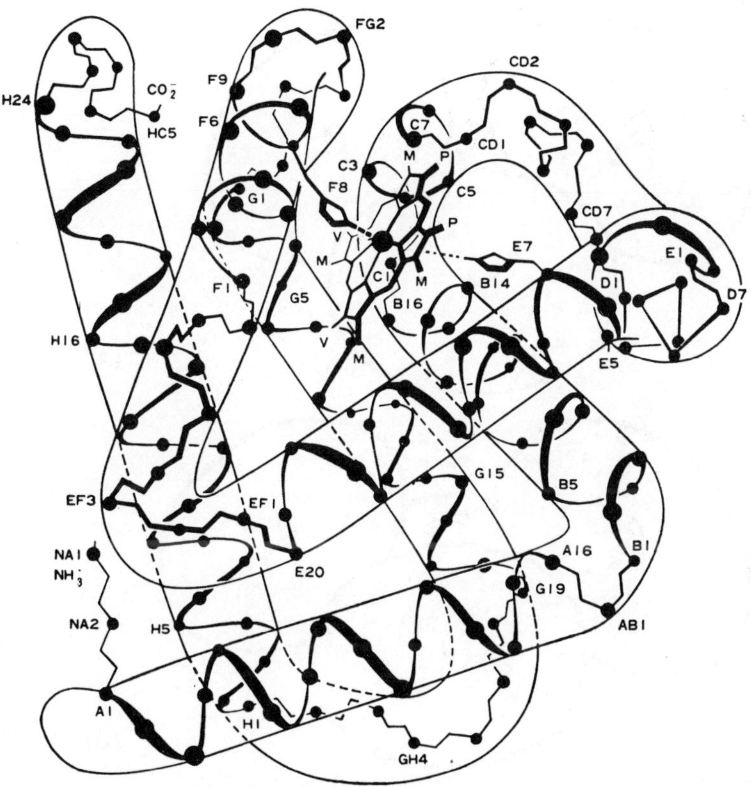

Figure 3-1. The conformation of sperm-whale myoglobin.

terminology introduced by Linderstrom-Lang in which the amino-acid sequence is called the *primary structure,* the local conformation of the polypeptide chain (here predominantly α-helical) is called the *secondary structure,* and the general folding of the coiled chain is called the *tertiary structure.* For larger molecules composed of more than one chain there is a higher level of organisation again, the *quaternary structure* which refers to the way in which folded chains or sub-units pack together to form larger aggregates and this, as it happens, is well illustrated by haemoglobin, the oxygen-transport protein. Perutz has shown, again by crystallographic analysis, that this molecule is made of four sub-units which are

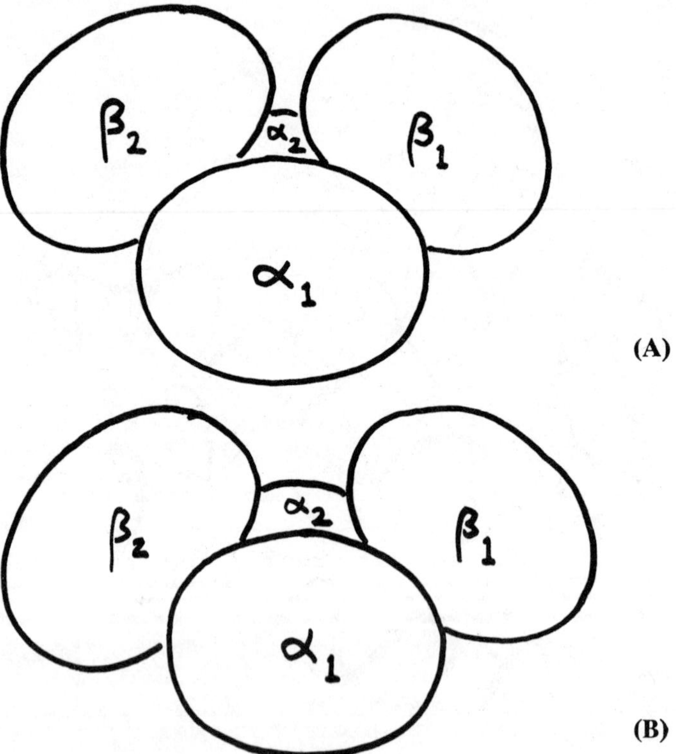

Figure 3-2. The arrangement of myoglobin-like sub-units in (a) oxy-haemoglobin and (b) deoxy-haemoglobin.

identical in pairs and closely similar to individual myoglobin molecules. Their arrangement is illustrated in Figure 3-2. Haemoglobin has more subtle properties than myoglobin and these arise from the rearrangement of the subunits as the molecule takes up oxygen: in a way that is not yet quite understood the uptake of one molecule of oxygen facilitates the binding of a second molecule, and so on, so that the haemoglobin molecules tend to be fully loaded with oxygen or oxygen-free. The general difference between the oxygenated and deoxygenated forms of the molecule are shown in Figure 2 but further details are not included here.

Enzymes

Similar analyses have revealed the structures of several more proteins—nearly all of them enzymes, the molecules which control individual chemical reactions by acting as catalysts. The first of these to be analysed in detail was lysozyme, an antibacterial agent discovered by Alexander Fleming. This molecule acts by breaking a chemical bond in the bacterial cell walls and its structure and probable mode of action were worked out at the Royal Institution in 1965. At this time we are concerned only with protein structure and the structure of hen egg white lysozyme is shown in Figure 3-3 which may be examined in comparison with the amino-acid sequence of Figure 2-4.

The three-dimensional structure is less regular than that of myoglobin but part of the chain is again in the α-helical conformation and there is also a small and somewhat irregular region of anti-parallel pleated sheet. Although the secondary structure appears to be rather irregular the conformations at each α-carbon atom lie within the allowed regions of the Ramachandran diagram *(Figure 2-8)* and rather more of the points are grouped near the angles required for the generation of α-helices than would be expected from the extent of the recognisable helices. This situation arises, of course, because helices are formed only when four or more residues in succession are in the appropriate conformation.

Similar analyses of other enzyme molecules have provided further examples of complex conformations in which the α- and β-structures play important parts. Just as in the examples of myoglobin and lysozyme, the balance between the standard secondary structures

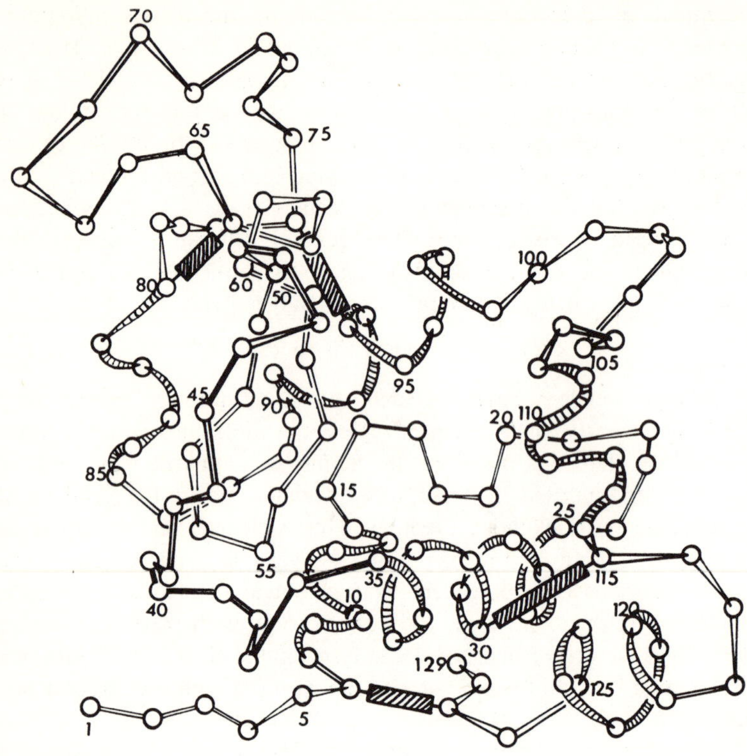

Figure 3-3. The conformation of the main polypeptide chain in hen egg-white lysozyme.

can vary quite dramatically. Thus, in the structure of α-chymotrypsin, worked out by Blow and his colleagues, there is very little α-helix but a large proportion of β-structure *(Figure 3-4)*. This enzyme is representative of a group of closely related enzymes that degrade proteins during digestion by hydrolysing certain peptide bonds. In carboxypeptidase, the function of which is again to degrade proteins this time by removing residues from the carboxyl ends of the polypeptide chains, Lipscomb has shown that there is a fascinating combination of α- and β-structure *(Figure 3-5)* with an extensive pleated sheet of eight parallel or anti-

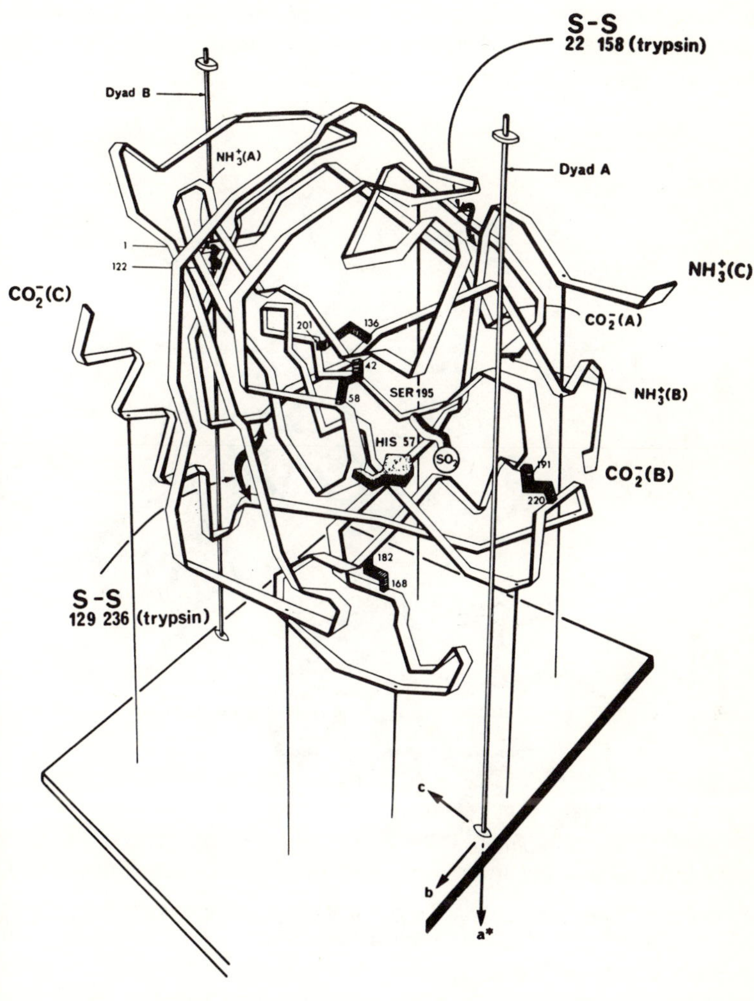

Figure 3-4. The conformation of the main polypeptide chain in α-chymotrypsin.

parallel chains running through the centre of the molecule and eight helices forming a cage-like structure around it.

These are enough examples to show the general complexity of globular proteins and before we enquire more deeply into their properties we can ask whether the ones we have considered so far have any general feature in common. The answer to this question

Figure 3-5. The conformation of the main polypeptide chain in carboxypeptidase A.

is that indeed they do; inspection of these structures shows at once that there is a common general pattern in the way in which the side chains of the amino-acid residues are distributed. The insides of these molecules, that is the regions that are shielded from contact with surrounding liquid by other parts of the molecule, are generally filled with non-polar side chains such as the aliphatic side-chains of leucine and isoleucine, or the aromatic side chains of phenylalanine and tryptophan. The polar side chains, and more particularly the ionisable side chains of residues such as aspartic acid or lysine, are distributed generally on the surfaces of the molecules where they can interact freely with the surrounding medium. For these proteins, which function in solution, the normal environment is provided by solution in water and this segregation of the side chains can be summarised by the statement that hydrophobic side chains tend to be buried and hydrophilic side chains tend to be exposed. There are exceptions: non-polar groups are found on the surfaces of these molecules and some polar—even ionizable—groups are buried, but this statement is a reasonable summary if interpreted cautiously and it brings out the importance of the interactions between protein molecules and their environments in determining their conformation.

The folding of globular proteins

There is compelling evidence that the overall three-dimensional structures of globular protein molecules are determined by the amino-acid sequences of these molecules and, as we shall see, their interactions with the surrounding liquid. Under certain conditions, such as high concentrations of urea, globular proteins are unfolded in solution and lose their biological activity. In very many examples it has been found, however, that the biological activity is recovered (implying that the molecules are re-folded to their active conformations) when the normal environment is restored. For this and for other reasons it seems clear that proteins fold to their active conformations during or after their synthesis on the ribosomes and that, in general, this is accomplished without the help of other molecules acting as patterns, moulds or templates.

Thermodynamic considerations suggest at first sight that a uniquely folded conformation of a protein molecule is unlikely to be found since, in comparison with the wide variety of unfolded states that are accessible to the molecule, it represents a state of very low entropy. The thermodynamic function that we have to consider is the Gibbs free energy

$$G = H - T.S$$

and in going from the unfolded to the folded state at constant temperature we have to suppose that the overall free energy change

$$\triangle G = \triangle H - T.\triangle S$$

is a decrease ($\triangle G$ negative) despite this large negative change in the entropy that results from the ordering of the protein conformation.

Following Kauzman in 1959 many workers have considered this problem (notably Klotz and Tanford) but a detailed quantitative understanding of the situation has not yet been achieved. Nevertheless in broad outline the explanation must be that the liquid surrounding the protein plays a critical part in the folding process. In assuming the overall free-energy change account must be taken of:

(a) the contributions to the enthalpy change arising from the favourable contacts that are made between non-bonded atoms in the protein in the folded state;

(b) the contributions to the enthalpy change arising from changes in the liquid structure and its interaction with the protein;

(c) the contributions to the entropy change due to the ordering of the protein conformation; and

(d) most particularly, the contribution to the entropy change arising from any changes in the order of the liquid molecules around the unfolded and folded protein molecules.

The explanation that is currently most favoured is that when a protein molecule is unfolded in solution, with side chains of all types exposed to the surrounding liquid, the molecules of the solvent (which we suppose to be essentially water in a living cell) are ordered to some considerable extent in cage-like structures around non-polar side chains such as leucine and phenylalanine.

When the protein molecule is folded, most of these non-polar side chains are buried in the middle of the molecule where they are shielded from contact with the surrounding water. Consequently, the liquid structure is less well ordered and this makes a large positive contribution to the entropy of folding. In other words, because of the more random distribution of solvent molecules when the protein is folded, there are more states accessible to the whole system with the protein folded and there is a higher probability, therefore, of a folded state being found.

The same considerations must, of course, apply to the folding of nucleic acids, including the formation of hydrogen-bonded pairs of bases and this raises an interesting question about the importance of hydrogen bonds. Why are hydrogen bonds considered important in determining the conformations of nucleic acids and proteins when the hydrogen bonding groups involved presumably make similar bonds with water molecules in the unfolded state? The answer appears to be that there is indeed a much smaller contribution from hydrogen bond formation to the enthalpy of folding than would be expected in the absence of water and that the important point is that potential hydrogen bond donors or acceptors should not be shielded from the water unless they are placed in an environment in which they can make appropriate hydrogen bonds, otherwise the enthalpy of folding would be unfavourably positive. When these conditions are satisfied hydrogen bonds are, of course, important in maintaining specific contacts in and between functional molecules because of their well defined properties of length and direction.

Species variations and evolution

In view of these rather general considerations underlying the folding of protein molecules and the uniqueness of the active conformations that are produced, it is surprising to find that proteins of the same kind from different species often differ very markedly in chemical structure while retaining essentially identical three-dimensional structures. For example, human lysozyme has very closely the same activity and three-dimensional structure as hen egg-white lysozyme but the amino-acid sequences of the two molecules differ in 53 of the 129 amino-acid residues *(Figure 3-6)*. In many of these changes the general character of the amino-acid

	1	2	3	4	5	6	7	8	9	10	11	12	13	14	15	16	17	18	19	20
HLL	LYS	VAL	PHE	GLU	ARG	CYS	GLU	LEU	ALA	ARG	THR	LEU	LYS	ARG	LEU	GLY	MET	[illegible]	GLY	TYR
HEL	LYS	VAL	PHE	GLY	ARG	CYS	GLU	LEU	ALA	ALA	ALA	MET	LYS	ARG	HIS	GLY	LEU	[illegible]	ASN	TYR
BAL	GLU	GLN	LEU	THR	LYS	CYS	GLU	VAL	PHE	ARG	GLU	LEU	LYS	...	...	ASP	LEU	LYS	GLY	TYR

	21	22	23	24	25	26	27	28	29	30	31	32	33	34	35		36	37	38	39
HLL	ARG	GLY	ILE	SER	LEU	ALA	ASN	TRP	MET	CYS	LEU	[illegible]	LYS	TRP	GLU	...	SER	GLY	TYR	ASN
HEL	ARG	GLY	TYR	SER	LEU	GLY	ASN	TRP	VAL	CYS	ALA	[illegible]	LYS	PHE	GLU	...	SER	ASN	PHE	ASN
BAL	GLY	GLY	VAL	SER	LEU	PRO	GLU	TRP	VAL	CYS	THR	THR	...	PHE	HIS	THR	SER	GLY	TYR	ASP

	40	41	42	43	44	45	46	47	48	49	50	51	52	53	54	55	56	57	58	59
HLL	THR	ARG	ALA	THR	ASN	TYR	ASN	ALA	GLY	ASP	ARG	SER	THR	ASP	TYR	GLY	ILE	PHE	GLN	ILE
HEL	THR	GLN	ALA	THR	ASN	ARG	ASN	THR	...	ASP	GLY	SER	THR	ASP	TYR	GLY	ILE	LEU	GLN	ILE
BAL	THR	GLU	ALA	ILE	VAL	GLU	ASN	...	...	ASN	GLN	SER	THR	ASP	TYR	GLY	LEU	PHE	GLN	ILE

	60	61	62	63	64	65	66	67	68	69	70	71	72	73	74	75	76	77	78	79
HLL	ASN	SER	ARG	TYR	TRP	CYS	ASN	ASP	GLY	LYS	THR	PRO	GLY	ALA	VAL	ASN	ALA	CYS	HIS	LEU
HEL	ASN	SER	ARG	TRP	TRP	CYS	ASN	ASP	GLY	ARG	THR	PRO	GLY	SER	ARG	ASN	LEU	CYS	ASN	ILE
BAL	ASN	ASN	LYS	ILE	TRP	CYS	LYS	ASN	ASP	GLN	ASP	PRO	HIS	SER	SER	ASN	ILE	CYS	ASN	ILE

	80	81	82	83	84	85	86	87	88	89	90	91	92	93	94	95	96	97	98	99
HLL	SER	CYS	SER	ALA	LEU	LEU	GLN	ASP	ASN	ILE	ALA	ASP	ALA	VAL	ALA	CYS	ALA	LYS	ARG	VAL
HEL	PRO	CYS	SER	ALA	LEU	LEU	SER	SER	ASP	ILE	THR	ALA	SER	VAL	ASN	CYS	ALA	LYS	LYS	[illegible]
BAL	SER	CYS	ASP	LYS	PHE	LEU	ASN	ASN	ASP	LEU	THR	ASN	ASN	ILE	MET	CYS	VAL	LYS	LYS	[illegible]

	100		101	102	103	104	105	106	107	108	109	110	111	112	113	114	115	116	117	118
HLL	ARG	...	ASP	PRO	GLN	GLY	ILE	ARG	ALA	TRP	VAL	ALA	TRP	ARG	ASN	ARG	CYS	GLN	ASN	ARG
HEL	VAL	SER	ASP	GLY	ASP	GLY	MET	ASN	ALA	TRP	VAL	ALA	TRP	ARG	ASN	ARG	CYS	LYS	GLY	THR
BAL	LEU	...	ASP	LYS	VAL	GLY	ILE	ASN	TYR	TRP	LEU	ALA	HIS	LYS	ALA	LEU	CYS	SER	GLU	LYS

	119	120	121	122	123	124	125	126	127	128		129
HLL	[illegible]	[illegible]	ARG	GLN	TYR	VAL	GLN	GLY	CYS	GLY	...	VAL
HEL	[illegible]	[illegible]	[illegible]	ALA	TRP	ILE	ARG	GLY	CYS	ARG	...	LEU
BAL	LEU	ASP	[illegible]	...	TRP	LEU	...	...	CYS	LYS	GLU	LEU

Figure 3-6. Amino-acid sequences in human leukemic lysozyme (HLL); hen egg-white lysozyme (HEL); and bovine α-lactalbumin (BAL).

residue is conserved so that, for example, a non-polar leucine (residue 12) in the human enzyme is replaced by methionine in the chicken enzyme. Furthermore, reference to the genetic code *(Figure 1-6)* shows that this change corresponds to a possibly minimal change in the nucleotide sequence coding for these amino-acid residues from CUG for leucine to AUG for methionine. Many examples could be produced of which this is true and the observation provides us with a clue about a possible evolutionary mechanism. We can suppose that from time to time mistakes may be made in copying the DNA in which one nucleotide is replaced by another. This will lead to a change in the amino-acid sequence of some protein, of the kind we have just examined, and provided that the overall change proves advantageous in some way the further generations that carry it will eventually outnumber those that do not.

A more complicated evolutionary mechanism is also suggested by Figure 6 which shows that the amino-acid sequence of the milk protein bovine α-lactalbumin is closely related to that of lysozyme even though these two molecules have different functions. This is supposed to have come about in the evolution of mammals by a process that began with gene-doubling: in a species producing lysozyme the DNA sequence coding for this enzyme was copied twice for some accidental reason during DNA replication. In later generations the second gene is then supposed to have changed, by the point-by-point mechanism just described, until a molecule with a useful new activity was produced. This process is likely to have been concerned in the evolution of many protein molecules, including haemoglobin *(Figure 3-2)* which is composed of sub-units of two kinds each of them probably derived from a common myoglobin-like ancestor.

CHAPTER FOUR

The Activity of Enzymes

For more than seventy years our general picture of enzyme action has been based on Fischer's famous lock and key hypothesis *(Figure 4-1)*. According to this one supposes that an enzyme has a structure that is complementary to that of its substrate, the molecule upon which it acts, so that the two fit together as a key fits into a lock. When the enzyme-substrate complex has been formed chemical reactions take place, which are promoted by the interaction between the enzyme and substrate, products are formed which diffuse away from the enzyme and the enzyme is left in its initial state ready for another round of catalytic activity. This scheme, analysed in terms of rate constants and concentrations by Henri, Michaelis and Menten and others, has provided a satisfactory basis for kinetic studies of enzyme catalysed reactions and now, thanks to X-ray structural studies, we can begin to see in detail the kinds of atomic interactions that are involved.

It is convenient to consider enzyme activity in two parts, dealing first with the specific interactions involved in the formation of enzyme-substrate complexes and secondly with the factors responsible for affecting the rate of the catalysed reaction. As we shall see, these two topics cannot really be separated completely but specificity is a general topic of importance in biology (for example, in the activity of antibodies towards antigens in the immune response) and we can consider later what part the formation of the enzyme-substrate complex plays in catalysis.

Lysozyme specificity

Most of the examples here are taken from our studies of lysozyme

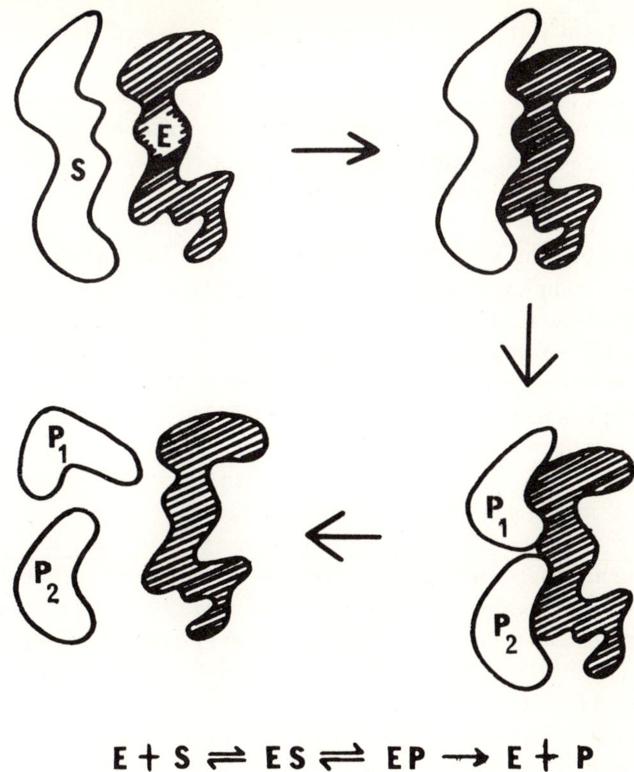

Figure 4-1. The interactions of enzymes with substrates and inhibitors.

at the Royal Institution and more recently at Oxford since they are well advanced and illustrate a number of general principles. One of these concerns the phenomenon of comparative inhibition of enzymes by molecules that are structurally similar to the substrate but are unaffected chemically by their interaction with the enzyme. This is shown schematically in Figure 4-1 where we see that such an inhibitor molecule, by binding to the enzyme in a part of its substrate binding site, presents the formation of an active enzyme-substrate complex and so inhibits the activity of the enzyme. Such a competitive inhibitor competes with the substrate for a binding site on the enzyme forming an inactive enzyme-inhibitor complex which may be quite stable and susceptible to detailed study.

This phenomenon of competitive inhibition has been exploited in our studies of lysozyme which have enabled us to build up a plausible picture of the nature of the enzyme-substrate complex involved in the activity of this enzyme.

Lysozyme was discovered by Alexander Fleming in 1921 in an experiment that was strangely similar to the one in which he discovered penicillin some six years later. He had a cold and for some reason it occurred to him to test the effect of placing some drops of his nasal mucus upon a bacterial culture. Imagine his delight when within a few hours the bacteria in the vicinity of the mucus were all dissolved. Quickly he tested the effect of other body fluids and secretions and he soon found that tears, sputum, plasma and many others all possessed this property of promoting the dissolution of bacteria.

All of these fluids were found to contain a substance which Fleming called lysozyme and for a time hopes were high that a generally useful anti-bacterial agent had been discovered. Stimulated by the methods used by Fleming to collect specimens there was even some fanciful discussion of commercial production *(Figure 4-2)* but the hopes were short lived. As Fleming himself said "Unfortunately the bacteria most strongly affected by lysozyme were bacteria which did not cause disease in animals so however interesting it was from the theoretical point of view its use in practical medicine was limited."

Happily lysozyme has proved very interesting and important in the development of our understanding of enzyme action since, as

Figure 4-2. Cartoon by Dowd based on Fleming's collection of tears from laboratory assistants for use in the preparation of lysozyme. (Reproduced by kind permission of Lady Fleming.)

we have noted previously, it was the first enzyme to have its structure and mechanism of activity described in detail as the result of chemical and X-ray crystallographic studies.

Investigations by Fleming himself and by Meyer, Chain, Salton and others have shown that the bacterial cell walls dissolved under the influence of lysozyme are composed in part of long chains of sugar molecules joined together by β (1-4) glycosidic linkages, the same links as those joining glucose units in the structure of cellulose. Two different but related amino-sugars, N-acetylglucosamine (NAG) and N-acetylmuramic acid (NAM), are found in the bacterial cell wall polysaccharide and they alternate as shown in Figure 4-3. Lysozyme acts by promoting the cleavage of bonds between N-acetylmuramic acid and N-acetylglucosamine (as shown in the Figure) and not, it is important to note, those between NAG and NAM.

Despite this specificity for particular glycosidic linkages in the cell wall, lysozyme is found to hydrolyse chitin, the simpler polymer in which all the sugar residues are N-acetylglucosamine and it was studies of chitin hydrolysis that provided the basis for one crystallographic work. Wenzel and others had shown that the mono-, di- and tri-saccharides derived from chitin, N-acetylglucosamine, di-N-acetylchitobiose and tri-N-acetylchitotriose, act as competitive

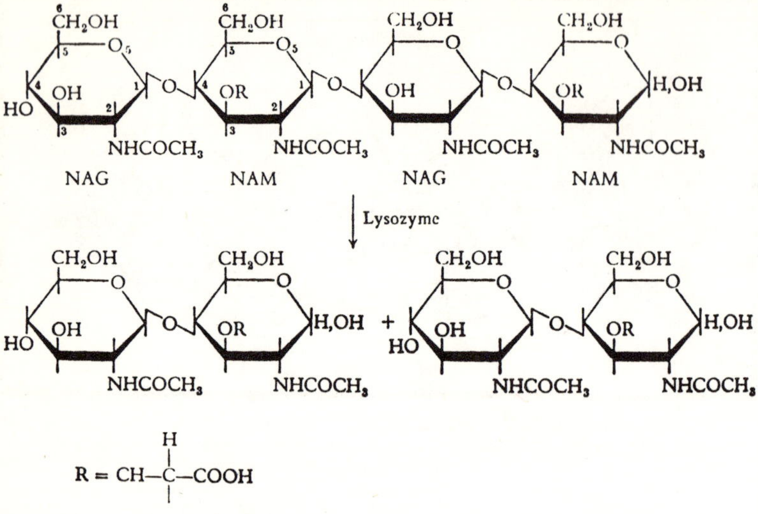

NAG NAM NAG NAM

Lysozyme

$$R = CH{-}\overset{H}{\underset{|}{\overset{|}{C}}}{-}COOH$$

Figure 4-3. The reaction catalysed by lysozyme.

inhibitors of the enzyme and we were able to study crystallographically the structures of these enzyme-inhibitor-complexes.

Molecules of N-acetylglucosamine change their configurations or mutarotate in solution to produce an equilibrium mixture of two anomeric α- and β-forms and molecules in these two configurations bind to lysozyme in different but related ways. The binding of α-NAG, in which the hydroxyl group bound to the C_1 carbon is in an axial position, is indicated in Figure 4-4 which for simplicity shows only the active site cleft in the enzyme molecule. The most specific interactions are hydrogen bonds between the NH and carbonyl oxygen of the acetamido side group of the sugar and the CO and NH groups of the main polypeptide chain belonging to amino-acid residues 107 and 59 respectively. A further hydrogen bond appears to be formed between the α-hydroxyl and another main-chain NH at residue 109. Particular emphasis is placed upon these hydrogen bond interactions because they are easy to pick out and contribute importantly to the specificity of the interaction. We must not forget, however, that in this example and generally there are many more non-polar interactions which have not been described

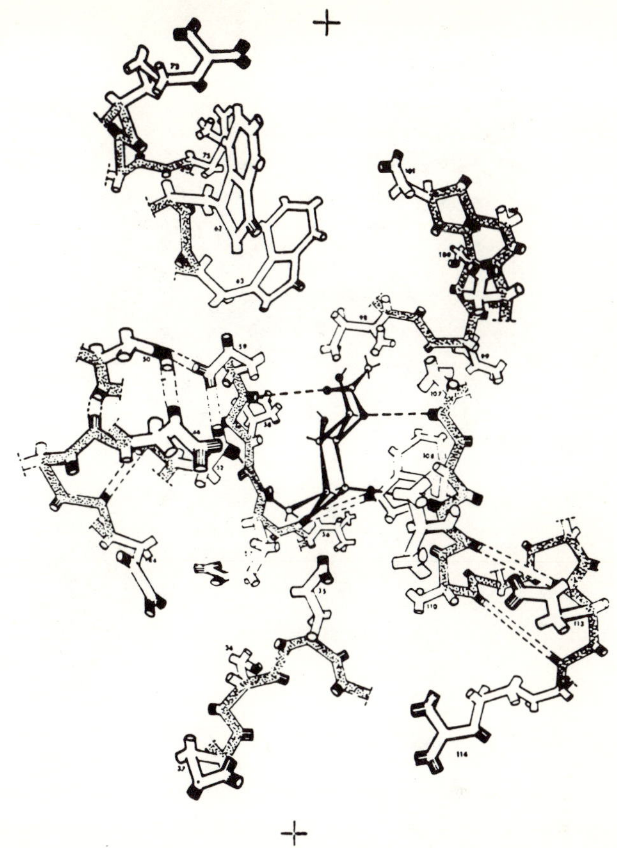

Figure 4-4. Atomic arrangement in the lysozyme molecule in the neighbourhood of the cleft where inhibitors are bound. The main polypeptide chain is shown speckled and N(H) and O atoms are indicated by line and full shading respectively. The superimposed line drawing shows the binding of α-NAG with hydrogen bonds indicated by broken lines. Carbon atoms in the inhibitor are represented by open circles and oxygen atoms by full circles. (Diagrams of this kind by Mrs. W. J. Browne.)

but which contribute essentially to the binding. The importance of the solvent interactions must also be remembered since the changes in the solvent structure play an important part in determining the nature and strength of the inhibitor binding just as they affect the folding of the protein molecule to its native conformation.

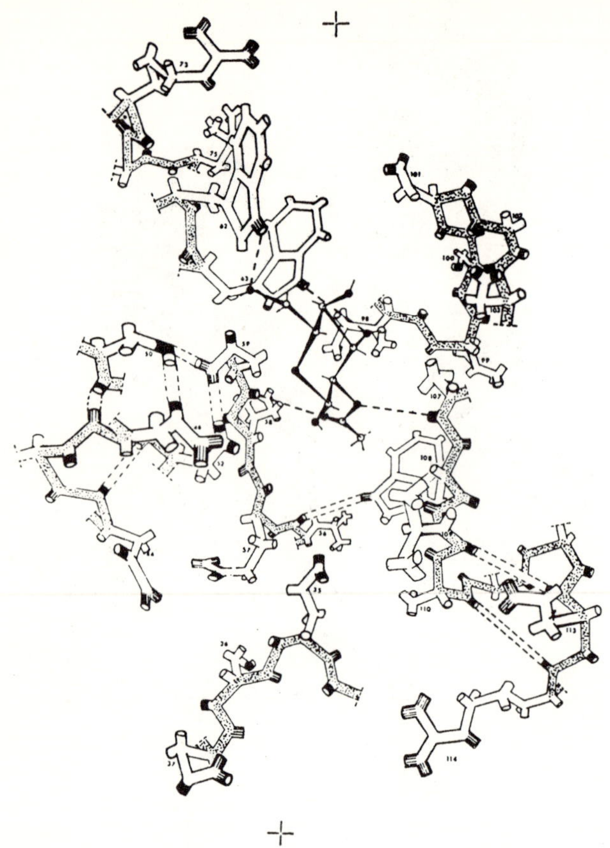

Figure 4-5. Binding to lysozyme of β-N-acetylglucosamine.

The related binding of β-N-acetylglucosamine is shown in Figure 4-5. Here the hydrogen bonds between the acetamido side group of the inhibitor and the enzyme molecule remain the same as those found in the binding of α-NAG but the sugar molecule is in effect rotated about the line of these hydrogen bonds so that it lies higher in the cleft and makes different contacts with the enzyme. Hydrogen bonds appear now to be formed between the 6- and 3-hydroxyls and the side chains of tryptophan residues numbers 62 and 63.

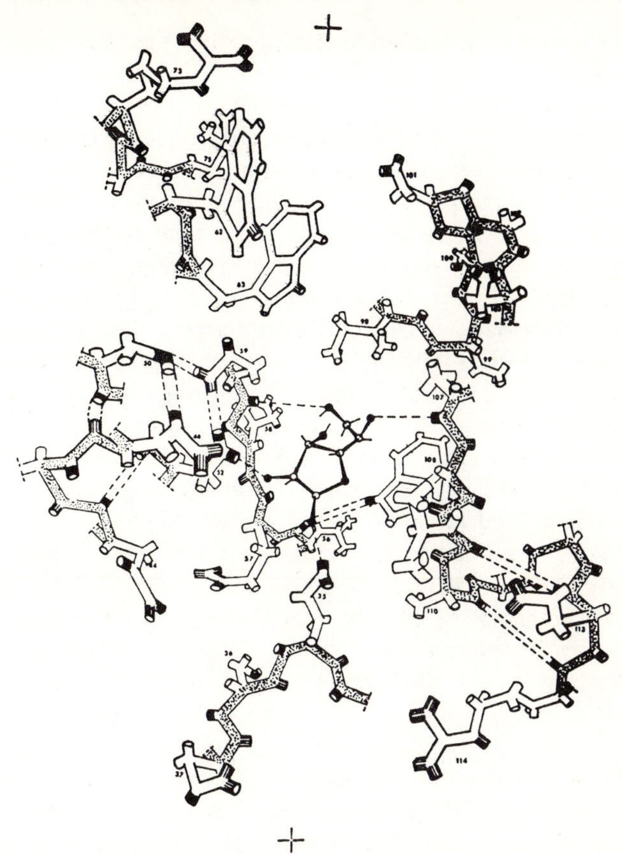

Figure 4-6. Binding to lysozyme of glucono-(1-4)-lactose.

This result underlines the important point that quite minor changes in the structure of an inhibitor may affect quite seriously the way in which it binds to an enzyme. It also suggests that lysozyme has a binding site that is specific for acetamido groups but we must not conclude from this that the enzyme cannot bind inhibitors with different structures even in this same site. Figure 4-6 shows that glucono-(1-4) lactose also binds to lysozyme in this position with its diol side chain making the same hydrogen bonds as the acetamido side chain of N-acetylglucosamine. Without detailed structural information this probably would not have been suspected.

Clearly the interactions between lysozyme and these small inhibitor molecules are less critically specific than might have been anticipated but, we might argue, this is because the lysozyme substrate is a large polymeric molecule and we should expect the specificity to increase with the size of the inhibitor molecule. This view is supported by our study of the binding of tri-NAG, shown in the top half of Figure 4-7, which occupies the site filled by β-NAG and two additional sugar binding sites above it. Various additional polar and non-polar interactions are found associated with these sites and the tri-saccharide forms a quite strong, specific and stable complex with the enzyme. Only weak interactions, hydrogen bonds, van der Waal's interactions and so on rather than covalent bonds, are involved and we can suppose that this lysozyme-tri-NAG complex is a good model for many stable complexes of biological importance. Antibody-antigen complexes for sugar antigens may well have similar structures.

Enzymes are not at all concerned, however, with making stable complexes with substrate molecules so that we must look beyond the structure of this lysozyme-tri NAG complex if we wish to examine a true enzyme-substrate complex. In doing so we may recall the suggestion of Pauling who pointed out in 1946 that enzyme-substrate complexes must satisfy some particular requirements. He wrote: "I think that enzymes are complementary in structure to the activated complexes of the reactions that they catalyse, that is to the molecular configuration that is intermediate between the reacting substances and the products of reaction for these catalysed processes. The attraction of the enzyme molecule for the activated complex would thus lead to a decrease in its energy and hence to a decrease in the energy of activation of the reaction and to an increase in the rate of the reaction."

The mechanism of action of lysozyme

Examination of the structure of the stable lysozyme-triNAG complex shows that it involves only the top half of the cleft in the enzyme surface and it seemed natural, therefore, to explore the possibility that additional sugar residues occupy the bottom half as

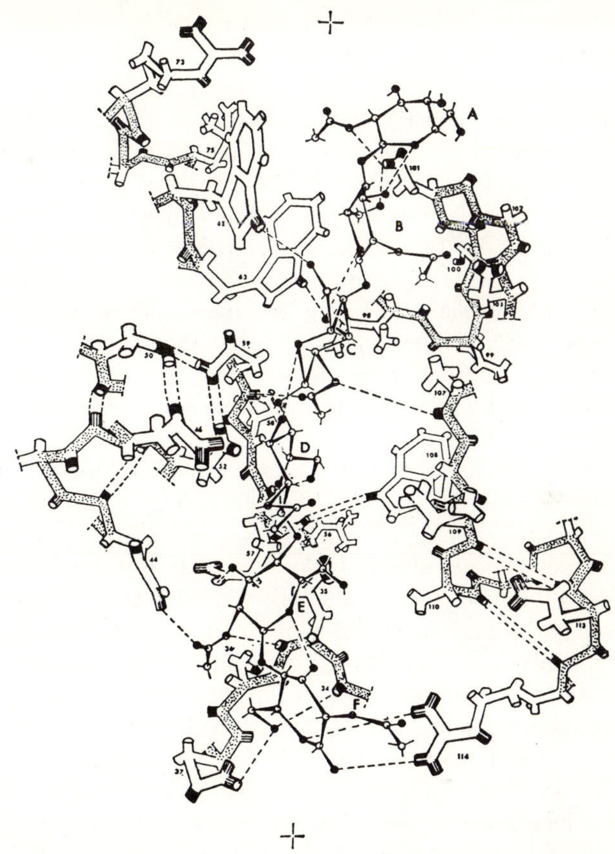

Figure 4-7. Proposed binding to lysozyme of hexa-N-acetyl-chitohexaose. Sites A, B, and C are as observed in the binding of tri-NAG. Sites D, E and F were inferred originally from model building.

well in the formation of the true enzyme substrate complex. At this stage there was no experimental evidence for the additional interactions but a stereochemical study, using scale models, showed straightaway that three additional sugar residues could be fitted into the cleft with very little difficulty. A number of satisfactory interactions between these sugar residues and the enzyme suggested themselves and the only problem was that the sugar residue immediately below the trisaccharide made some impossibly close contacts with the enzyme. These could be relieved by changing the conformation of this sugar from the normal chair towards the half-chair conformation and this, as we shall see, probably has functional significance in line with Pauling's proposal.

The structure of the lysozyme-substrate complex, according to this model, is shown in Figure 4-7. The enzyme is supposed to bind a hexasaccharide substrate in six sugar binding sites labelled A, B, C, D, E, F, the residue in site D being distorted. Our next problem was to locate the site of catalytic activity and this proved readily possible on the assumption that the binding of the cell-wall substrate is closely similar to that proposed for the chitin substrate. In the cell wall alternate sugar residues are N-acetylmuramic acid with an additional lactyl side chain in the 3-position. Which of the six binding sites for sugar residues could accommodate an NAM residue? Fortunately the answer is clearcut: in site C the 3-hydroxyl is pointing into the enzyme and involved in a hydrogen bond with the tryptophan 63 side chain. There is no room there for an additional lactyl side chain. It follows that, if the structure of Figure 4-7 represents also the binding of a cell-wall hexasaccharide to lysozyme, residues B, D and F must be N-acetylmuramic acid, with their lactyl side chains exposed to the surroundings, and residues A, C and E must be N-acetylglucosamine.

The glycosidic linkage that is cleaved in the cell wall is between NAM (contributing carbon-1) and NAG (contributing carbon-4) so that there are two candidates for the catalytic site. It is between either residues B and C or residues D and E. But the B-C linkage is part of the stable tri-saccharide complex and we are led, therefore, to consider only the vicinity of the D-E linkage in our search for catalytically active groups.

Near this linkage there are two carboxylic acid side chains; glutamic acid 35 has one of its oxygen atoms about 3.0 Å from the oxygen of the glycosidic linkage and aspartic acid 52 has one of its oxygen atoms rather less than 3.0 Å from the carbon-1 and oxygen-5 atoms of the distorted sugar residue in site D. Furthermore, these carboxyl side chains are in different environments so that we might expect glutamic acid 35 in a non-polar environment, to be protonated at pH 5 (at which the enzyme is active) while aspartic acid 52, in a polar environment is likely to be ionized.

These features suggest a catalytic mechanism according to which three factors are supposed to contribute to the hydrolysis: (a) glutamic acid 35 acts as a general acid catalyst transferring its proton to the glycosidic oxygen and hence promoting the breaking of the $C_1 — O$ bond with the formation of a positively charged carbonium ion at carbon-1 of sugar residue D; (b) aspartic acid 52 with its negative charge stabilizes this carbonium ion; and (c) this carbonium ion is further stabilized and its formation promoted by the binding of sugar D to the enzyme in the half-chair conformation.

This last factor (which is in accordance with Pauling's proposal) requires a little elaboration. Following the work of Lemieux and Huber, sugar chemists have long accepted that a sugar ring incorporating a carbonium ion at carbon-1 takes up the half-chair conformation because the positive charge is shared with the ring oxygen and the bond between carbon-1 and oxygen-5 acquires some double-bond character. Hence we suppose that the binding of this sugar residue to the enzyme in the half-chair conformation promotes the redistribution of electronic changes that takes place in the hydrolysis reaction.

The further course of the reaction is quickly described. On the cleavage of the glycosidic linkage between sugar residues D and E the sugars occupying sites E and F are supposed to diffuse away from the enzyme leaving a tetrasaccharide bound to the enzyme in an intermediate structure incorporating a carbonium ion at carbon-1 of residue D. The reaction now proceeds in reverse with water (or another sugar molecule) approaching site E, losing a proton to glutamic acid 35 (which acts in the ionized form as a general base catalyst) and neutralizing the carbonium ion. The tetrasaccharide then diffuses away, leaving the enzyme in its initial state ready for further cycles of operation.

Conclusion

This account of the probable mechanism of action of lysozyme, supported as it is by a great body of independent evidence, is intended to give some impression of the kinds of factors involved in the activities of enzymes generally. Lysozyme is a relatively simple enzyme and many more complex systems, particularly those concerned in the overall control of the metabolism, have already been characterised though their structures and modes of action are not yet known in detail. A quantitative understanding of the various catalytic factors affecting the enhancement of reaction rates still elude us but there is little reason to suppose that radically new principles remain to be discovered. It is more likely that further analyses of enzyme structures will contribute to the deeper understanding of reaction mechanisms generally and may lead eventually to the design of catalysts and other reagents of medicinal and industrial importance.

CHAPTER FIVE

Self-assembly and the Genesis of Higher Levels of Organisation

We have now seen something, albeit in very brief outline, of the structures and properties of individual nucleic acid and protein molecules and we have noted two things in particular, first that such molecules tend to adopt specific three dimensional structures in accordance with their chemical constitutions and the nature of the surrounding medium and, second, that evolution involves discrete and relatively small changes in these fundamental molecules. We must now consider the higher levels of organisation in living systems in order to discover the extent to which we understand how the properties of these molecules determine in turn the progressively more complex structures of organelles, cells, organs and whole organisms.

Protein folding and aggregation

Protein and nucleic acid molecules and the other constituent molecules of living systems, such as carbohydrates and lipids, which are synthesised under the control of enzymes can all be assumed to take up conformations that depend upon their chemistry and their environment. As we have seen, water may play a crucial part in this process but we must not lose sight of the fact that the environment in which such molecules are active may not always be aqueous within a cell. We might well find, for example, that the proteins involved in the structure of membranes have a somewhat different general nature than those that have been studied hitherto in relatively free solution since they may adopt conformations more suited to a non-polar lipid environment. The situation is probably closely analogous to the formation of soap micelles in aqueous and non-polar solvents *(Figure 5-1)* and such micelles do indeed provide good general models for the structures of these polyfunctional macromolecules.

Similar ideas clearly apply to the self-assembly of more complicated protein molecules. Single polypeptide chains rarely have molecular weights greater than about 100,000 and most of the

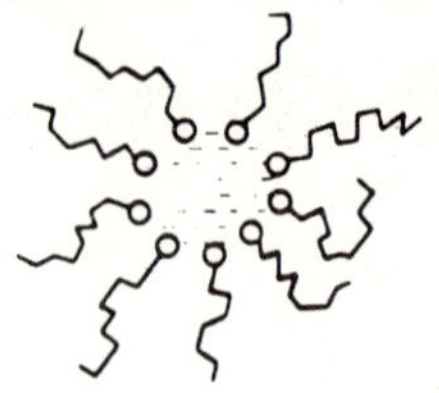

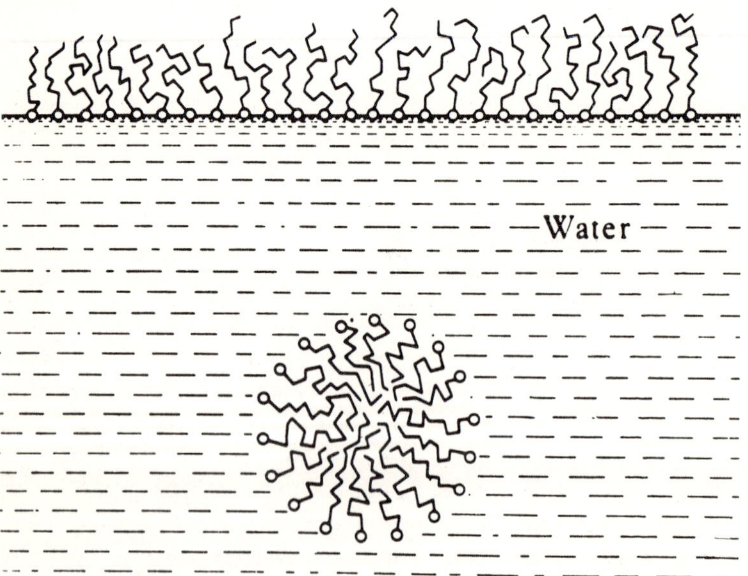

Figure 5-1. The formation of soap micelles in water and non-polar solvents.

larger protein molecules are oligomeric being composed of a number of sub-units within this size limitation. Haemoglobin *(Figure 3-2)* provides an example of such a molecule, being composed of four chains that are identical in pairs and each closely related to an individual myoglobin molecule. The forces holding these four sub-units together are all weak ones of the kind responsible for maintaining the folded structures of the individual chains, that is, they arise from the same polar and van de Waal's interactions and the entropic effects that we have already discussed. Clearly there is an element of specificity in the interactions, which depends to some extent again on the directional properties of hydrogen bonds, so that a closely defined aggregate structure is produced that can be disrupted or have its properties changed by critical changes in the amino-acid sequences of the constituent polypeptide chains. Incidentally it should be noted that Perutz has indeed made many studies of naturally occurring abnormal haemoglobins which have been altered in this way by chance mutations and that this work has led to the understanding of certain anaemias at the molecular level. These are among the first clinically important fruits of molecular biology and they have demonstrated the fundamentally molecular nature of some pathological conditions which could be remedied only by genetic surgery of a kind that is still beyond our capabilities.

Even larger molecules and molecular systems can be supposed to assemble themselves in the same way as haemoglobin. There are many enzymes for example, which have overall molecular weights in the range 100,000 to 1,000,000 and which are composed of various members of subunits. Many of them have complex properties which arise, as do the properties of haemoglobin, from their subunit structure and which depend upon the relative arrangements of the sub-units changing during their activity under the influence of various effector molecules in the environment which act as control agents. It may well be that this is broadly the mechanism of action of hormones and other chemical messengers.

Viruses and cell organelles

Self assembly and aggregation have been studied most intensively in work on viruses and this brings us to the consideration of functional units the size of cell organelles, such as the ribosomes on which proteins are synthesised. Like ribosomes viruses are composed partly of protein sub-units and partly of nucleic acids. There are two main types, the rod-shaped viruses, such as tobacco mosaic virus, and the spherical viruses like polio virus. Both kinds have been studied intensively by X-ray diffraction and electron microscopy and a detailed picture of their structures is emerging.

Tobacco mosaic virus was discovered by Imanowski in 1892 but it was not until 1931 that the virus was shown to be a solid particle rather than a poisonous fluid. A few years later, after extracting virus from a ton of diseased tobacco plants, Stanley showed that the particles are mainly protein and in 1937 Bowden and Pirie showed that they contain 5% RNA. Structural studies were also begun in 1937 by Bernal and their continuation by Rosalind Franklin, Klug and Holmes have provided us with the structure shown in Figure 5-2. Here we see that the protein sub-units form a helical structure which protects the helical RNA from contact with the surrounding medium which is likely to contain nucleases, enzymes which degrade nucleic acids. Electron microscope evidence shows that the TMV particles are all the same length in a native preparation but that the similar particles which can be made from protein alone have ranging lengths. This appears to be an example of one molecule acting as a template for the aggregation of another and it is a mechanism that we shall encounter again.

The self assembly of spherical and icosahedral viruses has been considered in great detail by Caspar and Klug who have shown how identical protein particles can aggregate to form icosahedral surfaces in such a way that each individual particle has very nearly the same interactions with its neighbours as every other. They were inspired in this work by the architect Buckminster Fuller whose domes are built on the same principles by dividing the surface into triangles which are as nearly equivalent to one another as possible. Over a full sphere this leads to icosahedral symmetry. A completed sphere might be subdivided into 720 triangles. If we imagine (with a change in scale!) that three protein subunits are arranged on each triangle we should have 2160 units in all having very similar but not strictly identical local environments. Caspar and Klug have shown, however, that these slight variations in environment require

Figure 5-2. The structure of tobacco mosaic virus. The loaf-like shapes represent protein sub-units and the helical coil within them is the RNA.

only tolerable variations in the lengths and angles of the interactions between the sub-units so that the principles of self assembly are readily applicable to aggregates of this complexity.

The nucleic acid of spherical viruses is confined within the protein shell but in ribosomes, which are generally similar in shape, the nucleic acid protrudes outside the shell where it is available for interactions with the tRNA and other molecules involved in protein synthesis. These particles are composed of a number of different kinds of protein sub-unit but again the work of Nomura and others shows that they can be assembled quite readily from their components. Thus, the ideas of self assembly, with provision under some circumstances for the enzymically controlled covalent coupling of sub-units, appear to explain the genesis of organelles adequately in general terms and we can turn to consider the factors that determine the overall shapes of cells.

The shapes of cells and organs

Single-celled organisms such as bacteria are bounded by a substantial cell wall which effectively determines their shape and which grows from within the organism during cell division and growth. As we have seen bacterial cell walls susceptible to lysozyme hydrolysis are composed partly of polysaccharide chains. In the complete cell wall these are linked together by short peptide chains (incorporating some D-amino acids) to form what Wiedel has called bag-shaped macromolecules.

Animal cells, on the other hand, are bounded from the chemical or physico-chemical point of view by a very thin layer of lipoprotein forming the plasma membrane. All cells have this but it is very thin indeed (of the order of 100 Angstroms) and can contribute very little to mechanical rigidity. John Pringle has described how the shapes of such cells are determined in two ways:

1) "by the production within the cell of substances which form aggregates there of definite shape which then affect the shape of the cell as a whole;" and
2) "by the production of substances which are secreted by the cells so that they form extracellular layers around them, constraining their surfaces. If these substances are produced evenly all around a cell and can withstand tension then a positive hydrostatic pressure inside the cell can maintain the shape (as with a rubber balloon). If the production of substance is unequal on different sides or if the external intracellular substances change their properties by crystallization or fibre formation after secretion, then the resulting

shape may become more spherical, like balloons that have different thicknesses of rubber in different parts."

"Both methods are used and, in addition, if the cells stick together because of the adhesive properties of the extracellular substance, organs and tissues result."

Many examples could be given of ways in which cell shapes are determined by means of the structures within them but I will mention only two. The first of these brings us back again to haemoglobin and the nature of genetic mutations. Perhaps the best known abnormal haemoglobin is the one known as sickle-cell haemoglobin because the red-blood cells of individuals carrying this mutant tend to adopt a sickle shape when the haemoglobin within them is in the deoxy-state *(Figure 5-3)*. It was in detailed studies of this haemoglobin that Ingram was able to show for the first time that a mutation leads to the replacement of a single amino-acid residue by another one. He demonstrated that in this example glutamic acid at position 6 in the β-chain of normal haemoglobin is replaced by valine in sickle-cell haemoglobin. How this leads to the sickling of the red-blood cells is not fully understood but Murayama has suggested plausibly that, as a result of this amino-acid substitution, deoxy-haemoglobin molecules aggregate to form long chain molecules which line up in such a way as to affect the external shape of the cells.

The second example concerns hair which is composed of cells that have differentiated so as to synthesize the protein keratin in large amounts. Intracellular fibres of keratin, in the α-helical conformation, ultimately fill and kill the cells which pack together to form the familiar fibres. As we have seen, the polypeptide chains are held in the helices by hydrogen bonds and these are cross-linked by di-sulphide bridges (of the sort seen in lysozyme). One consequence of this is that hair dissolves in strong reducing agents which break these disulphide bridges—a phenomenon that is exploited in permanent waving.

The shapes of cells are determined by extracellular tension in chemically different ways in different organisms. In most animals collagen is the most important extracellular fibrous element and it is produced within the cells in the form of tropocollagen molecules which are held together in the triple helices that we have encoun-

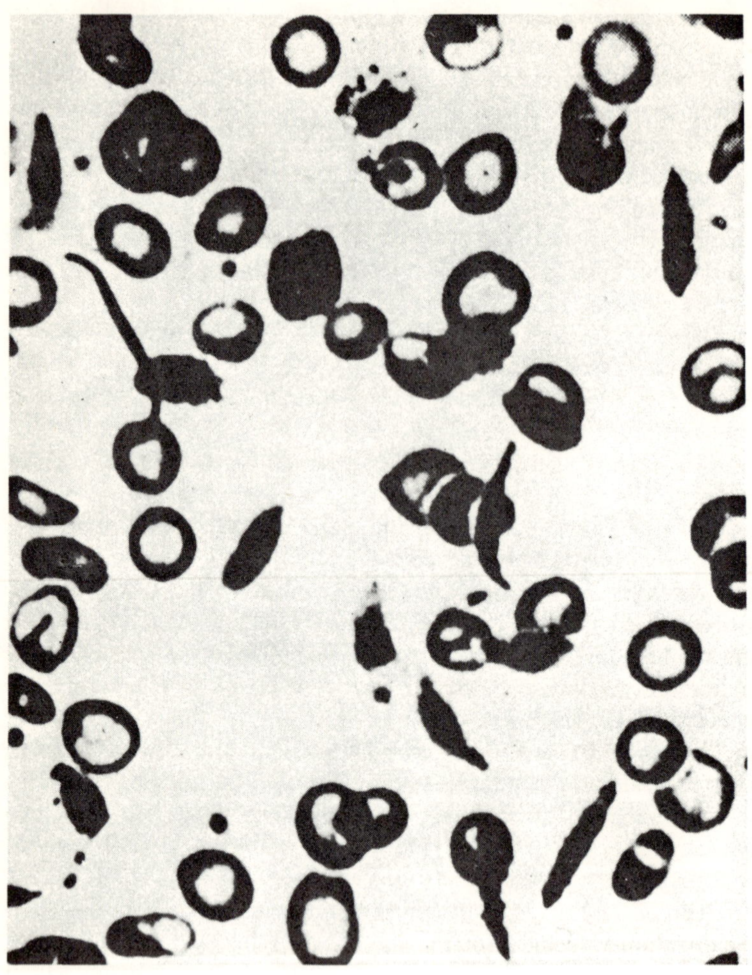

Figure 5-3. Sickle-cells.

tered already. Outside the cells these tropocollagen units aggregate spontaneously to form fibrils either in sheets that can withstand tension or in elements like tendon. Plant cells use cellulose in place of collagen.

In addition to these fibrous elements outside cells there has to be an intracellular cement. In metazoan animals these are complexes of polypeptides and amino-polysaccharides (like the lysozyme substrate) known generally as mucosubstances. They have various roles including the provision of lubrication, intracellular adhesion and sometimes, as with chitin, the basis for skeleton formation. This last brings us finally to the highest level of organisation, that of the whole organism.

The shapes of multicellular organisms

In order to create a viable animal not only must the cells have appropriate shapes but the assemblages of cells in the organs and, indeed, in the whole organism must also have the right shape. As with individual cells this is achieved in two ways, either:

1) by means of internal rigid elements; or
2) by means of a skin providing external tension.

Different animals use different methods and it is enough to note two examples. Among those that have an internal structure to maintain their shapes, vertebrates deposit crystals of calcium phosphate in the form of apatite in an orderly arrangement on preformed collagen fibres so producing bone. Insects, on the other hand, form a purely organic external element by tanning their cuticle, formed of chitin and protein, into a covalently cross-linked rigid structure.

Conclusion

This account may be enough to show that, superficially at least, we can go a long way towards understanding how the visible shapes and properties of living organisms are related to the properties of the molecules from which they are made. Closer examination reveals that great gaps still remain in our knowledge but we seem to be on the brink of understanding in great detail the chemical basis of our existence and, consequently, of being able to intervene when things go wrong with the assurance of putting them right. With so much achieved we must struggle also to ensure that this knowledge and the power it gives us are used aright.

FURTHER READING

In this account the achievements of many notable research workers have passed unacknowledged. Reference to the following books and papers will help to remedy this and provide a more balanced account of the subject.

Bernard, S. *The Structure and Function of Enzymes* (New York: W. A. Benjamin Inc., 1968).

Crick, F. H. C. *Nucleic Acids*. Scientific American Offprints (San Francisco: W. H. Freeman & Co., 1957).

Crick, F. H. C. *The Genetic Code III*. Scientific American Offprints (1966).

Dickerson, R. E. and Geis. I. *The Structure and Action of Proteins* (New York: Harper and Row Inc., 1969).

Gros, J. *Collagen*. Scientific American Offprints (1961).

Haynes, R. H. and Hanawalt, P. C. (eds.) *The Molecular Basis of Life* (San Francisco: W. H. Freeman & Co., 1968).

Holley, R. W. *The Nucleotide Sequence of a Nucleic Acid*. Scientific American Offprints (1966).

Kauzmann, W. *Advances in Protein Chemistry* (1959), *14*, 1.

Kendrew, J. C. "Myoglobin and the Structure of Protein Science", *Science (1963) 139*, 1259.

Kornberg, A. *The Synthesis of DNA*. Scientific American Offprints (1968).

Loewy, A. G. and Sickervitz, P. *Cell Structure and Function* (New York: Holt, Rinehart and Winston, Inc., 1969).

Mirsky, A. E. *The Discovery of DNA*. Scientific American Offprints (1968).

Perutz, M. F. *The Haemoglobin Molecule*. Scientific American Offprints (1964).

Perutz, M. F. "Stereochemistry of Co-operative Effects in Haemoglobin." *Nature* (1970), *228*, 726.

Rich, A. *Polyribosomes*. Scientific American Offprints (1963).

Watson, J. D. *The Molecular Biology of the Gene* (New York: W. A. Benjamin Inc., 2nd edn. 1970).

Watson, J. D. *The Double Helix* (New York: Atheneum; London: Weidenfeld & Nicolson, 1968).

Watson, J. D. *The Double Helix* (New York: Atheneum, 1968).

Watson, J. D. and Crick, F. H. C. *Nature*, 1953, *171*, 737.

Wolstenholme, G. E. W. and O'Connor, M. (eds.) *Principles of Biomolecular Organisation*. Ciba Foundation Symposia (London: Churchill, 1966).

Dolphins and Man

A Study in Mammalian Adaptation

by

RICHARD J. HARRISON

CHAPTER ONE

A Study in Mammalian Adaptation

The purpose of this account is to discuss the many anatomical and physiological adaptations which have evolved in the dolphin, a mammal that is successfully at home in the sea, and compare them with what has happened in Man, a mammal that has overrun our planet. Are there particular features which account for this success and are they present in both dolphin and Man? Both forms are the product of evolution, but for neither do we know a precise ancestral history: it is clear, however, that *Homo sapiens,* a member of the Order Primates, is far removed from the dolphins, members of the Order Cetacea.

Adaptation is an important concept when applied to comparative studies of structure and function. The term has been used in several senses, as for example when we consider adaptive growth, adaptive radiation, adaptive colouration, foetal and physiological adaptations. In the present sense we mean that adaptations are modifications in either structure or function that allow mammals to survive in particular environmental conditions. They develop, as generation follows generation, as responses to the changing environment in a continually altering succession of organisms which reproduce sexually. As the words adaptation and specialization both suggest, they are variations on previously existing structures or biological mechanisms; to adapt means to fit better under altered circumstances. Most adaptations are complex, some almost incredibly so, and to understand them at all we have, when we can, to look back at their origins, at their ancestry. Unfor-

---

Footnote: Illustrative material assembled by D. A. McBrearty, Assistant Technical Officer, University of Cambridge.

tunately, this is not possible to any great extent when we deal with the ancestry of some types of marine mammals. We just do not know very much about the ancestors of dolphins for example; maybe their remains are somewhere deep in the sea bed. We can, however, look at embryos and foetuses of the forms we wish to study. In the case of dolphins we can surmise something about what might have been their ancestral forms. We see, for example, that young dolphin embryos possess very small hind limbs, which retrogress during foetal life and disappear except for vestigial pelvic and occasionally limb bones found deep beneath the skin. This is not unlike what happens in human development when we had an embryonic tail and lost it early in development. A recent definition of adaptation states:

"Adaptations are those details which result in suitable and convenient morphological and functional correlation between parts of an organ, between the organs of a living organism, between individuals of the same species or of different species, and finally, between an organism and its organic environment." (G. Colosi: Encyclopedia of Biological Science.)

Obviously there are difficulties in deciding exactly what is an adaptation and what is not: there is always a subjective element in judging just how important are the characteristics of the organ or structure in question. It is always necessary to consider the relationships of the structure with others, with the organism as a whole and also with the environment. We would hope that an adaptation would exhibit in some way or other a marked adjustment to the part played, that it would provide increased efficiency, that it would assist in successfully maintaining the individual organism or the succession of organisms.

One of the great naturalists to dissect a type of dolphin, in fact a porpoise of the genus *Phocoena,* was John Ray (1628-1705) a Fellow of Trinity College, Cambridge. Whether it was because of what he saw in the porpoise, or in the many other creatures he dissected, he was among the first to comment on the adaptive characteristic of organisms. He gave eight general arguments as to why the structure of man was so perfect.

All the particulars he mentions he considers not to be proper only to the body of Man, but to be endowments and perfections conferred on Man, being adaptations to some extent common to other animals.

He remarks first on Man's upright posture, how it sustains the head which is full of brains and very heavy, enables him to look about, avoid dangers and discover whatever he searches after. An upright posture also frees the hand, that invaluable instrument, and allows full use and development of the legs for walking.

Secondly, Ray comments that in the body of Man, there is nothing deficient and nothing superfluous: if we pretend we do not know the use of a particular part, it is because we are ignorant. Thirdly, the parts and members of our bodies are conveniently situated and disposed, for use, for ornament and for mutual assistance. How, he asks, could the eye have been better placed either for beauty and ornament, or for guidance and direction of the whole body? How could the hands have been more conveniently placed for all sorts of exercises and works, for guarding and securing the head and the principal parts?

He notes, fourthly, the ample arrangements made for the defence and security of those principal parts, the heart, the brain, the lungs. Next he remarks on the abundant provision that is made against accidents and inconveniences, that there are two of so many parts, that there is an abundance of blood vessels so that if a branch is cut or obstructed, others may take over its function. There are many ways by which organs may be cleared, by sneezing, coughing, vomiting and sweating. His sixth point concerns the constancy that is seen in the number, shape, position and construction of the principal parts and the variety that exists in the lesser structures, so infinitely varying that no two men are alike. He then discusses the pleasure accorded to Man in those actions necessary to keep him alive and to perpetuate the species. Man actually delights in eating and drinking—and suffers the pains of hunger and thirst if he does not put his mind to these pursuits. As for the continuation of kind, I need not tell you, writes Ray, that the enjoyments which attend those actions are the highest gratifications of sense.

Finally, Ray draws attention in his eighth point to the very complexity of man's structure and how it all works harmoniously. For the muscles alone, he reckons there are no less than six thousand several ends or aims to be attended to. And a failing in any part would cause irregularity in the body, "and in many of them such as would be very notorious."

Ray also had a number of important observations to make about adaptations in dolphins. Although, like many before and after him,

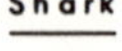

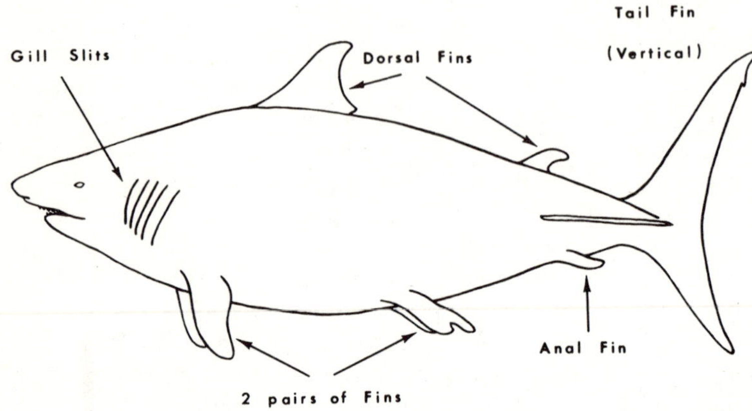

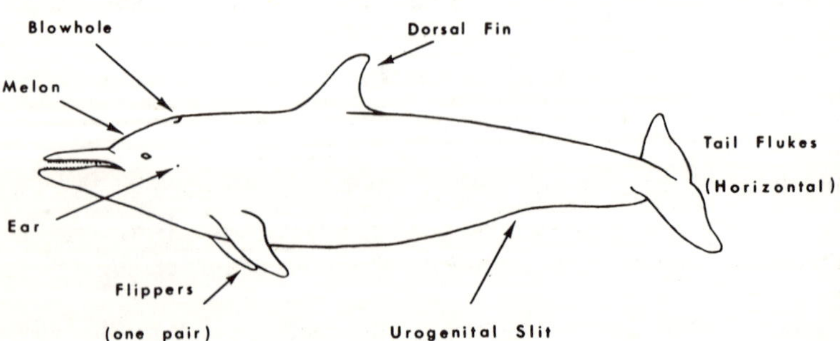

Figure 1. Diagrams comparing the external features of a shark (a fish) and a dolphin (a cetacean).

he called them fish of the cetaceous kind but he noted many of the differences which will be discussed in detail later in this work. He also remarked on the large size of the brain, considered it to have some correspondence to that of man and argued that the dolphin must be of more than ordinary wit and capacity.

Specialization is another word often used in relation to adaptation. In its true sense it means that an adaptation has reached such an extreme in shape, structure or function that it can only operate in a particular environment or in some limited sense or situation. Specialized features are thus those which have departed most from what is considered the ordinary, the normal or the general situation, structure or pattern: for these latter we use the term generalized features. However advantageous a specialized feature might be in a particular set of circumstances it implies a degree of limitation as to its usefulness. We can take two examples from cetaceans. First, there is the complicated blowhole apparatus with its several sacs, valves and cushion-like plugs and the numerous layers of muscle related to it. This is a specialized mechanism, derived from the generalized structure of the mammalian nasal region, that has evolved in relation to the problems of respiration, and possibly echolocation, in a diving mammal. Second, there is the filter made of numerous plates of whalebone found in the mysticete whales. There are horny specializations derived from palatal ridges in the roof of the mouth, they are limited in their use to filtering plankton.

Adaptations may be classified in various ways; a simple start can be made by considering those which are inherited and those which are not. Inherited adaptations are found in all members of a genus or species, and are even seen repeated in quite different groups. Some, such as fins, webbed feet, a torpedo-like form, and pointed teeth are repeated in many forms in the same or in similar environments. Others are concerned with the particular functioning of internal organs such as are found in parts of the alimentary canal and the vascular system (the ductus arteriosus and foramen ovale of mammalian foetuses). In yet other adaptations can be seen a specific relationship between a type of behaviour and the nature of the environment. This group includes such structures as copulatory organs, mammary glands and hair in relation to the evolution of terrestrial mammals.

Those adaptations which develop in individual animals and which are not inherited include any modifications that occur by way of hypertrophy, regeneration or as a result of injury (the opening up of collateral circulations). Others are associated with acclimatization, naturalization and even training to perform better in an unaccustomed environment.

There have been many suggestions to explain how adaptations are brought about, how they originated. John Ray, whom we have already quoted as one of the first to discuss adaptations, thought they were all manifestations of divine wisdom and that all were predetermined in a teleological sense. Subsequently there developed the concept that certain environmental factors bring about alterations in some germ cells of individual animals which when fertilized develop into new forms possessing favourable characteristics. Another concept was that the ability to produce adaptations is inborn in all living things and that it only needs the influence of the environment to bring forth an adaptation in all individuals exposed to that environment. Further, it could be postulated that some intrinsic morphogenetic mechanism is always ensuring during the development of shape in an animal, both in its individual development and in its evolution, that all the component parts are harmoniously interrelated in all structural and functional aspects. The incorporation of new characters into the system sets up the possibility of the animal encountering new conditions where its advantageous characters will fit. In this sense it follows that in order to invade a new environment there must have been some degree of preadaptation present at the moment of invasion.

There are several important adaptations associated with the evolution of mammals and of dolphins. Here, we mention only a few. Warm-bloodedness demands the ability to regulate body temperature and is related to numerous adaptations in connection with the vascular system, the distribution of blood and the control mechanisms of the circulation. A high oxygen content in the tissues of the body is required to maintain the high metabolic rate essential for life at 37 to 40°C. Heat is preserved by insulation and control mechanisms have evolved for heat production, heat dissipation and heat conservation. The porpoise, the smallest cetacean in the coldest seas, must generate heat by continuous activity: no wonder its whole physiology is staggered to allow it to keep warm.

One definitive feature of mammals is that they possess hair. Yet Man has been called a naked ape and dolphins possess no hair at all except occasionally when newborn a few foetal hairs are present on the upper lip. Most mammals have various cutaneous appendages, skin glands, nails or claws, and have a cornified external covering to the skin. Yet dolphin skin lacks any type of sweat or sebaceous gland or nails and its surface layer is microscopically thin and not keratinized.

Mammals reproduce by giving birth to young after an intrauterine period of gestation. They are viviparous. The number of eggs released from the ovary is reduced to one or a few at a time. The egg is fertilized internally after insemination at copulation. The fertilized egg is transported to a uterus where implantation allows the development of a placenta. Nutrition is supplied from maternal blood, allowing a prolonged gestation period in a warm environment. Foetal adaptations allow not only successful intra-uterine existence but also preparation for extra-uterine life. The hazard of birth means that such preparations must be developed sufficiently to be able to come into immediate operation. Newborn dolphins are not only faced with all such demands, they are faced with the additional risk of drowning! All mammals have mammary glands, a readily accessible source of a constant food type under hormonal control. They have evolved from sweat glands, indicating that a gland system originally subserving the function of temperature regulation has become adapted to a quite different function. Young dolphins are again faced with a problem, of sucking their milk while under water. Most female mammals exhibit varying periods of solicitude for their young: until the young can fend for themselves.

Many animals are seasonal breeders, all the females of the species reproduce at the same time of the year. Mammals exhibit varying reproductive patterns but in most the females are in oestrus, are receptive, at the time of ovulation. This may be seasonal or may be repeated at intervals throughout the year. Human females are potential breeders all through their reproductive lives: we are not yet quite sure but most dolphins appear only to ovulate on copulation (induced ovulation), during restricted periods. Some mammals display the phenomenon of delayed implantation where there is a pause in the development at an early stage of the blastocyst, which leads to birth taking place at a particular time of the year.

The success of mammals is associated with their excellent special senses and cutaneous sensibilities. Visual acuity and stereoscopic binocular vision have been essential to the advance of the primates. The most complex morphological changes have taken place in auditory mechanisms in the transition from aquatic to terrestrial life. The way in which hearing mechanisms have become adapted to the function of echolocation and navigation in dolphins is a fascinating recent discovery, only surpassed by the adaptations of a different sort which have accompanied the development of articulate speech and communication in Man. Many mammals have a marked sense of smell and a large portion of the brain, the olfactory system, is also well developed. Such mammals are called macrosmatic. Man is less well endowed, is microsmatic; dolphins have no sense of smell at all and are anosmatic. Mammals have many types of cutaneous sensory devices, sensitive vibrissae, tactile pads and various nerve endings in the skin subserving the modalities of touch, pressure, heat, cold and pain. In a way these adaptations can be related to the characteristic curiosity, the enquiring and adventurous nature of mammals. Strangely perhaps, it was at first thought that the skin of dolphins was devoid of much sensory innervation but recent investigations show that some parts at least are abundantly innervated.

The skeleton of land mammals shows two major adaptations, firstly in relation to the mechanical problems of supporting the body, secondly in relation to locomotion. The vertebral column was originally a strut preventing compression, in land mammals it lies dorsally in the body with the main mass of the animal suspended from it. Pectoral and pelvic girdles became strengthened and modified to bear weight and for locomotory purposes.

Of all the adaptations that have evolved in mammals, the most striking is that of the enlargement of the brain, seen most obviously in the primates. Cephalization and cerebralization have been linked with the evolution of upright posture, with the freeing of the hand from weight bearing, and with the elimination of sexual competition between males. Some authorities have also associated an increase in the importance of sensory information to a creature with a freely mobile upper limb with the expansion of the association areas in the brain. Both man and dolphin have high brain/body weight ratios and it is intriguing indeed to puzzle over the functions of the both absolutely and relatively large brain of dolphins.

Figure 2. A river dolphin (Pontoporia) *from the River Plate. Photo: R. L. Brownell.*

Before we consider all the numerous adaptations present in dolphins and compare them with the general situation in mammals, and to some extent with human structure, it is necessary to consider briefly what a dolphin is and how it fits into place in a classification of mammals which live in a marine environment.

A dolphin is a small toothed cetacean, an odontocete: it lives all its life in the sea and it is an example of what are collectively called marine mammals. These creatures are not all in one group and they are not all closely related, yet they are all in their various ways very well constructed for life in the sea. Indeed of all the mammals both extinct and living, the Cetacea, of which dolphins are members, are almost certainly the most highly modified in their structure. They are so remarkably adapted to a completely aquatic life, and are in general so aberrant among mammals that it is difficult to determine their precise position relative to the other groups of mammals.

It is always necessary to classify animals, in order that we may know exactly what the animal is with which we are dealing. The difficulty about classifying marine mammals is that they are not all that well known. Some live in far-away places, some are inaccessible, few can be kept in zoos or laboratories for reason of size or difficulties with feeding. Many are so large that examination is tedious, dirty and often unwholesome: in some instances bones only are available and the external appearances poorly described. Colouration may change as the animal ages so that there can be confusion of several species when there is really only one.

CETACEA

Order *MYSTICETI* (whalebone whales)
- Family Balaenidae
  - Genus *Balaena* Black right whale and bowhead whale
  - Genus *Caperea* Pygmy right whale
- Family Eschrichtidae
  - Genus *Eschrichtius* Gray whale
- Family Balaenopteridae
  - Genus *Balaenoptera* Blue, fin and sei whale etc.
  - Genus *Megaptera* Humpback whale

Order *ODONTOCETI* (toothed whales)
- Family Platanistidae
  - Genus *Platanista* Gangetic dolphin
  - Genus *Inia* Amazonian dolphin
  - Genus *Lipotes* Chinese river dolphin
  - Genus *Pontoporia* La Plata dolphin
- Family Delphinidae
  - Genus *Steno* Rough-toothed dolphin
  - Genus *Sousa* Humpbacked dolphin
  - Genus *Sotalia* Tookashee
  - Genus *Tursiops* Bottlenosed dolphin
  - Genus *Grampus* Risso's dolphin

Genus *Lagenorhynchus* White-beaked and white-sided dolphins
Genus *Lagenodelphis* Sarawak dolphin
Genus *Stenella* Long snouted and spotted dolphins
Genus *Delphinus* Common dolphin
Genus *Lissodelphis* Right whale dolphin
Genus *Cephalorhynchus* Commerson's dolphin
Genus *Peponocephala* Broad-beaked dolphin
Genus *Feresa* Pygmy killer whale
Genus *Globicephala* Pilot whale
Genus *Orcinus* Killer whale
Genus *Orcaella* Irrawaddy dolphin
Genus *Phocoena* Common porpoise
Genus *Neophocaena* Black finless porpoise
Genus *Phocoenoides* Dall porpoise

Family Monodontidae
Genus *Delphinapterus* White whale
Genus *Monodon* Narwhal

Family Physeteridae
Genus *Physeter* Sperm whale
Genus *Kogia* Pygmy sperm whale

Family Ziphiidae
Genus *Tasmacetus* Tasman beaked whale
Genus *Mesoplodon* Beaked whale
Genus *Ziphius* Cuvier's whale
Genus *Berardius* Giant bottle-nose whale
Genus *Hyperoodon* North Atlantic bottle-nose whale

PINNIPEDIA

Order *PINNIPEDIA*

Family Otariidae
Genus *Otaria* South American sea lion
Genus *Phocarctos* New Zealand sea lion
Genus *Neophoca* Australian sea lion
Genus *Zalophus* Californian sea lion
Genus *Eumetopias* Northern sea lion

Genus *Arctocephalus* South American, Australian fur seal etc.
Genus *Callorhinus* Northern fur seal
Family Odobenidae
Genus *Odobenus* Walrus
Family Phocidae
Genus *Phoca* Harbour seal
Genus *Pusa* Ringed seal, Baikal seal
Genus *Halichoerus* Gray seal
Genus *Histriophoca* Ribbon seal
Genus *Pagophilus* Harp seal
Genus *Erignathus* Bearded seal
Genus *Monachus* Monk seal
Genus *Lobodon* Crabeater seal
Genus *Ommatophoca* Ross seal
Genus *Hydrurga* Leopard seal
Genus *Leptonychotes* Weddell seal
Genus *Cystophora* Hooded seal
Genus *Mirounga* Elephant seal

SIRENIA

Order *SIRENIA*
Family Dugongidae
Genus *Dugong* Dugong
Genus *Trichechus* Manatee

The cetaceans can be considered as forming a single Order Cetacea with two Suborders of living forms, Odontoceti (toothed whales, dolphins, porpoises) and Mysticeti (whalebone whales). Some authorities consider cetaceans should be divided into two Orders, Odontoceti and Mysticeti, on the grounds of their evolutionary history and the differences in structure of the two groups. Pinnipeds are usually placed in the Order Pinnipedia with three main divisions; true or earless seals, the Phocidae; sea lions and fur seals, the Otariidae; walruses, the Odobenidae. Manatees and dugongs comprise the Order Sirenia.

Figure 3. A common porpoise (Phocoena phocoena) *jumping. Photo: Copenhagen Zoo.*

The term dolphin has been used both for freshwater or river forms as well as those of the open seas. The term porpoise has been applied generally to small cetaceans and in particular to the Common porpoise, *Phocoena*. Some of the cetaceans included for reasons of their anatomy in the Family Delphinidae are so large that they are called whales, such as the Killer whale and the Pilot whale. It is not possible to give descriptions here of all the species of marine mammal but as several types of dolphin and porpoise are mentioned frequently in the text a few comments will be made on a selection of them.

River dolphins

The living river dolphins are included in the Family Platanistidae. They have long, slender jaws. The Gangetic dolphin or Susu *(Platanista gangetica)* is found in the Ganges, Indus and Bramahputra Rivers: its name Susu is derived from the noise of its breathing.

The Amazonian Dolphin or Boutu *(Inia geoffrensis)* is a freshwater platanistid some 6-7 ft long found only in the upper Amazon. It has a long beak, with some bristles on it, an exceptional instance of a hair-bearing cetacean. The conical teeth number over 120. The flippers are large and fan-shaped and the flukes relatively large, a modification possibly related to the fast swimming it is said to accomplish. The La Plata dolphin or franciscana *(Pontoporia blainvillei)* is a pale brown short form from the estuary of the River Plate. The Chinese Lake or White Flag dolphin *(Lipotes vexillifer)* is restricted to the Tungt'ing Hu (lake) on the upper Yangtze River. It has a grey back, a white belly and has an upturned beak.

Bottlenosed dolphins

These are the members of the genus *Tursiops* of which two species are clearly recognizable: *T. truncatus* widely distributed in temperate and tropical waters and *T. gilli* off the coast of southern California. Several supposed races or species, such as *T. catalania,* have also been described. Bottlenosed dolphins are probably the best-known of all dolphins as a result of their exhibition in marinelands and appearances on television. They are easily netted in shallow water, will soon feed in captivity and can be taught to perform. They are good swimmers and jumpers and seem to enjoy being given various simple tasks especially when rewarded. Most of the *Tursiops* exhibited in the United States and now in Europe originally came from Florida waters. It is a large dolphin, up to 4 m long, with a short snout and slightly protruding lower jaw. The dorsal fin has a sharp apex and a recurved trailing edge. The

*Figure 4. Bottlenosed dolphins (*Tursiops truncatus*). Photo: Marineland of the Pacific.*

back is black to grey and the belly and lower jaw white. There are 20 pairs of teeth on both jaws.

White-sided dolphins

The several species in the genus *Lagenorhynchus* have a short, poorly marked beak or flask-shaped snout, a high dorsal fin with a concave posterior border, marked ridges behind the dorsal fin and anus and powerful flippers with pointed tips. There are

*Figure 5. Four white-sided Pacific dolphins (*Lagenorhynchus obliquidens*) in a display. Photo: Marineland of the Pacific.*

numerous distinguishing features in the skull, teeth and skeleton. The number of vertebrae can be as many as ninety. Quite marked colour distinctions are present as well, particularly on the flanks.

A number of species are little known, having been seen only by their original describers. More common are the Dusky dolphin of New Zealand *(L. obscurus),* the large schools of the northern Atlantic forms *(L. acutus* and *L. albirostris)* and that of the North Pacific *(L. obliquidens)* which is a fine jumper in captivity.

Common Dolphin

The Common Dolphin *(Delphinus delphis)* is widely distributed, in small groups and in very large schools, in temperate and warm seas the world over. It was known in the Mediterranean in classical times, often pictured on coins, vases and mosaics, and has long held the reputation of coming to the rescue of drowning men. The body is slender, hardly reaching 3 m in length and with a narrow but pronounced beak. The dorsal fin leans backwards, with a sharp, backwardly directed tip and with a concave caudal edge. The back is darkly pigmented, the belly white, and there are light grey or yellowish-brown stripes along the front flank. It is a powerful swimmer and probably the swiftest of all cetaceans through the water.

Pilot whales

The Pilot Whale *(Globicephala melaena),* also called the Ca'aing Whale or Blackfish of the North Atlantic, is common off the Faroes, Orkney and Shetlands. They are sometimes seen on calm, summer evenings off the north-east coasts of Scotland. They swim in schools of up to several hundred. The body is black, often with a white flash beneath the chin. The head bulges markedly (thus Pothead), the flippers are long and narrow. There are ten teeth on each side of both jaws. Adults can reach 9 m in length, but a few can be even longer. The schools have been hunted off the Faroe Islands and can be driven ashore by beating the water with oars. Large numbers have become stranded for unknown reasons in various parts of the world. Several species have been described but all are much alike but *G. macrorhyncha* has a shorter fin. Several have been kept at marinelands: one grew so large that it was able to burst a window of its tank and cause much flooding.

Figure 6. A Pilot whale (Globicephala). *Photo: Marineland of the Pacific.*

Killer Whales and False Killer Whales

Killer whales *(Orcinus orca)* are found in all oceans, but are more plentiful in northern waters. Adult males reach a length of ten metres, but the females only reach about five metres, one of

Figure 7. A young Killer whale (Orcinus orca). *Photo: Marineland of the Pacific.*

the few examples amongst cetaceans of females being smaller. The colouration is black but with several distinctive white areas on the belly, below the chin, above and behind the eye and on the flank behind the dorsal fin. The dorsal fin in older males becomes vertically triangular and up to two metres high. The flippers are rounded and not pointed. The shape of the animal reflects its speed and power and its jaws suggest a formidable predator. A beak is lacking. There are twenty teeth in both upper and lower jaws. They are large, oval in section and reach 2-5 cm in diameter. Not surprisingly killer whales have had an unrivalled reputation for ferocity; they are known to eat other dolphins, porpoises, seals, penguins, fish and squid. Usually they hunt in packs of a few to forty or more, even launching attacks on large whales, tearing at their tongues, lips and any projecting part. Observers have recounted how the large whale, overcome by terror, offers no resistance and submits to its attackers until it bleeds to death. Many young killer whales have been held in captivity, where their behaviour is far less mean; they can be trained to give rides to human beings, to jump for food and to allow close inspection of their formidable teeth. Whether they remain so placid and friendly as they grow older will be of interest!

Although closely related to the Killer, the False Killer Whale *(P. crassidens)* is not so large, is more slightly built, and is not as ferocious. The males do not grow that much larger than the females and the dorsal fin is small and recurved. The flippers are elongated and tapered. The colour is black almost all over. False Killer Whales are deep water, oceanic forms and are not often seen in coastal waters. They are almost world-wide in distribution in all temperate and tropical seas. They are gregarious and strandings may involve up to several hundred individuals. Alterations in the distribution of squid, cod and other fishes on which they feed have been thought to bring them inshore.

The Common Porpoise

Having recently been engaged in studying this small odontocete, the smallest cetacean species living in the coldest waters, I shall consider it in some detail. It also happens to be the cetacean most commonly encountered in British waters and most frequently stranded on British coasts. There has been controversy as to whether the animal called a porpoise by the British should be included in the family Delphinidae, as has been done in the classification given earlier, or whether it should be placed in a separate family, the Phocoenidae. In any event we give a short description of this interesting group, if only as a mark of respect for Dr. F. C. Fraser of the British Museum (Natural History) who has contributed so much to cetology.

The Family Phocoenidae comprises three genera, seven recent species. The genera are *Phocoena, Phocoenoides* and *Neophocaena.* The seven species are:

Phocoena spinipinnis	Burmeister's porpoise
Phocoena sinus	Gulf of California harbour porpoise.
Phocoena dioptrica	Southern harbour or South Atlantic porpoise. Spectacled porpoise
Phocoena phocoena	Common porpoise
Phocoenoides truei	True's porpoise (probably a colour phase of *P. dalli*)
Phocoenoides dalli	Dall's porpoise
Neophocaena phocaenoides	Black finless porpoise

The Phocoenidae have cranial features similar to those of the Delphinidae but can be distinguished by prominent bosses on the premaxillae in front of the nares. The teeth of *Phocoena* are spade-like with laterally compressed bilobed or trilobed crowns. The teeth of Phocoenoides are like needles. Three or more cervical vertebrae are ankylosed. The jaws are relatively short and there is no distinct beak as in many dolphins. The dorsal fin is triangular, obtuse and fairly low in *Phocoena* and *Phocoenoides*; it is absent in *Neophocaena* which exhibits a dorsal ridge in its place.

The common porpoise is found in the Arctic and North Atlantic Oceans and in the North Pacific. In the Atlantic it is distributed from the Barents Sea, Iceland, and Davis Strait to the Black, Azov and Mediterranean Seas and as far south as Senegal on the east, and on the west to Delaware. In the Pacific it is found from Alaska to Mexico on the east and to Japan on the west. English writers have referred to it as the Bucker, Herring Hog, Marsuins, Neessock, Pellock and Puffing Pig and the common or harbour porpoise (Porcpess). It is one of the smaller odontocetes, seldom exceeding 180 cm in length. In colour it is all over blackish or blackish above and whiter below. Between the years 1913-1966 out of a total of 1,547 identified cetaceans stranded on British coasts, 631 were *Phocoena phocoena*.

Phocoena may be seen at low water far up tidal rivers, as far as Emmerik on the River Rhine, Venlo on the River Maas, and near Teddington on the River Thames. Porpoises enter the Baltic in spring one by one or in pairs and in November they leave Danish waters to the North Sea: porpoises are found in the eastern Baltic only in summer. This migration can continue into December and January: the severity of the winter has much effect on the porpoise population in the Baltic. Porpoises also congregate in specific in-shore waters of the Canadian Atlantic coast during May to October. They are at peak numbers in August but are not observed after November. There is migration towards British coasts which reaches a peak from July to October: strandings are least common in December and February. The precise whereabouts of the north-western Atlantic porpoises from December to April is simply not known.

The Dall porpoise *(Phocoenoides dalli)* is not uncommon off south central California from October to June. It ranges from 39°N in Japanese waters, to the sea of Okhotsk and the western Aleutian Islands to as far south as at least 35°. Adults reach a length of 217 cm and a weight of 148 kg. The animal is black but with a conspicuous white region on the lower flank and belly and white patches on the trailing edges of fins and flukes. It is deep-bodied and has a characteristic dorsal keel producing a hump on the peduncle just anterior to the flukes. Dall porpoises are probably the fastest swimmers of all small cetaceans but they seldom leap clear of the water. They can ride a bow wave for considerable periods at speeds of 12 to 17 knots and will often accelerate far ahead of a boat travelling at such speeds.

The Finless Black Porpoise *(Neophocaena phocaenoides)* ranges through the Indian and Pacific oceans from the Cape of Good Hope to Japanese waters. It is closely allied to *Phocoena* but is distinguished by its lack of a dorsal fin and the degree to which the rostrum is bent down in relation to the basicranial axis. It inhabits bays and tidal channels, its colour resembles that of *Phocoena* and adults measure about 160 cm in length.

Several investigators have analyzed the tissue and organ weights of *Phocoena,* and have compared them on a basis of percentage of body weight with figures derived from large whales. It has been concluded that the whole physiology of the porpoise is staggered in order to allow it to keep warm. Indeed the resting metabolic rate

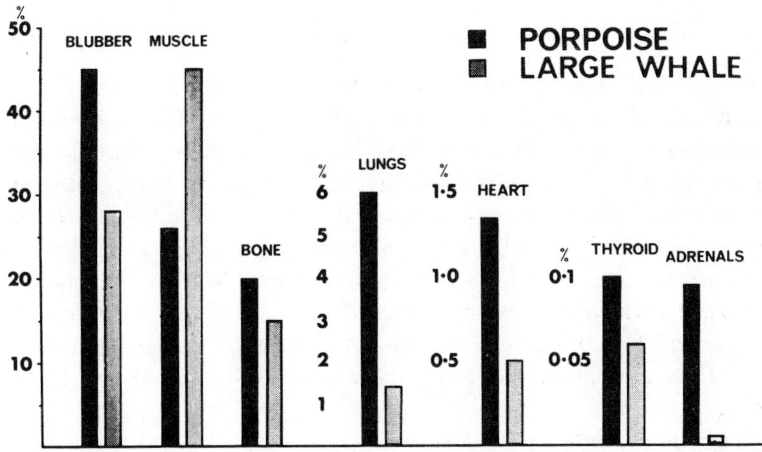

Figure 8. Comparison of the relative weights of the tissues in a small porpoise and a large whale.

(O_2 consumption 380 ml/min) is about three times higher than for other mammals of similar size. Our analysis of tissue and organ weights suggests that compared with large whales the porpoise has more blubber, less muscle, a relatively larger heart and lungs as well as more thyroid and adrenal tissue. Blood volumes are known for only a few cetacean species: they show however that *Phocoenoides* has a much higher blood volume (143 ml/kg) than *Tursiops* (71 ml/kg).

Weights of organs and tissues as percentages of total body weight

	Porpoise	*Large whale*
Blubber	32-45	18-27
Muscle	20-26	40-45
Bone	18-19.5	14-15
Heart	0.5-1.32	0.34-0.5
Kidney	0.5-1.5	0.4-0.5
Lungs	3.5-6.1	0.6-1.2
Thyroid	0.03-0.1	0.006-0.06
Adrenals	0.03-0.09	0.001-0.003
Brain	0.4-1.1	0.01

The maximum recorded length of a dive by *Phocoena* is 12 minutes and the maximum depth is 20 metres. The fastest recorded swimming speed is 9 knots. The porpoise has become adapted more to conserving heat, with its 32-45% of blubber, with an associated increase in respiratory capacity and a relatively large heart but has limited diving and swimming ability. There are a number of anatomical and physiological modifications in marine mammals which have been associated with diving and deep diving: these will be considered in detail in subsequent pages. Although some are present or occur in *Phocoena,* none is well developed.

Diving and related problems

The following account of diving by marine mammals, and the anatomical and physiological adaptations in relation to diving for prolonged periods to great depths, is based on work by Harrison, R. J. and Kooyman, G. L. in *The Behaviour and Physiology of Pinnipeds,* 1969, ed. Harrison, R. J. *et al.* Appleton-Century-Crofts and in *Diving in Marine Mammals,* (1971), Oxford University Press. The editors and publishers are thanked for permission to reproduce material and illustrations.

Marine mammals are skilled divers. They take their food while under water in the wild and some dive very deep but being mammals they must surface at intervals to breathe. How long does each type habitually stay submerged? How much longer could it remain underwater if it really had to? Is there a limit to the number and length of repeated dives, and must there be a recovery period after a very long dive? How deep do they dive, or could they dive if necessary, and why? If a whale dives to a great depth, why does it not suffer when surfacing from the "bends"? This is a painful condition when nitrogen bubbles out of solution in blood and body fluids and affects joints. If a man voluntarily holds his breath, the rising carbon dioxide tension in the blood affects the respiratory centres until eventually he has to start breathing again. Does such a mechanism operate in marine mammals? If not, what tells a seal it has been underwater long enough and must surface to breathe? The largest whale dives only with the air in its two lungs, so where does the extra oxygen it must need during its dive come from? Perhaps its lungs are different from those of other mammals. Could its blood have special properties, or are there special oxygen stores as there have to be when man dives for a long period?

Until quite recently all our information about the diving performance and general abilities of dolphins and other marine mammals had been obtained by observations from boats or from land and by deductions from cutting up dead specimens. Whaling and sealing industries, scientific expeditions and careful investigation of stranded specimens have all provided much valuable information and many indications as to what marine mammals might do. There are, however, obvious limitations to what can be gleaned by such methods and they seldom provide accurate answers to the questions we have been considering. When it comes to keeping marine mammals in captivity and carrying out experiments on them, many new problems arise. Marine mammals have to be caught; they live in places often not easy for man to get at them and are difficult to catch alive when they are so at home in the sea. An ancient Greek definition of the impossible is that it is like catching a dolphin by the tail. Special methods of catching have to be devised which will not damage the dolphin, cause it to enter a state of severe shock or otherwise reduce its chances of survival. There are also problems of transportation, of conveying the dolphin many hundreds or even thousands of miles to its destination. Then come the difficulties of maintaining the dolphins successfully, of feeding them and treating their diseases. When all this has been accomplished, we have still to devise the best methods of studying them, methods that will provide us with information about how dolphins behave in captive situations and also about their natural behaviour. Several excellent marinelands or oceanaria have now been established in North America, Australia, Japan and Europe. The photograph

Figure 9. The installation at Marineland of the Pacific where up to fifteen species of marine mammals have been kept successfully.

shows the first large marineland to have been built: it is the Marineland of the Pacific in California.

A distinguished French physiologist, Paul Bert (1830-1886) was the first to think seriously about the effects of submersion on mammals in his *Leçons sur la physiologie comparée de la réspiration*

(1870). He provided some of the earliest records of breath-holding under water and was also one of the first to investigate the "bends", or caisson disease at it is sometimes called. Sixty years later attention was again given to diving problems, when the Americans L. Irving and P. F. Scholander carried out numerous experiments on seals and some other diving mammals. They immersed the animals, having usually restrained them in various ways, in shallow baths or lowered them into the sea. Recordings were made of length and depth of dives, heart rate, blood pressure, oxygen consumption, carbon dioxide tension, lactic acid concentration, pH of blood and other biochemical changes during these experimental dives. There were few captive dolphins at that time and it was mainly seals that were investigated, as well as some aquatic rodents. It became apparent that when a seal dived, a dramatic *diving reflex* or *diving response* developed which lasted for the duration of the dive. This response involved a marked change in the circulation and also a slowing of the heart rate (bradycardia). Constriction of most of the peripheral blood vessels caused redistribution of blood to the central nervous system and the heart and thus it seemed that a diving mammal could be considered a sort of "heart/brain preparation".

Investigations were extended to several other types of marine mammals by other workers from 1950 onwards and were much stimulated by the invention of ways to catch seals and dolphins alive and by finding out with increasing success how to maintain them in captivity and for exhibition. Anatomical research has helped to reveal many adaptations in the vascular system and has suggested several functions which they might perform during diving. Techniques have also been developed for anaesthetizing seals and dolphins for long periods and for monitoring the levels of various substances in the body fluids. Diving has now been thoroughly investigated in many marine mammals, when they are restrained, when free-swimming in harnesses, and when in various types of pressure tanks. Diving activities by animals as large as elephant seals and killer whales have been recorded, as well as by coypu rats, beavers, otters and even by the hippopotamus. Diving has also been followed by attaching small packages containing recording instruments and then allowing animals to dive: the animal having been trained to return with the package to the point of release. Capillary manometers, depth-time recorders, ultrasonic depth transmitters, maximum depth recorders, miniaturized tape recorders and hydrophones and several types of heart beat recorder have been used, all under water. Development of several types of telemetering device has begun to give information about the behaviour of marine mammals when some way from land but there are many difficulties still to be overcome when using such techniques.

All these techniques have provided much information on normal or habitual diving habits and also on what happens when an animal is dived experimentally. Early records of diving capabilities were usually derived from estimating the time an animal remained submerged after harpooning and the depth it reached with a line attached. Observations of animals caught in nets or snared by cables on the floor of the ocean were also used as evidence, as well as the type of food found in the stomach. In some instances this sort of information is all we have, inaccurate though it well may be, to give us an idea of the abilities of a particular species. The normal diving capabilities of many species of marine mammal far surpass those able to be measured experimentally in pressure tanks and until we are able to attach recording devices to all of them we shall remain ignorant of the extreme depths they might reach. The Table on page 158 shows what is known about maximum duration of dives and the depth reached by various whales, dolphins and seals. Just how much longer and deeper they could dive is just not known. Some additional information is available about other species but is fragmentary or not well authenticated. Although they spend so much of their lives in water, sirenians do not dive for much longer than five minutes, though dives up to 16 minutes have been claimed. Nothing is known about how deep they can dive.

It can be asked for what reasons do marine mammals dive, and why so deep? All the evidence suggests that many species can, if needs be, dive to considerable depths, much deeper than one might think necessary. Obviously, there is the need for food and the type of food taken is found at varying depths in the oceans. Are there other reasons? Is a marine mammal so constructed that it has to dive beneath the surface at regular intervals, that it *has* to keep moving, to keep on diving? Could there be a metabolic necessity, or a thermoregulatory demand? Clearly there might be navigational requirements for long, exploratory dives but might a dolphin be so constructed, have become so adapted, that it only comes to the surface because, being a mammal, it has to in order to breathe? Perhaps, if it could, it would rather spend its entire existence completely under water.

In fact we know little of the natural diving habits of any one species. Diving physiology might well become more comprehensible if we did.

Studies on the diving patterns of Weddell seals *(Leptonychotes weddelli)* indicate how it survives successfully in the icy waters of its antarctic environment. These seals feed on fish which are taken beneath the ice for most of the year. They must return to the surface to breathe and have to depend on the existence of breaks in the ice caused by winds and currents. They can also use their curiously arranged incisor and canine teeth to prevent breathing holes from freezing up. The seals work at the ice from underneath using their teeth like a rasp. A Weddell seal can find its widely separated breathing holes from beneath the ice and in the dark. Swimming submerged, at about 5 knots, Weddell seals travel at least 2,000 m, and sometimes as far as 4,400 m (2.5 miles), in about half an hour. In this way the seals set out towards other breathing holes, all the time swimming under the ice. What is so puzzling is how does the seal know where the distant breathing hole is, in what direction, and that a particular breathing hole is within reach before it runs out of oxygen? Can the seal tell when it is nearing halfway in its range and that soon it must turn back or drown? In fact we can ask a simpler question:—what tells a seal under any circumstances that it *must* surface to breathe?

It is now suspected that Weddell seals display three basic patterns in their types of dive. There are short, shallow dives, which last for up to 5 min and which seldom reach depths of 100 m. These are thought to be exploratory dives for learning about the features on the undersurface of the ice. Short, but much deeper dives of up to 15 min duration and to depths of as much as 600 m, but usually straight down to 300-400 m below the breathing hole, appear to be used for hunting purposes. The dives of longest duration and in excess of 40 min, but usually at depths of less than

200 m, are directed horizontally towards adjacent breathing holes or in a quest for new ones. Restriction of the vertical type of dive for hunting purposes would seem to reduce the chances of getting

---

From: Harrison and Kooyman (1971) *Diving in Marine Mammals* Oxford University Press

Duration and depth of dives (maximum)

	Duration (min.)	Method	Depth (m)	Method
Cetacea				
Porpoise *(Phocoena phocoena)*	12	Exp.	20	Line
Dolphin *(Tursiops truncatus)*			170	Obs.
			200	Feeding habits
			300	Trained
Pilot whale *(Globicephala scammoni)*	15	Obs.	366	Feeding habits
Sperm whale *(Physeter catodon)*	75	Line	900	Caught in cables
			1134	
Fin whale *(Balaenoptera physalus)*	20	Line	355	Harpoon
			500	,,
Blue whale *(Balaenoptera musculus)*	49	Line	100	Feeding habits
Pinnipedia				
Northern fur seal *(Callorhinus ursinus)*	5	Exp.	55	Line
California sea lion *(Zalophus californianus)*			170	Trained
Walrus *(Odobenus rosmarus)*	10	Obs.	90	Feeding habits
Grey seal *(Halichoerus grypus)*	18	Exp.	146	Line
Hooded seal *(Cystophora cristata)*	18	Exp.	Pup: 75	Recorder
Northern elephant seal *(Mirounga angustirostris)*	40	Exp.	183	Line
			300	Exp.
Common seal *(Phoca vitulina)*	28	Exp.	91	Exp.
			200	Exp.
Weddell seal *(Leptonychotès weddelli)*	70	Obs.	600	Recorder

Techniques used to assess these records: Line = attached a line. Obs. = observed directly in some way. Harpooned = dive after harpooning estimated by amount of rope paid out and angle of descent allowed for. Recorder = see text. Trained = see text. Feeding habits = analysis of stomach contents for animals known to live at particular depths. Exp. = restrained (usually) in tanks of various types used for diving or to simulate diving.

lost beneath the ice and make it easier to return to the original breathing hole. The rates of descent and ascent during any dive are always of interest as they might indicate whether the seal pauses during a dive for any form of readjustment. The rates vary for short, shallow dives but are most rapid for hunting dives with maxima of 104 m/min (descent) and 120 m/min (ascent). This information is of particular relevance when we come to consider experimental dives.

Sea lions *(Zalophus)* have been trained to dive down to the end of a weighted line and have attained descent and ascent rates of 170 m/min. A bottlenose dolphin *(Tursiops),* trained in a somewhat similar fashion, was found to exhibit an average depth change rate of 180 m/min without difficulty.

It is perhaps hardly surprising that little is known for certain about diving patterns in the wild of most marine mammals. It is probable that one type of dive can be used for several purposes and that once embarked on a certain type of dive, circumstances could alter the nature of the dive. We have studied the diving behaviour of the Common seal *(Phoca)* in the relatively confined waters of the Wash in East Anglia. At high tide the Wash is a large expanse of water directly continuous with the North Sea: at low tide a complex system of mudbanks is exposed with many channels of varying depth leading out to sea. The seals haul out on the banks as the tide recedes or wallow at the edges of the sloping banks. Adults dive repeatedly for periods of up to 5 min when in the water and progressing to or from the banks: there are shallow dives of up to 10 m. Longer (15 min) and deeper dives (30 m) are spent in search of food. Experimentally, however, Common seals can dive for at least 30 min and have reached depths of 100 m. Young seals are not able to dive for as long or as deep as adults and are often carried on their mother's back at the surface or just beneath it. Experimental evidence indicates that seals may deliberately adjust before prolonged deep dives by a more rapid onset of the diving response and by emptying their lungs before or soon after a dive starts.

Another type of dive may be related to resting or sleeping. More than one species of seal and dolphin can go to sleep in a vertical position. Seals sink slowly downwards into a horizontal plane lying

*Figure 10. A Common seal pup (*Phoca vitulina*).*

on the bottom of a pool and return after some minutes equally slowly to the vertical at the surface with nostrils above water to breathe, asleep all the while. Large cetaceans can sleep while wallowing at the surface, or simply floating: they have been hit by ships while asleep. It is difficult to decide when a dolphin is asleep but studies on captive bottle-nosed dolphins *(Tursiops)* floating at the surface have shown them with conjugate eye closure for periods up to one hour: these animals were probably sleeping. On the other hand it has not so far been possible to detect any sort of sleep behaviour in Dall porpoises *(Phocoenoides)* while in captivity.

Trained dolphins, timed over known distances, can reach maximum speeds of from 15 knots *(Lagenorhynchus)* to over 21 knots

(Stenella). The Common Dolphin is reputed to be able to maintain speeds of over 20 knots for several hours. Large whales are thought to reach such speeds for periods of only a few minutes and they usually swim more slowly at 2-5 knots.

Adult Common seals *(Phoca)* have been timed by us and can reach 8-10 knots but they usually swim at slower speeds. Faster swimming speeds are of course attained by animals after diving into the water from a height.

As regards human swimming speeds, none of us can equal world record holders such as Michael Wenden of Australia who covered the 100 metres, free style, in 52.2 seconds, or Dawn Fraser of Australia who swam the same distance in 58.9 seconds. The highest speed reached by a human swimmer is 4.89 miles per hour (Stephen Clark, U.S.A.). Even so a dolphin could swim rings round all these record holders.

Increases, or decreases, of the pressure in the environment can affect living marine organisms in a number of ways. It is first necessary to differentiate a situation in which an animal is entirely immersed in the sea from one where the pressure is applied through the atmosphere or other gases. In the first situation it apparently is necessary for there to be a considerable change in pressure before there is any significant alteration in physiological activities. In the second, because of the altered tension of the gases dissolved in the cells, body fluids and cellular components, considerable damage may result from relatively slight pressure changes. Local application of pressure, as for example on a nerve where a pressure of some 200 mm of mercury soon stops conduction, can have drastic effects quite different to those observed with the whole animal having been immersed.

The effects of pressure are exerted primarily as a result of changes in the physical and chemical properties of water, seawater, and biological fluids. The changes in *volume* of water with pressure are commonly assumed to be unimportant for the majority of marine animals. It does indeed require considerable pressure to alter the volume of a mass of water. Pure water undergoes about a 20% reduction in its volume under a pressure of 12,000 kg/cm^2 (1 kg/cm = 0.97 atmospheres). As far as we know marine mammals are not exposed to such tremendous pressures and the problems are much more concerned with the effects of pressure on the gaseous components. In other words, what happens to the air taken down in the lungs and dead air spaces of a deep diving seal or dolphin?

*Figure 11. A Bottlenosed dolphin (*Tursiops*): note the recurved dorsal fin.*

Features of the body

The form of the body of all dolphins is adapted in various ways for a life that is spent wholly in the sea. An overall torpedo-like shape has evolved, streamlined and smooth surfaced, facilitating passage through water. Even so, a considerable degree of flexibility of the body has been retained and the thorax and abdomen are much less rigid than might be expected. The body form of pinnipeds is also torpedo-shaped but is not all that apparent when a seal is on land. Under water, however, the neck is extended forwards, the pelage becomes flattened and the elongated form is more obvious.

Cetaceans lack an obvious neck due to shortening and some

fusion of the cervical vertebrae. The rounded body tapers from the thoracic region to the tail stock. The tail is laterally expanded to form two pointed flukes lying in a horizontal plane. The caudal vertebrae extend almost to the end of the tail but the projecting flukes lack skeletal support. The flukes contain connective tissue which is arranged in stiff bundles. The tail flukes are the chief source of propulsive power when moved by the strong musculature extending backwards to the tail region. The arrangement of muscles and direct observation of rapidly swimming dolphins indicate that the up-stroke provides the driving force, and enough thrust to propel a dolphin weighing 150-200 kg at least 3 metres into the air. A dorsal fin is usually placed in the midline near the centre of the back or in some larger whales more caudally towards the tail. It varies in shape from a ridge-like structure in the Amazonian dolphin to a triangular fin in the Common Porpoise. It can be falcate with a recurved caudal edge in the white-sided Pacific dolphin, and in large male killer whales it is acutely triangular and

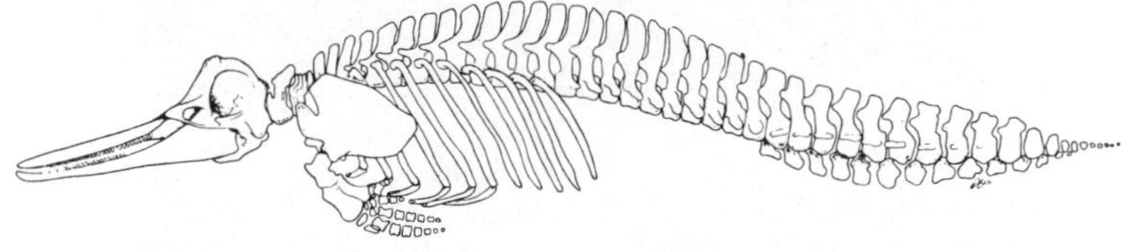

Figure 12. Diagram of the skeleton of a dolphin. Note the elongated jaws, compressed cervical vertebrae, reduction in size of sternum, increased number of floating ribs, increased number of lumbar vertebrae, and the chevron bones on the ventral surface of the caudal vertebrae.

may reach a height of 2 metres. The dorsal fin lacks any skeletal support, is devoid of muscle but is stiffened by connective tissue and contains vessels and nerves. It is absent in some species, such as the Narwhal, the Black Finless porpoise, the Right Whale Dolphins and the Greenland Right Whale. Its principal function has been said to be that of a stabilizer, helping to keep the animal upright on a straight and steady course. It has also been suggested that it plays a part in thermo-regulation and that it helps to guide amongst floating objects. Its absence in some northern forms has been ascribed as a negative adaptation to avoid the dorsal fin becoming trapped in the ice of freezing seas. The dorsal fin may flop over to one side, especially after capture, and also it is alleged, in sick animals. Tubercles and notches have been described on the dorsal fins of some forms, such as the Common Porpoise.

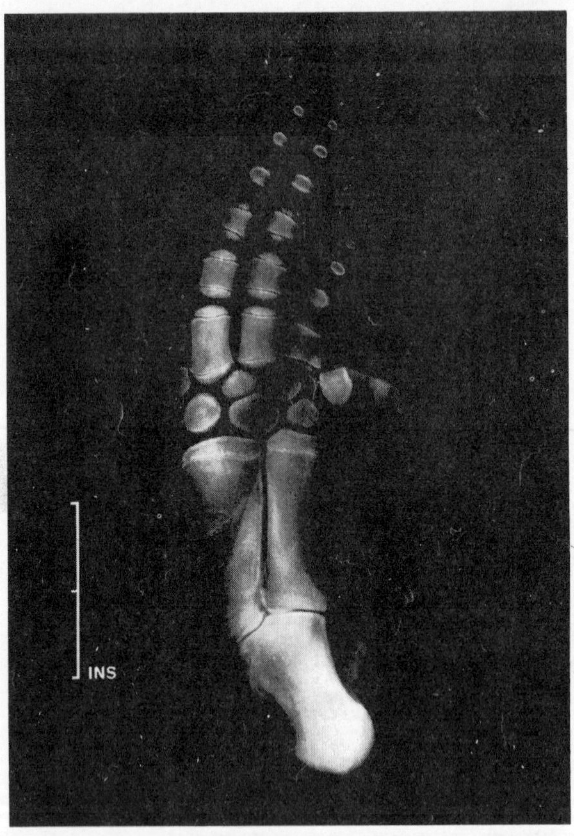

Figure 13. Radiograph of a flipper of the Common porpoise (Phocoena).

The fore-limbs have become modified flippers, with the usual basic arrangement of mammalian limb bones, but giving an appearance of a smooth paddle. There is no clavicle. The shoulder joint only is at all movable and even then the paddle cannot be brought forward to any degree towards the head. The bones of the five digits are embedded in the paddle-like flipper but the thumb is missing in some large whales. The flipper varies in length and the number of phalanges in the central digits may be increased to as many as fourteen. This condition is known as hyperphalangy and is unique to cetaceans amongst mammals. Very long flippers, almost a third of the length of the animal, are seen in the Humpback Whale which also has irregular protuberances on the lower margin of the flippers. Dolphins use their flippers mainly for balancing, steering and as lateral stabilizers: they provide little if any propulsive power. There is no evidence of a hind limb on the outside of a cetacean but it is often present transiently in early embryos as a primitive limb bud. Within the body, however, just anterior to the anus, are small elongated bones that represent the vestigial pelvis. They are not connected by bone to the vertebral column but give origin to muscles, especially those of the penis in males.

The shape of the head is characteristic for each type of dolphin and for marine mammals in general. The skull is large and heavy,

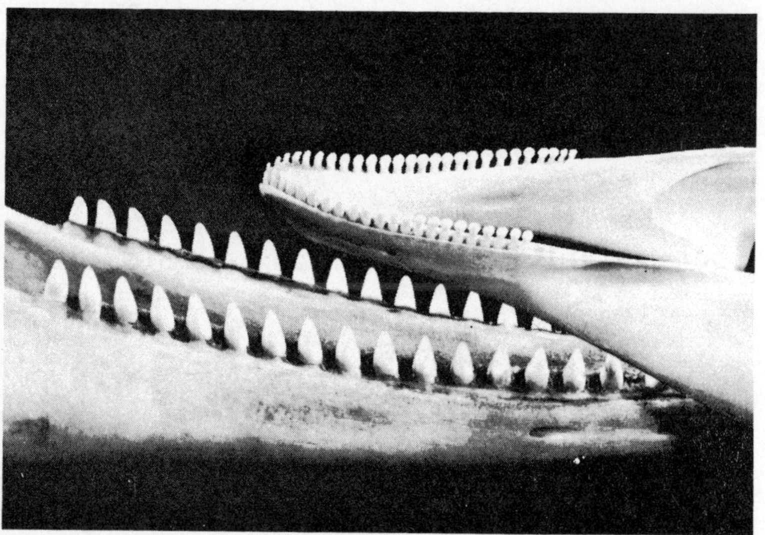

Figure 14. Photograph of the lower jaws and teeth of Phocoena *(top) and* Tursiops.

of dense bone; in dolphins it is telescoped from front to back and often exhibits asymmetry with the right side being larger than the left. The jaws of cetaceans carry baleen plates or teeth. There may be an elongated snout or beak in some dolphins, with the narrow jaws projecting for several inches like a pair of teeth-bearing pincers. Although called toothed whales, the number of teeth in odontocetes varies considerably. Female Beaked whales are virtually toothless, only one enlarged tusk-like tooth, an incisor is present in the Narwhal. The tusk is usually present only on the left side of males; sometimes a right one is also found; females lack a tusk. The Narwhal's tusk was thought before the seventeenth century to be the horn of a mythical beast, the unicorn. Dolphins have numerous similar (homodont) conical, wedge-shaped or peg-shaped teeth, usually surpassing in number the typical placental number of forty-four. In some porpoises there are as many as 300 teeth. The Common porpoise has characteristic spade-shaped teeth. In other cetaceans the upper part of the snout is heaped up, giving a blunter or more rounded appearance as in the Narwhal, the Pothead with its Latin name of *Globicephala,* and most characteristically in the Sperm whale which has a huge spermaceti organ in the snout. Many

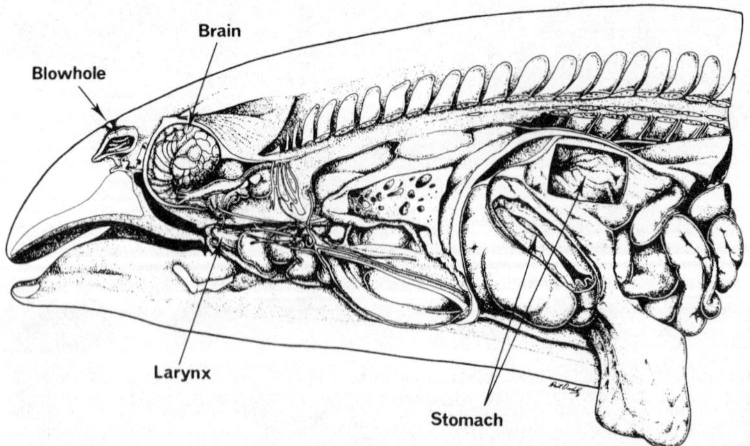

Figure 15. Diagram to show some of the organs in a Common porpoise (Phocoena)*: the larynx has been disengaged from the posterior nares.*

mysticetes have a lower jaw slightly longer than the high arched upper jaw which has rows of flattened baleen plates projecting down from the sides of the palate. Parallel grooves or pleats on the surface skin extend longitudinally beneath the chin and chest region of some baleen whales. These may possibly act as numerous "keels" stabilising the whale when it swims, or may allow distension on swallowing or even play some part in thermoregulation.

The external nasal opening, or blowhole, is usually situated on the highest point of the head region posterior to the snout and just in front of the brain case. The opening is either single or double and may be crescentic, slit-like, oval or with a double curve like the letter S. Air mixed with water vapour and oily secretions from the cranial air sacs is expelled on expiration to cause a "blow" or spout. If atmospheric conditions are right the spout can be seen from some distance and its characteristics are typical for each species. The position of the blowhole is important in that it has

Figure 16. A Long-beaked dolphin (Stenella).

to be thrust above the water when the animal breaks the surface to breathe. The blowhole is so placed that only a small part of the head region and back are momentarily exposed during the respiratory act. A dolphin does not have to expend energy by thrusting itself unnecessarily far above the surface and can thus exhale and inhale while swimming at speed. How a dolphin is able to tell that its blowhole has broken the surface of the water and that it can now breathe is difficult to explain.

That part of the facial region of a dolphin which lies between the snout or base of the beak and the blowhole is raised up into a smoothly curved mass, covered by skin and called the melon. It consists mainly of a fatty substance contained in fibrous tissue and there is also some muscle. At first it was argued that the melon provided buoyancy or helped to balance the dolphin as it swam. We shall see that other functions are ascribed to it nowadays.

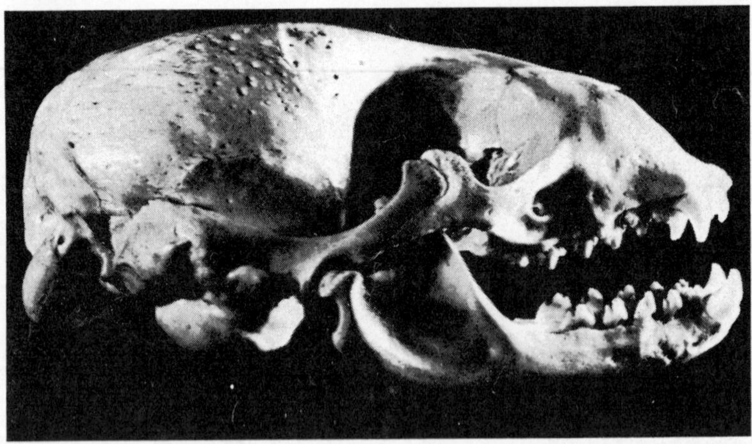

Figure 17. Photograph of the skull of a young Common seal (Phoca).

Dolphins lack an external pinna to the ear. A small pit or dimple lying between the angle of the eye and the anterior attachment of the flippers, and not always easy to discern, marks the position of the external ear. The visible part of the relatively large eye is small and there are stiff eyelids which can be closed. There are no eyelashes. The female urethra and the opening of the vagina lie in an elongated urogenital slit just in front of the anus. The nipples are found in shorter slit-like recesses on each side. The nipple is protruded during suckling; this takes place while both cow and calf are under water. When flaccid the penis is concealed in a

*Figure 18. A Common seal (*Phoca vitulina*); note the opening of the external auditory meatus.*

sheath beneath the abdominal skin opening in a midline slit just caudal to the umbilicus. The relatively long penis is arranged in a coiled form but is capable of being straightened out and thus protruded by muscular action at the time of erection.

Pinnipeds also have a streamlined torpedo-like body and have both fore and hind flippers. The face is somewhat elongated with the closible external nares directed anteriorly. The skull resembles that of a dog but the cranium is large and interorbital region elongated. The orbits are large and the snout is relatively short. The teeth of pinnipeds are quite different both in arrangement and form from those of toothed cetaceans. There are incisors, canines and cheek teeth, all with varying characteristics. The recurved cusps of the cheek teeth (postcanines) of the Crabeater seal of the Antarctic

Figure 19. A young otariid: note the external pinna to the ear and the characteristic flipper.

are most complicated in form. The walrus has the most modified teeth of all pinnipeds. The canines are developed into long tusks: present in both sexes but longer in males and can weigh over eleven pounds.

The eyeball is large; the eyelids can be opened wide giving an appealing appearance. There is a nictitating membrane in several species. The upper lip is like a pad and bears numerous stout vibrissae which are profusely innervated at the base and can be used as sensitive brushes. The external ear is reduced to a small, furled, leaflike pinna in sea lions giving them the name otariid, or eared seals. In true seals only a small closible orifice caudal to the lateral canthus of the eye marks the entrance leading to a meatus that turns downwards for some distance beneath the skin to reach the tympanic membrane. The external auditory meatus is surrounded by a kinked tube of cartilage.

An obvious neck can be extended with strong thrusting movements to snap up fish or attack enemies. The fore-flippers of sea lions are oar-like structures with naked palmar skin and nails reduced to nodules. Strong muscles move the flippers for swimming and help to support the body on land. Flippers are used for hauling out and even for climbing. The naked skin plays an important part in thermoregulation. The seal fore flipper is much shorter with marked claws. It is kept close against the body when swimming but can be used to change direction. The seal's body tapers caudal to the thorax but is expanded at the pelvic region by the presence of a pelvic girdle which is attached to the vertebral column but which lacks a symphysis pubis. All pinnipeds have hind flippers. In sea lions they are very mobile and in swimming they are turned backwards, soles together, making a movable rudder. On land a sea lion can rotate its hind flippers forwards and laterally with the sole pressed down against the ground. Seals can use their hind flippers, assisted by strong, sinuous side-to-side movements of the body, to propel them rapidly through the water. The hind flipper is so constructed that it can fan out its propulsive stroke because the digits are connected by webs of thin skin. The two lateral digits are the longest and strongest. Even so a seal's hind flipper cannot be rotated forwards and seals are not able to progress on land using their hind flippers for support. They hump themselves along in a caterpillar-like fashion, carrying their weight on the chest and then the abdomen and pelvis. The fore flipper is used to help jerk the body along, the terminal phalanges being flexed to increase purchase on the ground.

The vulva is situated immediately in front of the anus and there

*Figure 20. A Common seal (*Phoca vitulina*) making for the sea: note the way the fore flippers are used for locomotion on land.*

is a common cloaca. Testes are inguinal in seals but there is a scrotum in sea lions. The penis is retracted in a cutaneous pouch within the abdominal wall and only projects when erect. The opening lies nearly midway between umbilicus and anus. The penis contains a bone, an os penis (baculum), which increases in size as the seal grows and is useful for ageing purposes. Sea lions have four nipples, seals have two: in sea lions two are in front of and two are behind the umbilicus, in seals the nipples correspond to

Figure 21. A manatee (Trichechus)*: note the large paddle-like tail.*

the caudal pair of sea lions. The nipples are retracted in pits covered over by fur but become erect when the pup sucks. Suckling takes place on land or in shallow water.

Sirenians, manatees and dugongs are bulky, heavy, awkward and lethargic. Their bodies are less streamlined and less elongated than those of dolphins and obviously not adapted for swimming at speed. The head is heavy, the jaws are massive and there is a short snout. There are characteristic, mobile upper lips, looking like a bloodhound's chops, and there is a curious-looking muzzle. The valvular openings of the nose lie on or behind the upper edge of the muzzle. Only skin creases mark the position of the neck. The large forelimbs are like paddles and are used for sculling or equilibration under water and for clasping the grasses which are eaten beneath the water. The body narrows only slightly in the region of the plump abdomen and caudally expands laterally to form a large paddle in manatees and flukes in dugongs. No hind limbs are present and the pelvic girdle is very undeveloped. The eye is smaller than that of a dolphin and only wrinkles indicate the position of eyelids. No external ear pinna is present; the external meatus is narrow and is occluded for part of its course. The genitalia are considerably modified in dugongs although a basic pattern is discernible. The nipples in all sirenians are pectoral in position.

Physiological considerations

A superficial explanation as to how marine mammals hold their breath and dive so long and so deep could be that they must have exceptionally large lungs in which to store oxygen. This suggestion is simply not true; compared with body size the lungs are about the same size as in Man. Many seals usually exhale a large volume of air before diving or early in the dive and thus spend the period of the dive with a quantity of air in the lungs smaller than that which would be present at the surface during normal ventilation. In any event the amount of oxygen in two lungs, even of the largest size possible, would be insufficient to maintain the animal for much longer than a minute and a half without it taking another breath, assuming of course that no other mechanism were in operation. Could there be special oxygen depots, or instant sources of oxygen, almost as there are in modern submarines? Or are there other ways by which marine mammals can perform what seems the impossible?

Oxygen stores are in fact appreciably larger in diving than in terrestrial mammals. The increases are, however, related primarily to the vascular system rather than in respiratory tissues. Blood volume of marine mammals generally is increased to about 10-15 per cent of body weight compared with that of about 7 per cent for Man. The oxygen capacity of the blood is again raised to more than 35 ml/100 ml of blood in some seals and whales. This figure is almost twice as much as that in Man. This is due to a raised packed cell volume and to increased haemoglobin concentration in the blood. The normal value of 63 per cent packed cell volume in some seals would have to be considered pathological were it encountered in a human subject. Such a condition is called polycythaemia and causes trouble in extreme forms due to circulatory problems arising from increased viscosity of the blood and other physiological disturbances. Muscle haemoglobin, or myoglobin, is increased to about ten times that in terrestrial mammals and is more concentrated in diving mammals and gives the muscles a characteristic dark colour. It has high oxygen storing capacity and thus the muscles carry, as it were, their own local stores of oxygen with them on a dive.

Even if all the oxygen is added up that could be stored in a diving dolphin or a seal, considering lung oxygen, blood oxygen and muscle stores, and then divided by the known resting oxygen consumption, it is soon obvious that the oxygen stores are nothing like adequate for the known durations of dives. Some other mechanisms must operate during the dive otherwise the animal would soon become dangerously short of oxygen and would have to come to the surface after a minute or so.

The solution of this fascinating problem began to be uncovered by the French physiologist Paul Bert, in 1870. He observed that when ducks were submerged a profound slowing of the heart rate, known as bradycardia, occurred almost as the dive began. Sixty years later Laurence Irving and Per Scholander working in the United States began to provide more detailed information from experiments on diving various marine mammals on the mechanisms of the prolonged asphyxial tolerance which occurs on diving. During a long dive in seals they found that the heart rate falls to a much slower rate than its resting rate on the surface. They also demonstrated that oxygen dissociated from haemoglobin at a rate different to that from myoglobin. The dissociation of oxygen from myoglobin was exponential and was near zero in a few minutes, whereas the blood depletion of oxygen was linear and never reached zero. At the end of the dive there was always a considerable increase in lactic acid in the blood. Lactate is an end product of anaerobic metabolism. During a dive it appeared that a major redistribution of blood took place. Blood was shunted away from tissues tolerant of oxygen lack such as muscle and viscera and strictly conserved for use by the heart and central nervous system. Metabolic activity was also somewhat reduced during the period of the dive and this helped to conserve oxygen. To some considerable extent, therefore, the seal became a form of heart/brain system.

It was only after World War II that techniques were developed to assess what happened during the period of vascular shunting and how the redistribution of blood might be controlled. Cardiac output has been shown to drop dramatically during a dive but the amount of blood pumped through the heart at each stroke, the stroke volume, remains constant as does the arterial blood pressure. Blood flow in the mesenteric and renal arteries is much reduced by arterial constriction. If a seal is surfaced for a few seconds and, then again submerged, the heart rate rises and falls correspondingly and blood flow through the renal artery starts and stops in harmony. The bradycardia is procured through the vagus nerves and the venous return from abdominal reservoirs is modulated by the degree of contraction of the caval sphincter. The results of all experiments draw attention to delicate refinements and subtlety in control of what has often been called a diving reflex but which, because of more than suggestions of some voluntary element in its control, is perhaps better considered a diving response. In more than one species it seems as if the seal sets the response for dealing with a particular type of dive, again as if in particular circumstances the animal predicts the type of dive which is needed. There is still much to be explained; for example, how does a seal appreciate that it has been diving for long enough and that it must surface for breath? How do muscles continue to contract when they may be without oxygen for very long periods? How does an increased hydrostatic pressure of as much as sixty atmospheres influence muscle conduction or even nerve conduction? Can a deep-diving seal alter its buoyancy in any way so as to facilitate descent and ascent?

These and many other problems would seem to become more cogent when we consider really deep dives. To understand such problems we must consider the physical laws about gases under

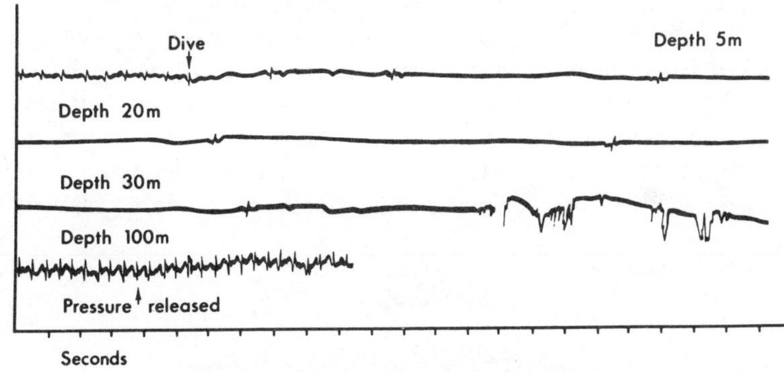

Figure 22. Tracing of an electrocardiograph recording during a dive by a Common seal (Phoca) *in a pressure tank to 100 m: note the bradycardia to as little as 5 beats a minute.*

pressure and gases in solution. Boyle's law states that PV = C, where P = pressure, V = volume of the gas and C = a constant; volume is thus inversely related to pressure. Dalton's law states that the partial pressure of a gas is equivalent to that pressure it would exert if it alone occupied a given volume. The gas pressure in the atmosphere at sea level is about 760 mm Hg of which 21 per cent is oxygen which is a pressure of about 159 mm Hg. If one considers, however, the situation up on a mountain where the atmospheric pressure was only 500 mm Hg, the oxygen concentration would still be 21 per cent, but the pressure exerted by oxygen alone would now be only 105 mm Hg. Henry's law states that at a given temperature the amount of gas dissolved in a solvent is directly proportional to the partial pressure of the gas, provided that there is no chemical combination.

In diving deep as the depth increases, the hydrostatic pressure also increases at a rate of about one atmosphere for every ten metres sea water. Spaces within a diving mammal containing air, such as the lungs, trachea and the middle ear spaces, are naturally liable to collapse under compression. If these structures were not able to be compressed, a pressure differential would build up between each air space and the tissue and blood vessels lining its walls. Only a small pressure difference would be required to cause extravasation of fluid and eventual rupture of vessels. A similar disequilibrium is appreciated by a human being who dives to several feet below the surface when his paranasal sinuses are blocked by swollen mucous membranes and mucus after a heavy cold in the upper respiratory tract. Diving mammals possess numerous modifications which allow the internal air spaces to remain in equilibrium with the external ambient pressure. The thoracic wall is more flexible and the diaphragm is more obliquely

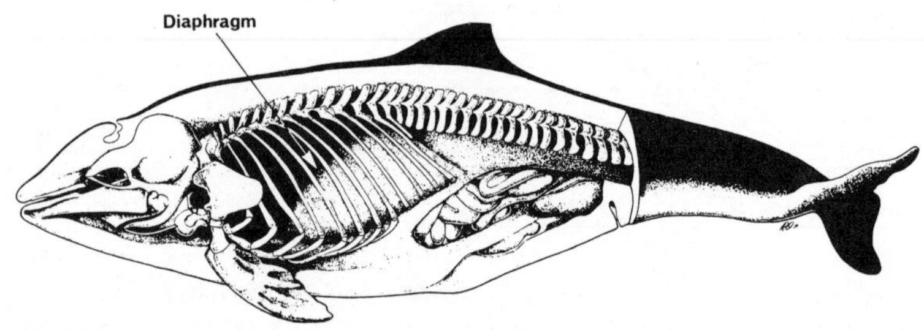

Figure 23. Diagram to show the oblique arrangement of the diaphragm in a porpoise: the lung is shown in black and can be seen to extend beyond the level of the last rib.

placed than in terrestrial forms, thus permitting blood and abdominal viscera to displace the diaphragm into the thorax. The lining of the middle ear spaces contains abundant vascular sinuses which become engorged and so help to reduce the volume of a particular air space when it is under increased pressure.

As the animal dives deeper, the pressure increases in the lungs, the partial pressure of nitrogen also rises, consequently more nitrogen is absorbed by the blood according to Henry's law. This is a potential danger to the diving animal in at least two important respects. High partial pressures of nitrogen are in any event known to have a narcotic effect on the central nervous system of human subjects. In the language of divers the effect has been called the rapture of the deep or in the slang of the Royal Navy, "narks". It can be dangerous indeed, causing disorientation, forgetfulness and a casual lack of concern, in all resembling drunkenness. The condition appears at depths of around 50 m and progresses to unconsciousness at about 100 m. Then again, too rapid a decompression may cause the excess nitrogen dissolved in the blood and tissues to come out of solution and form bubbles. These can cause local damage and pain and can also become air emboli which enter and occlude small vessels in vital regions. Symptoms in human divers develop on return to the surface: the severity depends on the depth of the dive, the time spent there and the time taken over the ascent to the surface. Most common are joint pains, known as "the bends". More serious are "the staggers", an unsteady walk resulting from a spastic paresis, and more serious still are "the chokes", with coughing, shortness of breath and chest pain together with signs of inadequate oxygenation such as pallor and cyanosis. Development of these distressing features may occur some hours after return to the surface.

It is these factors that make a low lung volume in marine mammals advantageous when diving, even though a sacrifice is made of a small part of the oxygen store. If a seal dives with a small volume of air, its lungs are collapsed even at shallow depths and the gas is compressed into non-respiratory spaces. Consequently nitrogen is not absorbed in large amounts by marine mammals and presents no danger. In any event, the volume of nitrogen would be only that from two partially filled lungs and not that derived from the continuous respiration of human divers. Clearly it is thus advantageous in human divers to replace the nitrogen in the gases breathed with some other less harmful gas. Helium is used because less of the gas is absorbed by fatty tissues and by the central nervous tissue and also because the gas leaves the body tissues more rapidly than nitrogen. As far as is known marine mammals do not normally experience the bends or the chokes for reasons explained above. What happens if a seal dives with its lungs full? Experiments suggest that it attempts forcibly to exhale if it cannot at once return to the surface. Sea lions and dolphins are believed to dive always with lungs nearly full but we know only little about their diving to extreme depths. Elsewhere in this text it is argued that it could be essential for all marine mammals to retain some gas in the respiratory pathway in order to be able to produce sounds.

Anatomical adaptations

Every anatomist, ancient or modern, who has dissected a marine mammal has immediately been struck by the curious modifications in structure which his knife or his microscope reveal. Many of the anatomical adaptations seem to be related to life spent wholly or partially in the sea, to swimming and diving. Not a few are found, in varying extent, in all forms of marine mammal, others are more developed in pinnipeds than in cetaceans, and vice versa. Some modifications in structure are even present in terrestrial mammals though usually in not such marked or exactly similar form and the function subserved is in many animals quite different or unknown. On the first approach to a dissection of a marine mammal one is confronted by the strangely hairless, smooth surfaced skin of a cetacean or by the thick hair or fur of a pinniped. There seems to be a contradiction: if a marine existence is better tolerated by a hairless cetacean, why has hair persisted in a seal and for that matter why has man apparently lost so much hair that he has been referred to as a naked ape.

The skin

A striking feature of the bodies of some dolphins and porpoises is

Figure 24. Diagram to illustrate the distribution of pigment in the skin of a Common dolphin (Delphinus).

their complex pattern of skin pigmentation: mainly patterns in black and white. One authority has recently devised a terminology, using for convenience the common dolphin, *Delphinus,* as the most complexly pigmented delphinid. Four basic patterns have been identified: in ascending order of complexity they are called: saddled, spotted, striped and crisscross. The first is so-called because of the large saddle or cape of pigmentation in front of the dorsal fin with the ventral surface having a lighter colour. The saddled pattern is thought to be the most generalized and probably the most primitive. It may have adaptive significance in its use for concealment and even for making the dolphin seem to be something which it is not, disruptive colouration. The striped form exhibits a pattern of pigmented stripes on the sides of the lower trunk. These act as a disguise, breaking up the shape of the dolphin and providing camouflage. They may deceive prey as to which is the dangerous tooth-bearing end, and which the tail end. Stripes masking the eye, or an eye-patch of pigmentation, as well as a lip patch, could be of value in concealment, and, in the case of an eye mask, for reducing glare. Spotting or mottling over the body gives the dolphin camouflage against certain types of background or even making it resemble shoals of smaller organisms. The spotted pattern is found as an overlay upon other patterns. The crisscross pattern is the most complex and is presented as two lines crossing diagonally on the flank with dark pigmentation in the dorsal quadrant, light below, yellowish or brownish pigmentation anteriorly and grey colouration posteriorly. This pattern would undoubtedly help to break up the shape of the dolphin and to disguise and mask its telltale features. It would provide camouflage through a countershading effect and shadow mimicry. The brownish-yellow anterior colouration is more difficult to assess and several suggestions have been made as to its significance. It could be a warning signal to other cetaceans, in a way that the black and yellow markings on certain insects exhibit their dangerous properties. It might be an orientation patch, an identification mark rather like flying a similar flag to indicate membership of the same squadron. Most patterns seem to be slightly different so it could aid in telling one individual from another. On the other hand the patches of yellowish colouration could be quite accidental or a feature carried into adulthood from the juvenile condition. As well as the four types of pigmentation outlined above there are others called uniform, lobate and disruptive: the terms are self-explanatory.

In view of the high swimming speeds of some dolphins, the question as to whether there is anything peculiar about the skin surface might be of great interest. Suggestions have been made that the surface cells of the epidermis of dolphins might be shed and influence the degree of hydrodynamic drag. Suspensions of scrapings from the epidermis have been tested experimentally but have been found to have a negligible effect on drag. It has also been postulated that secretion of high molecular weight liquids into the boundary layer of water close to the dolphin's skin could reduce drag. Unlike seals and sea lions, dolphins lack cutaneous glands so

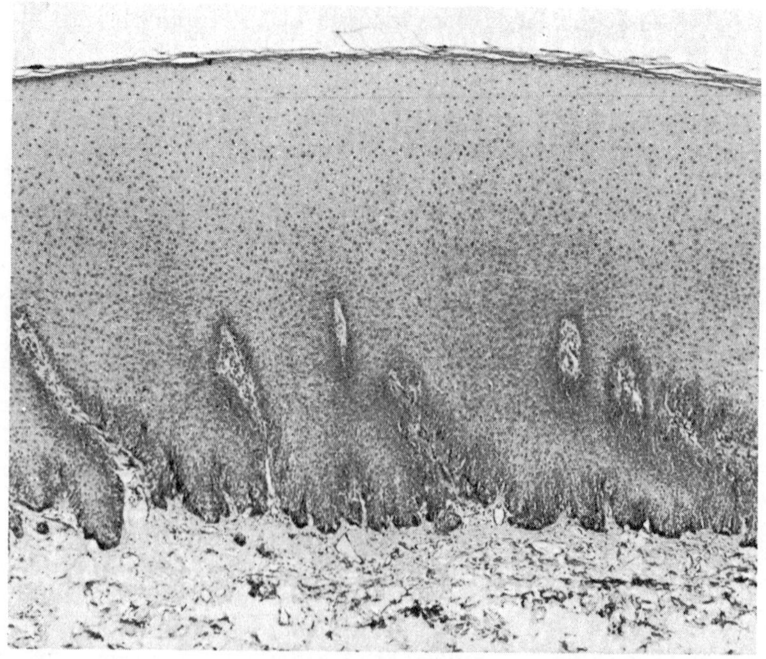

Figure 25. Photograph of the skin of a Common dolphin (Delphinus) *from in front of the flipper: note the thin stratum externum.*

cannot sweat anything on to the skin surface. Another suggestion has been that the lacrimal gland might produce some special substance that could flow back over the dolphin's body and help to reduce turbulence and drag. Experiments with tears have not provided evidence of any such substance and the structure of the small lacrimal gland is not remarkable.

The epidermis of dolphins is from 0.5 to 1.5 mm thick. The surface is smooth and the superficial layer of stratum externum is very thin indeed and is not keratinized. The epidermal cells possess healthy nuclei right to the surface of the skin and thus the epidermis would seem to be alive all through and to lack the relatively thick dead covering of terrestrial mammals and man. No stratum lucidum is present. There are no signs of hairs or of hair follicles except in the newly born when a few foetal hairs protrude from pits on the edges of the upper lip: these are soon lost. The epidermis grows rapidly from the basal region where division of the basal cells occurs along the sides of the dermal papillae. The epidermis is easily scratched or scarred but even quite deep scratches have been seen to disappear after about two weeks, again implying a rapid loss of surface cells and steady replacement from below.

The dermis of dolphin skin is thin, being made of loose connective tissue, and blends imperceptibly with the subjacent fat-containing hypodermis or blubber. It is raised up into numerous narrow dermal papillae which can occur at intervals of 0.1 mm and which project upwards almost halfway through the thickness of the epidermis. The slender papillae are composed of connective tissue and carry vascular loops high up into the epidermis. They are said to be arranged in more or less definite lines and it has been argued that the arrangement of these lines could indicate the shearing forces developed at the surface of the animal. On anatomical grounds it seems more likely that the arrangement of the dermal papillae would be such as to allow easy flexion, extension and bending of the skin mass. They could also anchor the epidermis to the underlying tissue yet at the same time allowing slight changes in shape. Crinkling of the skin, and even quite large skin folds have been observed in some types of dolphin but there is little evidence that these help to reduce turbulence or increase speed. The

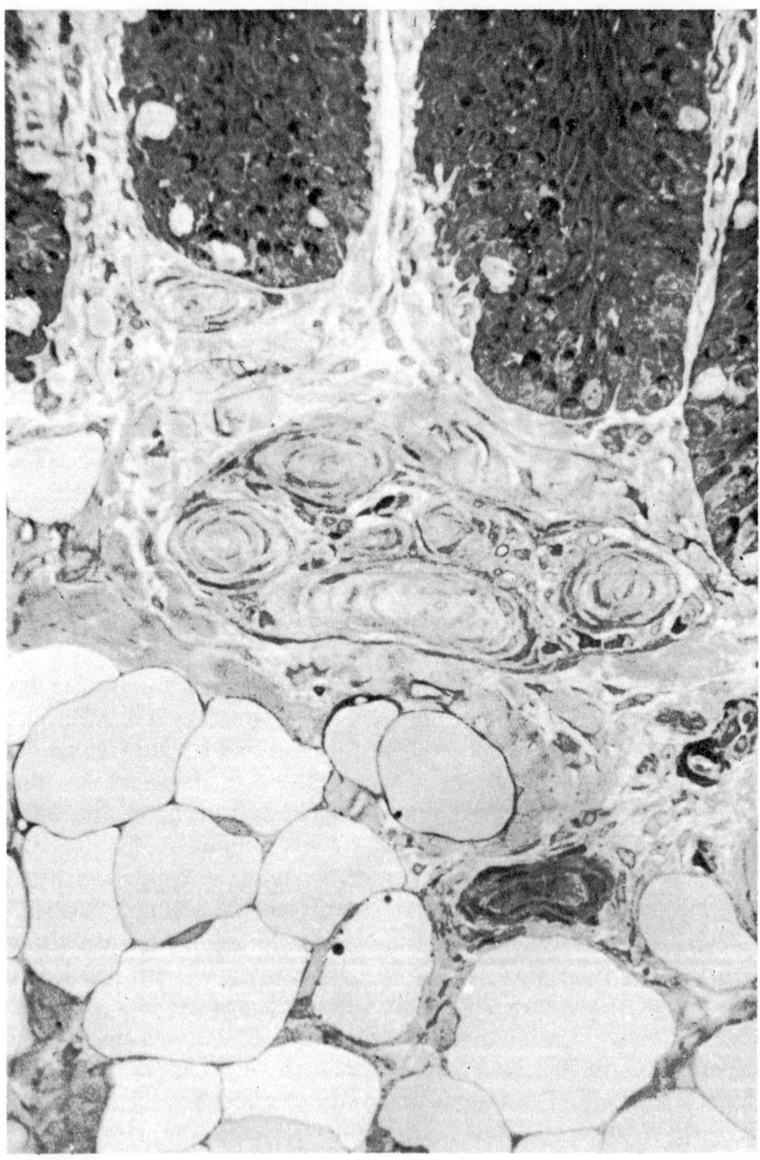

Figure 26. Photomicrograph of a sensory receptor under the epidermis of the snout of a Common dolphin (Delphinus).

dermal papillae contain small arteries which extend to the tips of the papillae as capillaries and blood returns through small veins lying close to the arteries in the dermis. In certain places, such as the upper lip, chin, round the blowhole region and about the urogenital slit there are organized, corpuscular type nerve endings arranged singly or in groups just below the epidermis and at the bases of the dermal papillae. They are ovoid structures made up

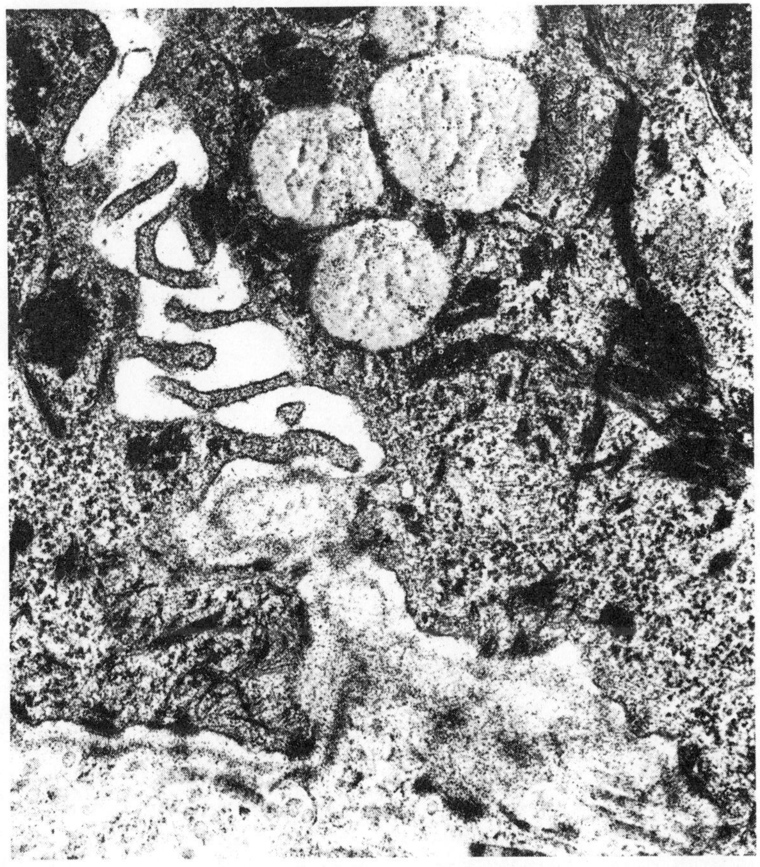

Figure 27. Electronmicrograph of the basal region of the skin of Delphinus*: note the intercellular channel.*

of central cells with a many layered capsule of connective tissue elements, the whole being penetrated by a curving or almost coiled nerve fibre. They resemble the structures in human skin known as Meissner's corpuscles, and although it is tempting to consider them as endings subserving touch, it seems not impossible that they could also respond to vibrations in the water.

The hypodermis, or blubber, varies in thickness over the various parts of the body. It is virtually absent over certain regions in the head, round the blowhole and on much of the flukes, fins and flippers. The fat is accumulated in numerous fat cells held in a loose meshwork of connective tissue. In large whales it can account for up to 45% of the body weight. The composition of the fat is such that it sets solid at low temperatures which is said to be due to the presence of numerous unsaturated fatty acids. This is clearly advantageous to cetaceans frequenting cold waters. Calculations of heat loss suggest that the blubber is usually not thick enough and its conducting properties such that more heat would be lost than is created by metabolic activities. This means that the animal has to keep moving all the time to generate enough heat. Besides being an insulating blanket, the blubber helps to streamline the shape of the animal. Only when a cetacean has become severely emaciated from disease or starvation do we see any of the subjacent hard structures outlined beneath the skin. The blubber fat is a source of food, of energy and water for metabolic purposes as the result of oxidation. It is important during migration, when the available food in the sea is low, and in the females especially during lactation.

Electronmicrographs show that the greater part of the epidermis consists of cells of the stratum spinosum. They are characterised by a remarkably large number of desmosomes in fine processes that connect one cell with its neighbours across complex intercellular spaces. Nuclei are present even in the sub-surface regions. The cells of the epidermis all contain some droplets of fatty material, and mucopolysaccharide is present in the intercellular spaces in the outer region of the epidermis. The cells are flattened in the superficial zone but the intercellular spaces, processes, desmosomes and other characteristic elements can still be seen. Every cell is surrounded by these intercellular spaces and they extend all the way to the surface. They are narrower near the surface and are

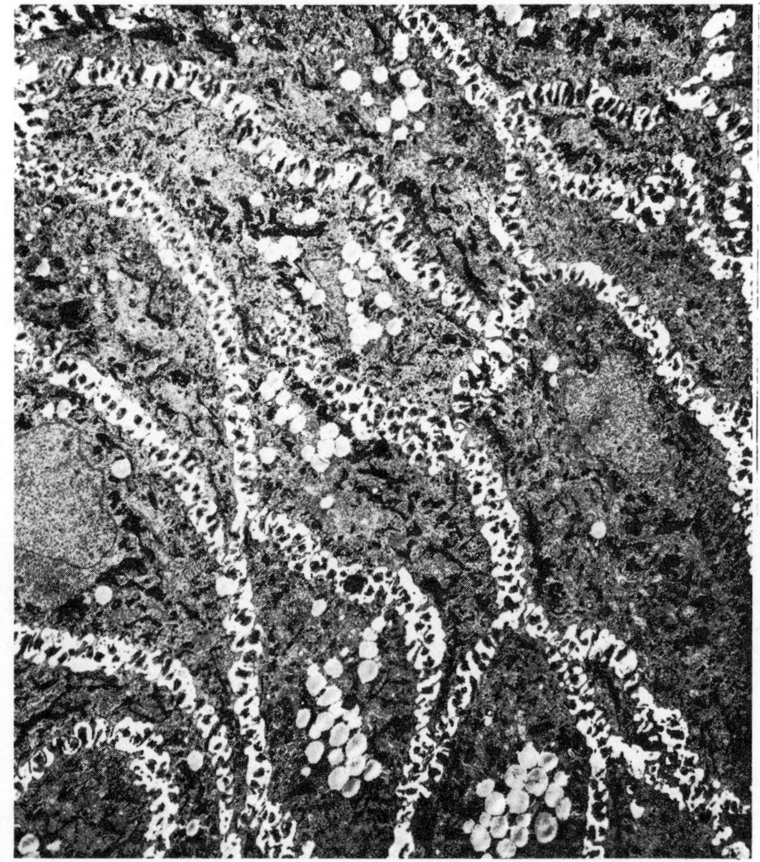

Figure 28. Electronmicrograph of the deep part of the epidermis of Delphinus*: note the extensive intercellular channels.*

arranged parallel to it. In places the droplets of fat-containing material actually protrude from the surface and in others the intercellular space seems to erupt on to the surface so that its contents could ooze or be squeezed out to the exterior. These features may play a part in facilitating intrinsic movements in the skin, in thermoregulation, and in the transport of nutriments and other materials through an epidermis that is alive almost to its surface. In view of the way in which captive dolphins habitually rub or scrub their skin against objects and brushes provided for the purpose, like being scratched or rubbed down, it would seem that they avoid piling up of the surface of the epidermis. If the surface layer is continuously being lost, it seems inevitable that some material in the superficial regions is liberated on the skin surface.

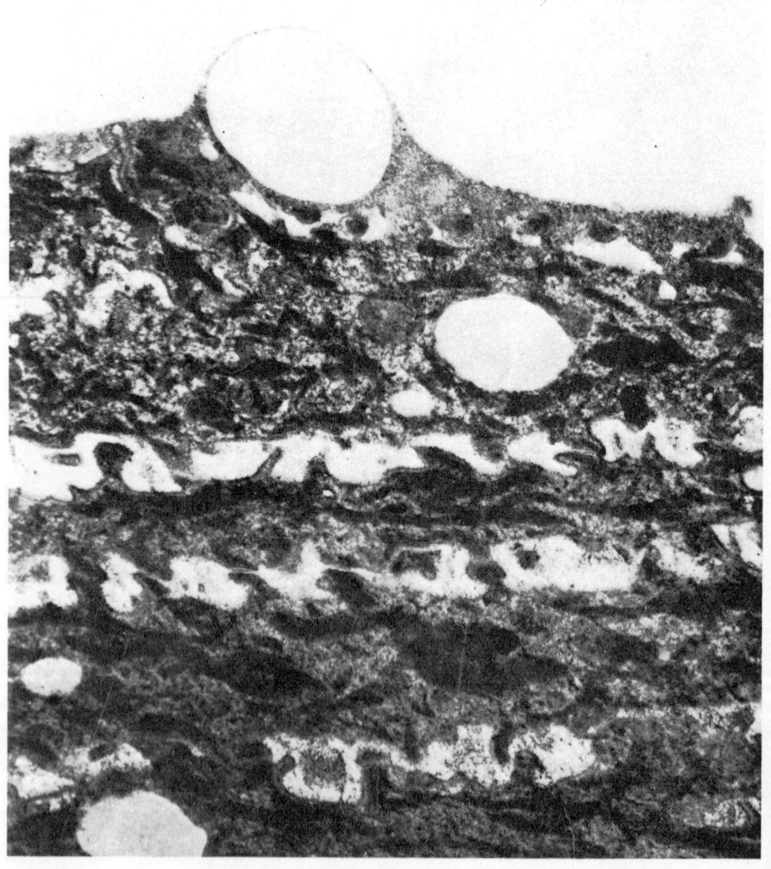

Figure 29. Electronmicrograph of the superficial region of the epidermis of Delphinus*: note the intercellular channels and surface droplets.*

Seal skin possesses several interesting adaptations and although several may be related to a marine existence not all are understood.

The epidermis is relatively thick, measuring up to 1.0 mm over the entire body. The stratum corneum is deep and compact and composed of nucleated cells which extend almost to the surface. The entire layer seems to be a relatively permanent one as there are few mitotic figures in the epidermis. A stratum granulosum is absent and there is little keratinization. The stratum corneum is not unlike that found in mucous regions or in parakeratotic lesions. This would seem to be an excellent adaptation as any superficial layer of scaly keratin could very well become waterlogged and slough. The thickest region of the epidermis is probably on the chest. The epidermis of all otariids and the walrus is whitish or light grey and in phocids it is brown or black: the epidermis exhibits a scaly relief.

Pigment is abundant in the epidermal cells, as it is in pigmented dolphin skin, and is always localized as a cap over the nucleus in the superficial region of each cell. Pigment is more concentrated in the outer cells of the stratum corneum. The functions of the pigmentation may be to absorb heat, and also to shield cells from sunlight when the seal is lying out on a bank.

The dermis is very vascular and, in the papillary layer of many regions, many enlarged venules can become so dilated that they look like cavernous sinuses. It has been argued that this increased vascularity of the integument may perform the function of another blood reservoir. Blood could accumulate in the dermis during dives, kept there as a result of peripheral vasoconstriction and a sluggish venous return. The insulating capacities of the pelage and blubber could then be important factors in preventing heat loss. The strong fibrous layer of the dermis is like a sieve of connective tissue with masses of fatty tissue in its meshes. The fatty tissue is continuous with the underlying hypodermis or blubber. The insulating capacity of some types of seal blubber has been shown to resemble that of certain widely used insulating materials such as asbestos fibres. It has been calculated that the insulating effect of the blubber is sufficient to prevent loss of heat from the seal even when the temperature of the seal is 36° and the external temperature is as low as 0°C. This assumes that the seal is at rest: if it were to be active for prolonged periods then it would seem that the problem is the dissipation of heat rather than its conservation. Seal pups at birth are to some extent poikilothermic. Newborn pups often display shivering and are restless in warm sun. The elephant seal cow sometimes flicks sand over her pup until it is partially covered and protected. Blubber accumulates rapidly with suckling and efficient thermoregulatory mechanisms have become established by the time of the first moult or within several weeks of birth. A fall

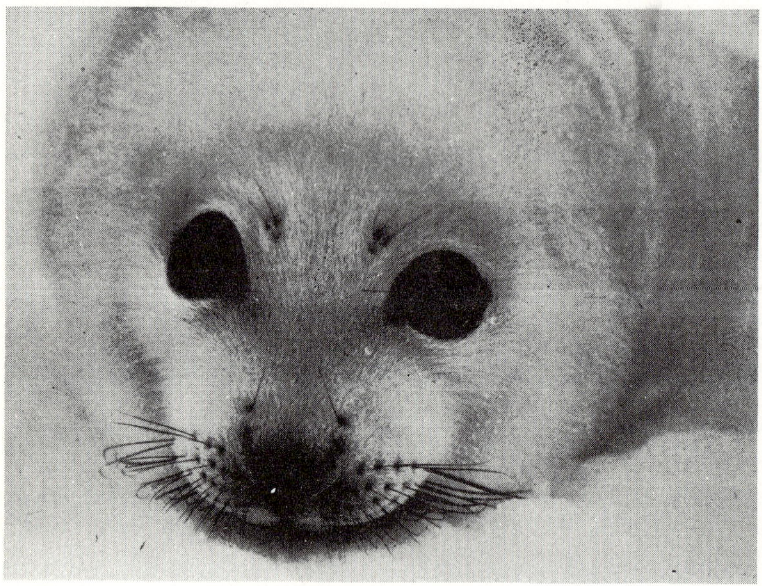

*Figure 30. A Harp seal pup (*Pagophilus groenlandicus*): the thick white coat is much in demand by commercial hunters. Photo: Professor K. Ronald.*

in body temperature of up to 2.5°C, has been observed in young seals used for experimental dives and shivering during such dives is not infrequent even in older animals. If seals are kept in captivity for long periods without exercise and are well fed, the accumulation of much blubber can cause difficulty in dissipating heat especially in warm weather. If a seal is chased over land it can generate so much heat, which it cannot dissipate, even by a type of panting, that it becomes exhausted and can even die. One can conclude that within a certain range of external temperatures and under appropriate activity states the seal manages very well in balancing heat conservation, heat production and heat loss. It can also be asked whether seals can sweat; we have already shown that dolphins are unable to, as they lack sweat glands of any sort.

The cutaneous glands of seals are of the apocrine sweat and sebaceous types. The sweat glands of sea lions open into the pilary canal above the sebaceous gland openings, but in true seals the situation is the other way round. Sweat glands in seals are simple coiled tubes; they are much larger in sea lions. Previously thought to be absent from the hairless flippers of sea lions, they have now been found in this region. Evidence has also been produced of definite sweating from the flipper skin in hot weather but the composition of the sweat is not known.

Sweat glands in the walrus, an almost hairless animal, are large but are said to exhibit seasonal activity. Sebaceous glands are large in seals and sea lions, but smaller in walruses. Sebaceous secretion may waterproof the hair and epidermis and help the hairs to pack closely and sleekly together.

Hair pattern in seals is much more uniform than in terrestrial fur-bearing mammals. A unit construction is the basis of the hair pattern; a unit contains one coarse, primary or overhair and several slender, secondary or fur hairs. Units are equivalent in size but are spaced more irregularly. Primary and secondary hairs all protrude from a common pilary canal and the primary hair lies on the distal or anterior side of the bundle of secondary hairs. The primary hair becomes rounder and smaller with growth and is more deeply rooted than secondary hairs. The primary hairs lack a medulla in seals but it is present in sea lions.

Primary hairs have persisted in all seals and exhibit little variation. Hair units are arranged regularly in sea lions but are spaced in groups, or arranged in groups and rows in other forms. The number of primary hairs for each 10 mm^2 of skin is from about 18 (walrus) to 115 (fur seal). Male otariids have the longest hairs as long as 30 mm or more: this is perhaps a defence against biting during the breeding season. Water temperature does not seem to influence hair length; in fact, ice-frequenting seals have the shortest hairs. Again, there is no precise relationship between the size of the animal and the diameter of its primary hairs: in other words, giants do not necessarily have hairs of the thickest sort. The seals with coarser hair do, however, have primary hairs of a more uniform diameter.

The number of secondary hairs varies from none in some seals to 50 or more in fur seals. The average number of secondary hair follicles is, however, only about 15: old fibres remain unshed and are accumulated. Walruses, which are almost hairless, have only one secondary hair for every 50 primary hair bearing pilary canals. Secondary hairs accumulate with age in some seals but there is much individual variation. The diameter of secondary hairs varies from 7 μ in fur seals to over 30 μ in antarctic seals. Sea lions and fur seals possess fine underhairs whilst in true seals they are usually coarse: all secondary hairs are nonmedullated.

Some experts argue that the pelage of pinnipeds has lost the primitive ability to vary its insulating capacity. Arrectores pilorum muscles are absent and except in the fur seals the pelage no longer has much significance as a body covering. The thick fur and blubber of fur seals can be considered such an effective insulation that they are adaptations related to occupancy of cold waters. Such a mechanism for heat conservation can be a difficult physiological problem for fur seals during terrestrial activities and once again could make sustained vigorous activity on land impossible. Major heat loss must be procured from the hairless flippers, and through other pathways such as passage of urine and faeces and through the respiratory system. Once again, fur seals can be killed from heat stroke while being driven across land: their body temperature can rise to 44°C. Pelts obtained from deaths from this cause are known as "road skins".

There has been some argument as to whether fur seal pelage (especially) is so dense that in some regions it never becomes properly wetted. Numerous minute bubbles of air have been seen to be trapped in the secondary hairs of seals when they have been dived experimentally. Such an insulating blanket of air bubbles trapped in the fur might provide a temporary barrier to heat loss during the first few dives or even for quite a time during a long dive. It has even been suggested that this mechanism, as well as the sebaceous coating to the epidermis and the hairs, could provide a thin but effective heat carrying layer on the surface of the seal which might affect the properties of the boundary layer and thus be of advantage in progress through the water, especially ice cold water.

An important characteristic of the pelage of pinnipeds is that it undergoes the phenomenon of moulting in relation to the season. The significance of moulting is still obscure and the cause of the moult has been sought in relation to more than one factor. The seasonal type of moulting starts just before or at the commencement of the breeding season and is thus annual. The degree of

*Figure 31. A Harp seal (*Pagophilus groenlandicus*) showing its vibrissae. Photo: Professor K. Ronald.*

loss of moult varies from individual hairs, though some may remain fast, to a widespread sloughing of hair together with extensive sheets of epidermis. Hair loss may be patchy in many species. Most of the primary hairs are moulted within a few days of the loss of the underhairs. It has been argued that moulting is a replacement phenomenon, if so it is strange that it may be so severe, as in elephant seals, that the animal is virtually incapacitated and must clearly suffer profound physiological re-adjustment during the moult. Hair replacement and growth can occur in waves or in localised patches. This suggests some basic regional difference in timing of the replacement response. In view of the long-held suspicion that no hair grows without some basic hormonal stimulus, it has been tempting to incriminate adjustments in the level of sex hormones, and/or pituitary hormones, as the primary factors in causing the moult. A seasonal shedding of the epidermis has not, to our knowledge, been described in dolphins.

Yet another interesting feature of the integument of pinnipeds is the facial vibrissae: these can be mystacial, supraorbital and facial. The degree of development varies amongst pinnipeds and their functions are still unknown. The follicles of the vibrissae are abundantly innervated and the vibrissae seem to be elaborated in terms of amplification of sensory stimuli. It is possible that the vibrissae could help in establishing the existence of underwater disturbances, tides and currents, even objects in the water. An irregular surface to the vibrissae ought to be of assistance in this respect but by no means all pinnipeds exhibit such a feature to their vibrissae.

An appreciation of immersion would seem to be essential to any pinniped but just how it appreciates such a sensory modality of 'wetness' is unknown. It might involve detection by the skin of changes in temperature, or of increased resistance exerted by water against the vibrissae as opposed to their free movement in air. There might even be some specific receptor, but any evidence of its existence is lacking. The face of a seal is profusely innervated by a very large trigeminal nerve of which the infraorbital branch is a major component. The trigeminal nuclei are more developed than in other carnivores, and all this suggests that the facial region is particularly sensitive.

Heart and vascular system

It might at first seem obvious that a huge whale, or a fast swimming dolphin ought to have a very large heart, both absolutely and relatively. In fact the heart has been shown to differ little in relative shape, size and weight from that in other mammals: in large whales it would seem smaller and perhaps less powerful than might be expected. It is much more in the peripheral vessels that striking modifications are found which are undoubtedly related to vascular participation in the diving response. Many arteries are thick walled and have increased elastic tissue. In cetacea the arteries in certain regions are formed into massive plexuses or retia mirabilia. These structures are present in substantial masses in the thorax between the ribs and vertebral column, about the spinal medulla, at the base of the brain, in relation to the eyeball and about some arteries, especially those of the appendages. Veins and lymphatics may also be present in these vascular conglomerations, which astounded those who first saw them and who, then as now, sought to think up some explanation for their existence. Pinnipeds do not possess arterial retia to anything like the same extent as cetaceans. They do, however, exhibit numerous and marked venous plexuses, some in situations where retia exist in cetaceans and others in quite different sites.

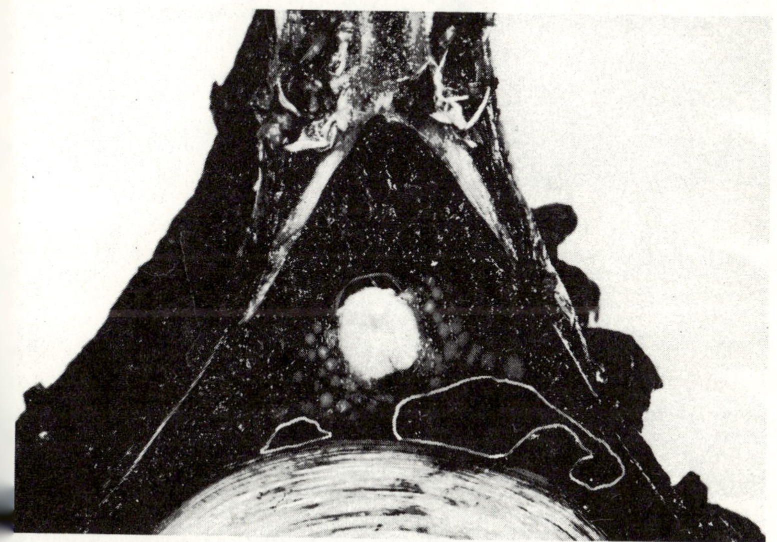

Figure 32. Photograph of the retial tissue, spinal cord and spinal nerves in Globicephala. *One large and one small ventral extradural vein are also present.*

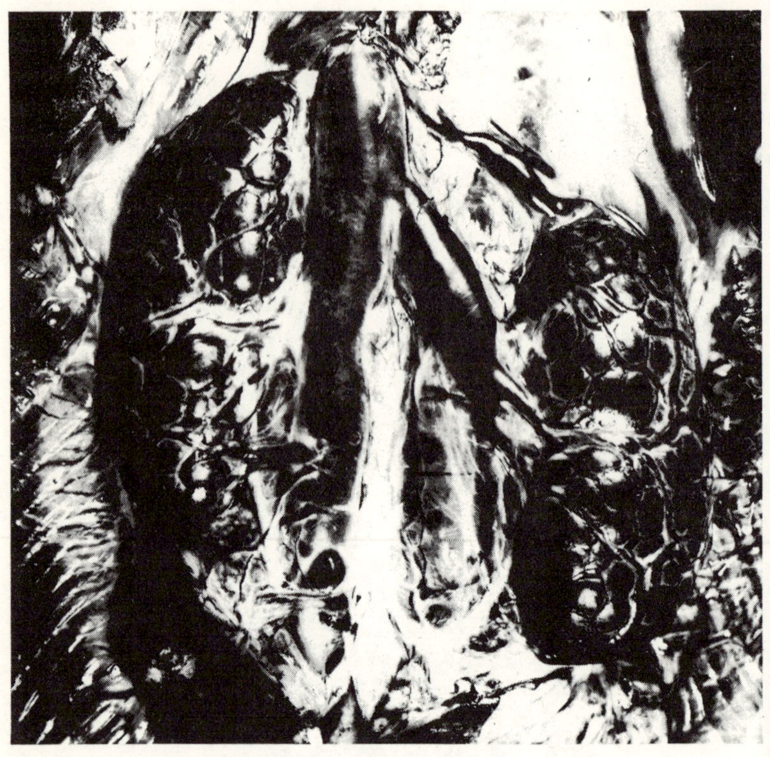

Figure 33. Photograph of the kidneys of Phoca *showing the stellate renal plexus and duplicated posterior vena cava.*

The precise functions of retia in marine mammals are still obscure. They could be involved in maintaining a steady blood flow, in equalizing pressure differences, during diving, as temporary reservoirs of oxygenated blood, as space fillers or cushions, again to act during diving, as thermoregulatory devices in counter-current heat-conserving structures or simply as stores available during the re-distribution of blood that must occur during the diving response. Retia mirabilia are, in fact, widely distributed in mammals and have been classified under several headings:

1. Diffuse plexiform anastomoses of arteries and veins in the limbs of certain mammals.
2. Vascular bundles of small arteries, with some veins, enclosed in a common fibrous sheath, as in cetacean limbs and body wall.
3. The thoracic masses of retial tissue as found in cetaceans.
4. Arterial plexuses or convolutions not associated with any modification of adjacent veins, as in the carotid rete of ungulates.

There is always marked enlargement in marine mammals of all the many thin-walled veins. This is associated with a total blood volume that is relatively greater than that of terrestrial mammals. The posterior vena cava is often duplicated and in pinnipeds there is usually a large hepatic sinus or considerable enlargement of the hepatic veins. The vena cava alone may be large enough to hold one fifth of the total blood volume. These enlarged vessels are clearly blood reservoirs and before a dive they contain blood which is relatively well oxygenated. Another aspect of this considerable venous reservoir is that it provides a space where part of the blood volume can be isolated safely when the limited type of circulation develops during the diving response.

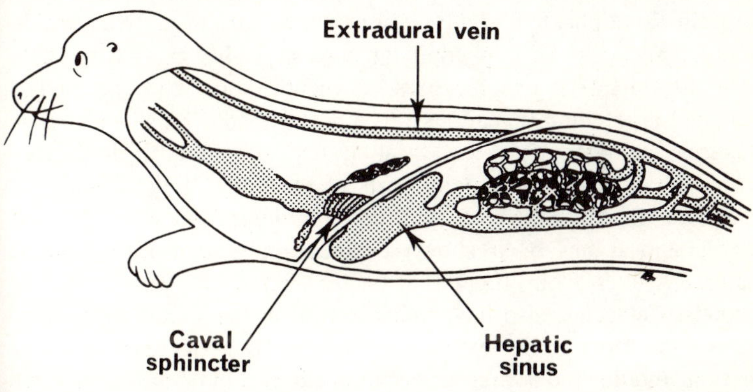

Figure 34. Diagram to show the position of the caval sphincter, hepatic sinus and extradural vein in a seal.

In most pinnipeds, especially in deep or prolonged divers, but not in cetaceans, there is a thick sphincter of striated muscle about the intrathoracic posterior vena cava just anterior to and connected with the diaphragm. Supplied by the right phrenic nerve, it is capable of occluding the abdominal venous return and has been shown under experimental conditions at least to contract when a seal starts a dive. It does not seem to contract when the diaphragm contracts during respiration on land. This caval sphincter can prevent venous engorgement of the heart in bradycardia and probably regulates the blood flow returning to the right atrium and thus to the lungs. The sphincter may also come into action when a seal ascends vertically to the surface to take a breath: its closure at the moment the seal took in air would prevent blood flooding the pulmonary system. A true caval sphincter is lacking in cetaceans but in a few there is a sling of muscle passing dorsal to the vena cava.

Another modification of the venous system is found in relation to the spinal cord. In pinnipeds there is a large thin-walled vessel lying dorsal to the spinal cord, it has been called an extradural intravertebral vein because of its anatomical relations. Anteriorly, the vein is duplicated and receives venous drainage from inside the skull through the hypocondylar canals. It also gains tributaries from the dorsal venous plexuses, segmental veins and the azygos system. Large communications are made with the renal and pelvic plexuses and thus the extradural system provides an important venous channel between many parts of the body, both cranial and caudal to the diaphragm. It has free communication with all these vessels and it has been shown experimentally that radio-opaque material passes from the heart to the extradural vein in a few seconds. Cetaceans also have veins ventral to the spinal cord which is also for most of its extent surrounded by retial tissue. Those vascular modifications present in pinnipeds and in certain cetaceans, and which may or may not be related to diving can be summarised as follows.

Large, distensible, thin-walled veins.
Retia mirabilia of various types.
Venous plexuses often of large capacity.
Enlarged, duplicated abdominal posterior venae cavae.
Renal stellate venous plexuses.
Caval sphincter or caval "sling".

Hepatic sinus with enlarged hepatic veins.
Enlarged intravertebral extradural veins.
Modified azygos system.
Reduction of the internal jugular system.

The reasons why one hesitates to associate all these characteristics with a diving capability is that they are not always present together in every aquatic form and also because some are found in terrestrial mammals not known voluntarily even to take a bath. Furthermore, although several features are especially well developed in deep diving pinnipeds such as the antarctic seals, some are absent in cetaceans known to reach considerable depths. It would seem that there are several ways in which the venous system at least has become modified to deal with the re-distribution of blood that occurs when the diving response develops. Many electrocardiographic recordings are now available to show what happens to the heart rate and the cardiac cycle during the diving bradycardia in a wide variety of marine mammals. More is known about pinniped responses than about those in cetaceans and certain generalizations can be made, always bearing in mind that the very complexity of the vascular system and its delicately balanced control mechanisms through the autonomic nervous system will allow greater plasticity in the type of response than we realize.

The bradycardia is not an isolated vascular phenomenon but is associated with decreased cardiac output, a widespread peripheral arterial vasoconstriction and continued maintenance of blood pressure. There is a reduced blood flow in viscera, skeletal muscle and much of the skin but cerebral and coronary artery flow is maintained throughout the dive. The bradycardia is brought about through the vagus nerves. If one vagus is cut the degree of bradycardia is influenced; if both vagi are severed the heart rate increases and the bradycardial response is abolished. Heart slowing during experimental dives develops rapidly in smaller phocid pinnipeds *(Phoca, Halichoerus, Cystophora)* but takes longer to develop in larger forms *(Mirounga)*. The diving bradycardia is more profound and it develops more abruptly in adults than in pups at the start of a dive. There is some evidence that the slowed heart rate, in young phocid seals at least, increases as the dive progresses beyond what may be a critical length of time. One possible explanation is that after the oxygen tension in the blood has fallen to a certain level, the available blood has to be circulated more rapidly to keep the brain and heart functioning.

The bradycardia appears to be most marked when a seal is dived experimentally to great depths. It is less marked in unrestrained seals than in those secured for experimental dives but we have little information about unrestrained seals diving to great depths. In those seals known to dive deepest and longest and in which a diving bradycardia occurs, there is marked enlargement of abdominal and some other venous pathways, a large hepatic sinus and a well developed caval sphincter. Section of the right phrenic nerve does not completely abolish the diving bradycardia but significantly delays its onset after the start of a dive. In those sea lions which dive for relatively short periods the true diving bradycardia may not develop for up to half a minute. This could mean that the sea lion is hesitating over what sort of a dive is needed; its response might therefore be optional rather than obligatory. A temporary bradycardia, probably a type of cardiac arrhythmia, can be induced by forced breath-holding and by other means.

Properties of the blood of some marine mammals
(Most figures represent mean values from several individuals)

Species	Packed cell volume %	HG (g%)	Oxygen capacity (vol. %)	Blood Volume (ml/kg)
CETACEA				
Bottlenose dolphin *(Tursiops truncatus)*	45	18.2	24.1	71
White-sided dolphin *(Lagenorhynchus obliquidens)*	53	18.8	25.1	108
Dall's porpoise *(Phocoenoides dalli)*	57	20.3	—	143
PINNIPEDIA				
Californian sea lion *(Zalophus californianus)*	45	14.7	17.5	92
Walrus *(Odobenus rosmarus)*	42	16.2	23.4	106
Harbour seal *(Phoca vitulina)*	52	20.0	26.4	132
SIRENIA				
Manatee *(Trichechus manatus)*		17.2		
Man, male	40-54	13.5-18.0	18-20.4	70-72

Respiratory system

The respiratory passages and the lungs of marine mammals are modified in numerous ways that can be related to diving with the breath held or which allow control of the amount of air retained in the lungs at the start of a dive or during its progress. The basic mammalian pattern of a nasal component of soft and hard tissues, a larynx and windpipe or trachea, a bronchial tree and lungs together with a thoracic cage and diaphragm are all present but there are differing features in all of them.

The nasal region of all cetaceans has evolved in remarkable complexity to provide a series of sinuses, nasal passages, valves and plugs. All these are found in the region between the posterior nares behind the palate, and the blowhole. Numerous sheets of the complex maxillofacial, nasal and labial muscles overlie and surround these spaces and structures. They are able to open or close them or modify their arrangement. Water can and does enter the most superficial parts of the nasal apparatus but various factors prevent water passing farther into the passages. The nasal sacs may be concerned in some way in the production or modification of sounds emitted under water as will be discussed in a later section. The elongated larynx, modified from the epiglottic and arytenoid cartilages, is inserted and probably held permanently in place by a strong muscle sphincter, into the posterior part of the nasal cavity. There are no vocal cords but it is still possible that the larynx is the source of certain underwater noises (see later). The intranarial position of the larynx, with channels passing on each side into the oesophagus, allows food to be swallowed without disengagement of the larynx. Cetaceans are thus able to breathe with the mouth open and while swallowing. The situation is very different in pinnipeds: they have no complex nasal apparatus and the general arrangement resembles that found in dogs. The larynx is some way from the posterior nares and is covered by a typical epiglottis when swallowing occurs. The external nares can be closed tightly to prevent the entry of water. There is also much development of the conchae, or turbinate bones, which allows warming of inspired air.

The trachea is short but wide in cetaceans but is relatively longer in pinnipeds. The rings of cartilage are completed, or almost so, except in antarctic seals which are known to dive deep. In them, the rings are incomplete and flattened and allow the trachea to collapse completely. Observations on seals dived experimentally show that the diameter of the trachea from a lateral view is reduced to less than half its original dimension at pressures of about 31 atmospheres absolute which equals a depth of just over 300 m. It appears to be important that the trachea should not collapse completely for as long as it remains open it is able to function in sound production and this could well be of value at great depths. The structures for closing the glottis are well developed in deep diving seals, implying that retention of air caudal to the glottis could be required under certain circumstances.

It is also known that the thorax of a dolphin trained to dive to depths of 300 metres collapses markedly. A 200 kg dolphin has a lung volume of some 10 to 11 litres and it is argued that at a depth of 300 m, the compression of gases could cause a difference in buoyancy of about 10 kg.

Many marine mammals possess an increased number of floating ribs. These are ribs with no connection to the sternum; and because of this and for other reasons the thorax is remarkably flexible. Together with the extreme obliquity of the diaphragm, this flexibility can result in almost complete collapse (atelectasis) of the lung at the end of expiration. Physiologically, the implication is that the lung volume is almost entirely tidal air on inspiration and that hardly any residual air is to be found in the lungs, but only in the dead air spaces. Lung volume appears to be less in deep divers: ability to expel air almost completely from the lungs could have several advantages when embarking on a deep dive. It seems, too, that there are volumetric differences in the amount of air in the lungs depending to some extent on whether the seal anticipates embarking on a shallow or a deep dive.

The lungs of marine mammals display much increase in the extent to which cartilaginous rings extend along the bronchial tree, even as far as the alveolar ducts or respiratory sacs. This means that the bronchial tree is less compressible than the trachea or the alveoli. It is able to retain its dimensions under pressure and so to accommodate gases which are squeezed out of other regions. There is reduction in the length of bronchioles in odontocetes and there is a remarkable series of myoelastic sphincters along the terminal bronchiolar segments. Much elastic tissue is also found elsewhere in the lung substance. Powerful muscle sphincters are present in the lungs of large baleen whales about the mouths of the alveolar sacs. These arrangements could control distribution of air in the alveoli and bronchi, regulate pressure, trap air in specific regions, and modulate its delivery to dead air spaces. Some could also be concerned in providing both elasticity and flexibility in a respiratory system which needs suddenly to be re-inflated after virtual collapse.

Respiratory function in some marine mammals

Species	Tidal volume (L/100kg)	Tidal volume (L)	Lung volume (L/100kg)	Lung volume (L)	Oxygen utilisation (%)
CETACEA					
Bottlenose dolphin *(Tursiops truncatus)*	5.9		6.6		9-10
Pacific pilot whale *(Globicephala scammoni)*	2-8.8	9-39.5	10	45.1	
Killer whale *(Orcinus orca)*	2.7-5.6	30-61			
PINNIPEDIA					
Harbour seal *(Phoca vitulina)*	1.8		9		5-7
Californian sea lion *(Zalophus californianus)*			11	4.2	
SIRENIA					
Manatee *(Trichechus manatus)*	2.9		5.1		7-10
Man	0.8		6		4-5

Production of sounds under water

Who it was who first discovered that dolphins emit noises under water is not known. Their sound producing abilities were definitely commented on by ancient Greek writers and Pliny the Elder remarked that a dolphin's voice had a moaning or wailing quality not unlike that of a man's. There are many charming ancient fables and anecdotes about dolphins and their habits and modern investigations have shown that many of them are not all that far from the truth.

Despite the claims of sailors and harpooners that delphinids squealed and that the White Whale or Beluga *(Delphinapterus leucas)* 'sang' so loudly under water that it was called the sea canary, it was only the use of underwater listening apparatus during World War II that revealed that many cetaceans could make noises. It had been generally believed that cetaceans were mute, or that if they did make noises it was the result of forced exhalations. Cetaceans lacked vocal cords in the larynx and so could not be expected to be especially loquacious: though we now know it is not necessary to possess even a larynx to be able to talk. Human patients after laryngectomy can learn quite quickly to speak understandably by using the oesophagus.

It was not really until Arthur McBride noted about 1947 at the Marine Studios in St. Augustine, Florida that dolphins were able to avoid fine mesh nets under difficult visual conditions and postulated that they did so using a sound echolocating mechanism that any notice was taken of the possible meaning of their sounds, and then only after some years. In the early 1950s recordings were made of bottlenosed dolphins *(Tursiops truncatus)* swimming freely in the sea and it became possible to elicit their auditory responses to frequencies above 100,000 cycles per second. The human ear is sensitive to sounds of frequencies between 30 cps and 15,000 cps: musical instruments use a range of between 25 (organ) and about 5,000 cps (piccolo flute).

Many species of cetacean are now known to make sounds under water. The noises made by whalebone whales are in general of low frequency, typically low moans or screams of complex type ranging from 20 to about 1,000 cps. The Californian Grey Whale *(Eschrichtius),* the Finback *(Balaenoptera),* the Greenland Right Whale *(Balaena)* and Biscayan Right Whale *(Eubalaena),* and the Humpback Whale *(Megaptera)* have all been heard. Of the odontocetes the delphinids are particularly vocal and have been studied the most. The knowledge of how to keep dolphins in captivity has resulted in more refined studies involving expensive and complex listening and recording devices.

A catalogue of the noises produced by bottlenosed dolphins made by various investigators includes several classes of sounds. They are firstly high pitched whistles of rising, or falling or undulating form; the fundamental frequency is usually continuous and modulated. First and second harmonics are often produced, sometimes as many as seven; they are usually discontinuous. The frequencies of the sounds emitted extend from about 8 to over 64 kcps and some harmonics of up to 150 kcps have been detected.

The second type of sound produced by dolphins, in fact by all odontocetes that have been studied so far, is described as a click. The clicks may be single, emitted at intervals in a desultory manner, or with increased rate up to about 600 clicks per second. The repeated clicks follow each other so fast that they produce a 'click train' which at slow rates of repetition sounds like a rusty hinge or a creaking gate opening. Noises that sound like squawks or barks underwater in fact consist of clicks emitted with the astonishing rapidity of up to 1200 per second. These sonic clicks are of between 6 and about 30 kcps, but more than one investigator has demonstrated that ultrasonic clicks of over 100 kcps are also produced. These clicks appear to be directed forwards straight in front of the animal in the direction its rostrum is pointing. Although the evidence is not conclusive it seems that the Common Porpoise *(Phocoena phocoena)* and the Dall porpoise *(Phocoenoides dalli)* produce clicks only and no pure tone signals.

Dolphins also produce other raucous noises under water which are quite inaudible to man and can be described as short barks or snaps. They may be accompanied by sudden closing movements of the jaw or gagging or gawking postures. These may be directed at an intruding dolphin or at a human being approaching the animal. The noises appear to be formed by a rapid series of clicks emitted at great intensity. They indicate that the dolphin is disturbed, displeased or even angry.

Dolphins can also make noises in air and can be trained to 'creak' or emit sounds with the blowhole open. After some weeks in captivity the animals learn that their underwater sounds are not heard by man and so they emit sounds out of the water.

The problem of just where the clicks and whistles of dolphins are produced anatomically is not easy to solve. Some authors have maintained that clicks and whistles can be produced at the same time; clicks on the left side and whistles on the right together or separately in one and the same dolphin. Sounds can also be produced with the blowhole open or closed and bubbles of air can sometimes but not always be seen escaping from the blowhole during vocalization.

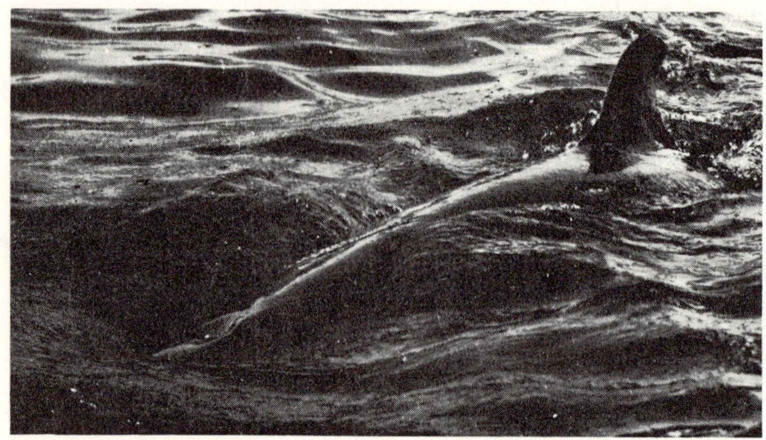

*Figure 35. Photograph of a False killer whale (*Pseudorca*) taking a breath.*

The nasal apparatus, that is the blowhole, nasal passages, valves and related sacs, of odontocetes has long been known to be an anatomical wonder and its functions only poorly understood, if at all. The muscles that open and close the blowhole of the Common Porpoise and that act upon the various sacs have been said to form one of the "most complicated yet more exquisitely adjusted pieces of machinery that either nature or art presents" (F. Sibson, 1848, whose original drawings may still be seen at the Library of the Royal Society). Several anatomists have attempted the almost impossible task of making homologies between the nasal arrangements of several cetacean species and those of other mammals including man. Earlier workers just could not fathom the complexities of this highly adapted region. One of them wrote that the nasal sacs or bye-cavities had no special function, they were nothing but the vestiges of a widely expanded nasal cavity which ceased to have any useful function because of lack of cartilaginous and bony support. More recently, two enlightened investigators have written more aptly that no anatomical description can really describe the breathing cycle of a swimming dolphin, or convey the "sense

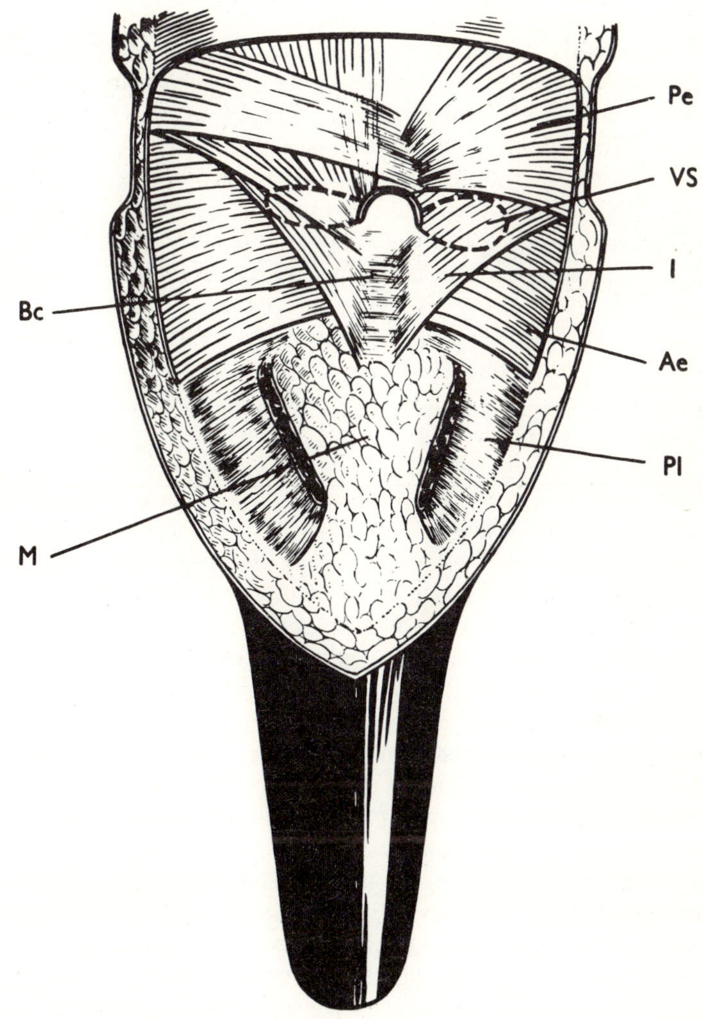

Figure 36. Diagram to show some of the components of the complex blowhole muscles of Delphinus. *M = melon, Bc = blowhole commissure, VS = vestibular sac, Pl = labial musculature, Ae, I and Pe = antero-externus, intermedius and postero-externus.*

of urgent purpose behind the grace and power with which it rolls to the surface and the split-second timing when it goes down".

The pattern of the sacs in relation to the nasal passage is basically similar in many species of delphinid but there are some striking differences in the arrangements of the sacs in *Phocoena, Delphinapterus, Hyperoodon* and especially in *Physeter* and *Kogia.* The delphinid blowhole is single and crescentic, the concavity facing rostrally: it usually lies just to the left of the midline. No muscles are present immediately about or inserted into the walls of the opening of the blowhole. The superficial layers of the complex maxillonasolabialis muscle converge from their broad origins on the facial bones to be inserted into the walls of the blowhole below the opening, into fibrous commissures in front of and behind the blowhole and onto the walls of the vestibular sacs. These paired sacs project, like waterwings, on each side of the nasal passage; they have pleated walls, very marked in *Phocoena,* and can be distended. Deeper beneath the skin and the vestibular sacs lie further muscle layers and a pair of fleshy, cushion-like plugs, anteriorly placed which can occlude the bony nares. Numerous muscle bundles in the fleshy nasal plugs are presumed to pull them rostrally and thus disengage them from the bony nares. Air can pass round the edges of the nasal plugs to enter a pair of tubular sacs which encircle the nasal passage almost completely in a horizontal plane. From the lateral aspect of the broad nasal passage a further small connecting sac projects on each side at a level just below the opening to the tubular sac. In dead dolphins a small tongue-like extension of the nasal plug projects into the opening of the tubular sac and appears to occlude it except perhaps at its lateral edge where air could pass into the connecting sac from the tubular sac. Beneath the nasal plugs lie the large pre-maxillary sacs, bounded by bone ventrally, a continuation of the nasal septum medially and by powerful muscles laterally. Air can enter the sac from the nasal passage through the posteriorly placed entrance and can pass round the lateral edge of the nasal plug extension to enter the connecting sac. All the sacs, and the muscles moving them and the nasal walls and fibrous commissures are larger and heavier on the right. The bony nasal passages are double; the right is again the larger. At the supero-lateral edges of each bony opening, where the nasal plug occludes, is a thin diagonal membrane with a sharp edge, often it leaves an imprint on the nasal plug in a dead animal.

The problem of just where the sounds produced by cetaceans are made is still unsolved. There are several possible sites and one

Figure 37. Photograph of the anterior bony nares and nasal plugs (turned forward) of Tursiops.

or more might be implicated. It is known that noises can be made by slapping the water with the flukes, the flippers and even with the body. Snapping of the jaws has already been mentioned as a form of aggressive or threatening display. Fluttering movements of the edges of the blowhole have been observed but it seems unlikely that this region is an important source: most authorities maintain that the source is deeper within the forehead region.

The vestibular sacs are true diverticula in their lateral extent, the slit-like entrances in the medial part of the floor can be narrowed by muscular action. These sacs would seem to act as reservoirs of air or water traps rather than sound producers. Vibration of the thin septum between the two nares has been postulated but there is no experimental evidence that vibration is set up here though the septum might conduct sound vibrations initiated elsewhere.

The nasal sac system has, not surprisingly, attracted attention from several points of view; control of respiration, pressure regulation of the contained air in a diving dolphin and also in the

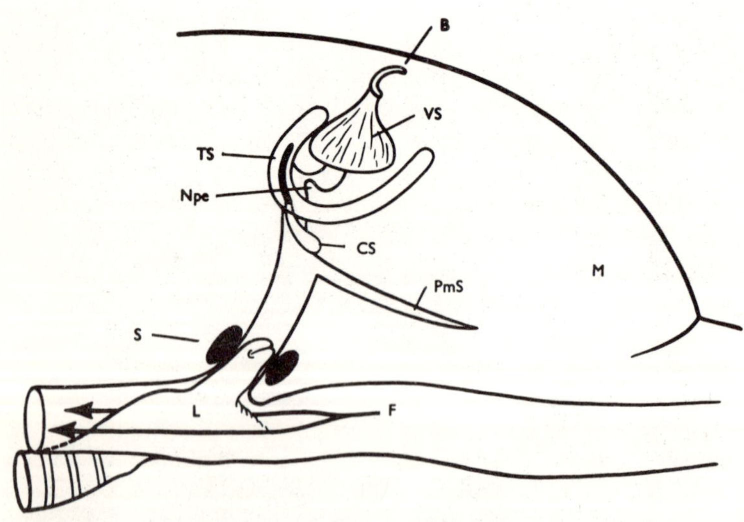

Figure 38. Nasal sacs in a dolphin. B = blowhole, VS = vestibular sac, TS = tubular sac, CS = connecting sac, PmS = premaxillary sac, M = melon, Npe = nasal plug and its extension, S = palato-pharyngeal sphincter, L = larynx, F = food pathway.

pharyngo-tympanic air sinuses which may well play an important part in sound reception, and also in connection with sound production. Early suggestions were that click trains were produced by "shock excitation of the resonant frequencies and harmonics of air-filled cavities in the head". The small semilunar membranous flanges, the diagonal membranes, on the supero-lateral aspects of the bony nares have also been implicated as a possible sound source. In particular, the tubular sacs have received attention as a source of at least some sounds. There are lip-like extensions of the nasal plugs which could be inserted into the entrances of the tubular sacs to effect a sort of "raspberry" noise.

More than one investigator has argued that the tubular sacs could be the site of sound production and has suggested that the following sequence of events might occur. First the blowhole closes: then increased intrathoracic pressure drives air up into the upper nasal passages. The nasal plugs are open, the vestibular sacs fill, with the result that the top of the dolphin's forehead becomes more swollen. The small postero-lateral extensions of the nasal plugs move caudally into the slit-like openings of the tubular sacs as the nasal plugs begin to occlude the nasal channels. The musculature around the vestibular sacs contracts and upward pressure is exerted by the muscle of the nasal plugs. This increases the pressure in the upper part of the nasal passages. Muscles of the extensions of the nasal plugs and of the slits of the tubular sacs allow controlled streams of air bubbles to enter under high pressure. The passage of these controlled streams produces pops or clicks. The tension of surrounding muscles, aided possibly by water pressure on the dolphin's head, may together control the frequency composition of the train of clicks produced. Air entering the tubular sacs after the clicks have been produced passes into the connecting sacs and back into the nasal channels below the extension of the nasal plugs. The air pressure in the various sacs is controlled by surrounding muscles, that in the tubular being kept somewhat below that in the upper nasal passage. As the air is passed back through the system the vestibular sacs gradually empty, but their muscles keep the contained air under pressure as they shrink in size. Eventually the contained volume of air becomes insufficient for noise production and the clicks cease, the plugs open, more air is forced up from the thorax and the sequence recommences. The frequency and repetition rate of the clicks are possibly controlled by increases or decreases of muscle tone thus affecting the pressure within the sac system. If this sequence of events does in fact occur, it would suggest that the larynx acts as another valve, controlling the entry of air into the system.

Despite the suggestions that the nasal sacs and nasal plugs may be implicated as sound sources other investigators still look to the larynx as the originator of the whistles, and even the clicks produced by dolphins. The larynx is interestingly modified in its position in that it is inserted into the back of the nasal passage above the soft palate and held in place by a powerful sphincter, a modified palatopharyngeus muscle. There has been some argument as to whether the larynx is ever retracted from this position, even when a large fish is swallowed past the larynx along one or other lateral food channel. Since the larynx can be disengaged by hand and can be returned by the animal without assistance, and because of an observation on the possible inhalation of a stone by a captive Pilot Whale, it has been suggested that the larynx can be voluntarily withdrawn from the nasopharynx. There must be some doubt that the dolphin has this ability on anatomical grounds, in that there are no muscles to disengage the larynx, that its orifice is slit-like and can hardly be enlarged, and that as there is no need for dolphins to do so it seems unlikely that the patent utility of an intracranial larynx would be undone and by-passed when it is most needed. It is agreed that it would seem disadvantageous to disengage the larynx but the organ could be ejected partially from the nasopharynx by sealing the nasal passage with the nasal plug and 'blowing back' the larynx by a forceful expiratory movement.

There has been active investigation of the usefulness to dolphins of their clicks and vocalizations and whether there is any meaning in a communication sense. The clicks have been definitely implicated in the ability of dolphins to echolocate in a manner similar to that employed by bats and some shrews. In its simplest terms this phenomenon involves the emission of a sound by the animal, the reflection of the sound from the target, a rock, obstacle, net or fish, back to the ear. Experiments have shown clearly that the dolphin's 'sonar' system (sound navigation and ranging) is complex

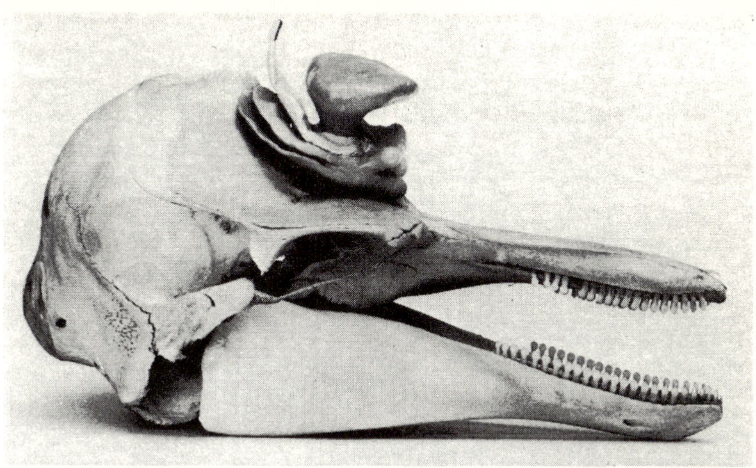

Figure 39. Photograph of a reconstruction (enlarged) of a vestibular and other nasal sacs in Phocoena. *Specimen prepared by A. Greenwood.*

and capable of remarkable discrimination. Dolphins swimming in muddy water or in the dark can navigate successfully through complicated arrangements of obstacles or poles. They can detect and find small objects thrown in the water and even discriminate between different sizes and, possibly, different species of fish. Echolocation can be demonstrated by placing rubber suction cups over the eyes and sending the dolphins off to collect plastic rings floating vertically in the pool. Undoubtedly they make use of their visual powers where and when possible in association with sonar emissions. The range at which dolphin sonar can or could operate is difficult to assess and experiments are vitiated by interference of one sort and another. There is some indication that in shallow water clicks can be heard over 800 yards away and possibly much farther.

Other experiments have strongly suggested that there is a distinct directionality in the way sound is emitted from the dolphin. It is not emitted from the head equally in all directions. Dolphins are apparently very poor at detecting objects in water which are not directly in front of them. This suggests that they are 'sonar blind' in other directions and may explain why they occasionally become stranded, though of course they are quite capable of turning their heads, rotating their bodies with great agility, or swimming upside down. Ingenious experiments with heads of dead animals and skulls have suggested that if the sound source is situated above the skull and in the region of the nasal sacs it is projected in a definite beam about 15° upwards from the rostrum and between 30-40° to the left and right of the rostrum. The width of the beam is related to the frequency, it being narrower the higher the frequency. It is tempting to speculate that the low frequency occasional clicks are involved in casual navigation associated with, or perhaps checking, visual perception. Clicks are repeated more often when an interesting object or unusual obstruction is encountered. Their frequency in cycles per second increases as the dolphin requires more discriminative information at shorter range. There is more attenuation of high frequency sound waves than those of lower frequency during their passage through water. It is also reasonable to suppose that whatever is the mechanism producing the clicks, it would be less efficient at very high frequencies. There is evidence, however, that interference and diffraction phenomena are almost certainly involved in the propagation of sound through the anatomical structure in the pharynx, nasal sacs and head region and that these might well account for an increase in intensity that has been recorded of some components of clicks emitted at 100 kcps. There have been suggestions that the melon (the fatty mass enclosed in fibrous tissue lying in front of the nasal passages) acts as an acoustic lens. Dolphins often swim towards objects close to them weaving their heads repeatedly from side to side. Many delphinids turn on one side under water, flex the head and again nod it from side to side. This behaviour could alter the timing and intensity of echoes returning to the two ears and give information about distance and movement of objects (acoustic scanning). It could also be related to an attempt to analyse frequency differences in echoes returning from a narrow directional beam of sounds of multiple frequencies.

It is sometimes stated that the whistles of cetaceans are used for communication purposes though some authors consider they too are involved in echolocation. Numerous attempts have been made to associate particular sound emissions with certain situations, but with a few exceptions little progress has been made so far with such analyses, intriguing possibilities though they present. Some evidence that individual dolphins do warble or whistle differently has been provided by slowing down and overlapping recordings from several animals. Although the contours of each whistle or warble were much the same, they did vary enough in their rise and fall to be distinguishable. Regrettably it has not yet been found at all rewarding to play back the whistles of one dolphin to another. Perhaps it is unfortunate that there is virtually no change in the expression on a dolphin's face when it is whistled at, except under situations charged with danger or emotion, when it makes sharp snapping movements of the jaws associated with high-energy cracking sounds.

It is naturally very tempting to suggest that dolphins are communicating something to one another. The question of whether there is a dolphin language is another matter, as it has been considered that a meaningful speech language is only possessed by man. Do the discrete whistles of dolphins comprise some type of language? First, it seems certain that there are particular, definite, discrete and repeated whistles. Second, some at least of these whistles provoke a response in other dolphins, either by action or by 'replying', particularly if the signals are of a distress type. Third, there is a tendency for certain types of whistle to be repeated under repeated circumstances. Fourth, dolphins isolated in separate tanks display interest in what noises are being made if the two tanks are connected electronically with hydrophones. Fifth, the occurrence of the whistles when analysed show some of the statistical relationships common to human language. Sixth, there is accumulating evidence that each dolphin's voice is different and individuals can be recognized by their vocalizations.

All this has aroused much interest and has stimulated the investigation of the dolphin's noises using sensitive apparatus to record their sounds. Several species have been shown to emit many different whistle contours both at sea and in tanks. Recent research has shown that the brain areas for the reception of whistles are isolated from the click responsive areas. The whistles are of low frequency and evoke slow responses (far too low to be involved in echolocation) in the posterior cerebral cortex. Whistles produced no responses in the cochlear nucleus or posterior colliculus (these areas are considered to be extremely sensitive to clicks of the echolocation type). It thus seems reasonable to argue that a dolphin may have two separate systems in the brain for analysing sound. One is for low frequency whistles, possibly for some type of social communication, and another for high frequency and broad band clicks used in echolocation.

Food and alimentary canal

Dolphins feed principally on nektonic fishes in the epipelagic zone of the open sea but some forms also take benthic fishes of inshore bays: the fish taken by small dolphins are small and are swallowed whole. Squids are also taken by many species, especially Pilot whales. Common porpoises eat fish, mostly whiting and herring and usually less than 25 cm in length. False killer whales feed on larger fish and the notorious, predatory killer whales take a wide variety of prey such as small dolphins, seals, diving birds, sharks and large bony fishes: they also attack large whales.

The stomachs of several species of dolphin, *Tursiops truncatus, Delphinus delphis* and *Stenella longirostris* and *S. roseiventris* have a similar construction. The oesophagus leads into a dilated sac, the forestomach, which is really a continuation of that tube and is lined by a thick stratified epithelium. At the cranial end of the forestomach a small orifice leads into an almost spherical second compartment which is broken up by large folds of remarkably thick and spongy mucous membrane containing long gastric glands. This compartment has been called the second, the main and the glandular stomach. From its right hand upper wall a narrow connecting channel curves downwards for some centimetres to enter the elongated, thinner walled pyloric compartment of the complex delphinid stomach. The connecting channel has a varying form amongst dolphins, sometimes having valves or folds or short recesses leading off it. Both the entrance to and the exit from the channel are very small in size. The channel has been referred to as a third stomach: it has a glandular lining that resembles the lining of the fourth, or pyloric, compartment. At the distal end of the usually kinked pyloric stomach is the pyloric sphincter. Beyond this is the duodenum, the first part of which is so dilated that it has been mistaken for yet another gastric compartment.

The arrangement of the gastric compartments in *Phocoena* is not dissimilar but there are certain characteristic differences. The forestomach is more elongated, smaller and more finely rugose than in *Tursiops*. The opening to the sac-like second compartment is also relatively larger but would still restrict the passage of an object over 1.5 cm in diameter. The second compartment is lined by mucous membrane thrown into numerous longitudinal folds

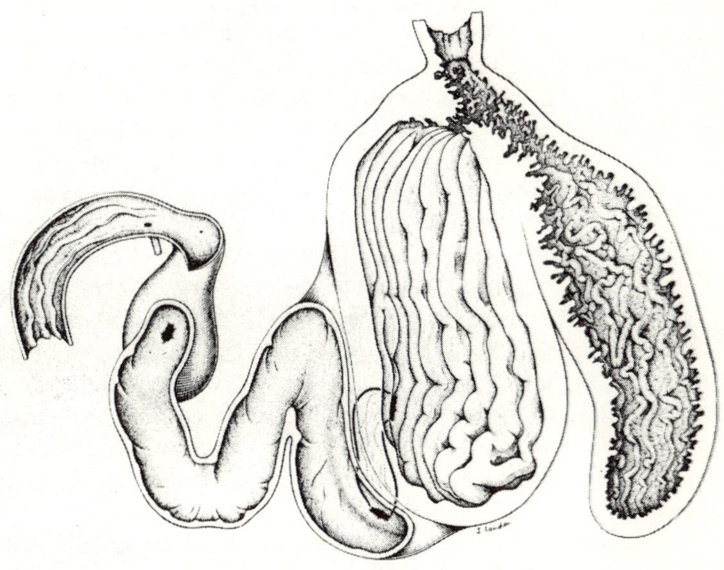

Figure 40. Drawing to show the compartments of the stomach of Phocoena.

that extend almost to its tip. The opening into the third compartment is far more caudally situated on the right side of the second chamber than in other forms. The third compartment is the smallest and it seldom exceeds 5 cm in length: it leads by a small nozzle-like opening into the proximal limb of the pyloric stomach.

There has been much discussion as to how the process of digestion is accomplished in the complex delphinid stomach. Fish are swallowed whole but only the very smallest could pass through the narrow opening leading from the forestomach (which lacks glands) to the second stomach (which possesses abundant glands).

Many authors have conjectured whether gastric juice is regurgitated into the forestomach and whether preliminary digestion occurs there. A mechanical action could be exerted by the muscular forestomach, aided by the presence of ingested sand or stones which are often found there in dolphins, and then there is the added effect on cold fish of the warmer stomach. Ultrastructural studies of the forestomach lining do not indicate any particular adaptation for resisting chemical ulceration nor is the turnover of lining epithelial cells high. This suggests that there is very likely a lubricant of some sort assisting the passage of fish into the forestomach. The salivary glands are absent or poorly developed in dolphins, and if mucus is the lubricant, it must originate from the pharyngeal glands or by regurgitation up from the main stomach. There is also the possibility that seawater is the lubricant. In the wild, dolphins and porpoises take their fish under water and although it has been suggested that the tongue forces water out of the fish and out of the mouth, like a squeegee, and prevents it entering the oesophagus, it is difficult to accept that no seawater is ever swallowed. Dolphins have been transferred to freshwater for a period and then returned to seawater, and the urine has been analysed before, during and after. Urine electrolytes fell during the period in freshwater but rose above the pre-test level after return to seawater.

The forestomach may be found after feeding to be packed with several kilogrammes of fish in varying stages of decomposition. It may also contain "scales, vertebrae, eye lenses, teeth, bits of smelly fish, and other fish fragmentia, making up the soupy, often worm-ridden mixture". The forestomach may also be found empty on other occasions except for a little mucus, sand and small stones. No fish fragmentia have been found by us in the second or main stomach or in any other compartments: only fluid contents were present. Whatever the mechanisms involved in this preliminary trituration of food in the forestomach, they are powerful enough to separate bones and free otoliths from the skull and even possibly to dissolve the entire skeleton of a fish. The rate of removal of material from the forestomach is not known but it must be performed with some rapidity in view of the daily consumption by a dolphin or a porpoise of up to 10 kg fish. It is therefore not surprising that many authors have been forced to conclude that digestive juice of strength sufficient to accomplish such rapid dissolution can only be provided by regurgitation from the second stomach.

The structure of the epithelium of the forestomach, as revealed with electron microscopy, resembles that of the rumen of the ruminant stomach across which the products of microbial digestion are absorbed and large amounts of sodium are transported. If the functional aspects were also similar, then it would seem potentially disadvantageous to a dolphin to swallow excessive amounts of seawater and that any method of preventing its entry or of regurgitation would be advantageous.

The intestinal canal in marine mammals is long, probably longest in sperm whales. The intestine of *Phocoena* measures from 13 to 16 metres in length from the pyloric sphincter to the anus; in a full grown bottlenosed dolphin, *Tursiops,* it may reach 25 metres in length, nearly three times the total length of the intestine in Man. There is no clear distinction between small and large intestine: a caecum is lacking. A similar arrangement is present in pinnipeds. Again the intestine is remarkably long. An Elephant seal 5 m in length has an intestine of over 20 m: a caecum is absent but is present in sirenians.

Kidneys and water metabolism

Diving mammals of one sort or another inhabit almost every type of aquatic environment from polar to tropical seas, fresh water lakes and some of the major river systems of the world. Those living in the wide seas might seem to be in a vast desert where drinking water is non-existent. There is indeed plenty of water everywhere, as the Ancient Mariner declared, but if man's experience of drinking seawater is anything to go by, it could well be dangerous also to dolphins. Seawater has a higher salinity than the blood of dolphins and as far as is known to date dolphins lack special salt-excreting glands such as are found in some fishes and birds. Whether cetaceans and pinnipeds can drink seawater even in small quantities without ill effects, or whether adequate water for metabolic and other purposes is available from ingested food are important questions. Dolphins eat fish and their diet contains much fat. High oxidation rate and high fat intake would yield much metabolic water and a dolphin has been calculated to consume enough fish to provide its own metabolic water. Fish do not, however, have a very high salt content and therefore it may be necessary to imbibe small quantities of seawater to obtain adequate sodium chloride. Problems of maintenance and difficulties in devising good experiments have prevented the accumulation of data on water metabolism. Even a simple adding of a radioactive isotope to the pool water hardly helps when dolphins are able to take water into the mouth and also the oesophagus and forestomach and then squirt it out. Many substances which might be tested are also taken up rapidly through mucous membranes.

We know that blood and other fluids have the same components as in other mammals and in approximately similar amounts. Urine concentration is not exceptional and although the urinary bladder is small, urine appears to be passed to the exterior frequently. Little body water is lost in evaporative cooling from skin or lungs. More would seem to be lost from the small lacrimal glands than anywhere else. The kidneys are subdivided into numerous lobules (over 3,000 in the largest cetacean kidney) but their significance is not clear unless it be that the number of reniculi in action at once can be varied from few to all. There is a relatively high kidney/blood weight ratio in marine mammals (*Tursiops,* 1.1 per cent; *Phoca,* 1.05 per cent). Freshwater dolphins appear to have somewhat smaller kidneys and fewer reniculi. The degree of separation of the reniculi varies. In otariids the furrows between them are shallow: the reniculation is more marked in phocids and in *Ommatophoca* it resembles the more extreme separation seen in most cetaceans. In dolphins the ureter is found on the ventral aspect of the caudal pole of the kidney. Reniculation has been considered to be directed towards securing the largest amount of renal cortex without the existence of a disproportionately large organ. Diving is known both to reduce renal blood flow and urine production and it may be that the large kidneys act both maximally and most efficiently at irregular intervals. It also seems that dolphins, and some other cetaceans, possess a perimedullary muscular arrangement (a sporta) which could have a compressing or even a milking action on the excretory ducts of Bellini and so positively express urine into the calyces. If so, then a renicular arrangement would be more advantageous than having such a mechanism at work in only a few papillae.

The most demanding period for water in all marine mammals is probably during lactation. A young seal pup gains several times its birth weight in a short period and the suckling mother seal does not appear to eat for days or weeks on end. She loses weight rapidly and uses up blubber fat for her own needs and those of lactation. The milk of both pinnipeds and cetaceans has a high fat content. Milk of Weddell seals contains only 27 per cent water, the rest being fat and protein. In seals of several species, lactation is short, and is therefore a heavy demand on maternal supplies. In dolphins, however, the mother takes food, at least in captivity, during a lactation period which may be extended for as long as a year.

Adaptations related to thermoregulation

Marine mammals of all types have problems with regulating their body temperature. Many have to move from warm to cold seas in a relatively short time and even during dives they may be exposed to water varying in temperature from 27° to 4°C. They generate heat during muscular activity and their recorded internal temperatures range from 33.8 to 38.7°C under varying circumstances. They have at one time a need to conserve heat and at other times they must dissipate heat. Heat can be lost to a small degree by expiration of warm air, through excretion of body fluids, and through the vascular mucous membrane lining the mouth but almost certainly the major avenues are through skin, fins and appendages. Thermoregulatory problems are, therefore, faced by marine mammals all their lives, day and night, at the surface or during a dive. On cold days, or nights, they must continue to be active to generate heat: in hot climates it could well be necessary for them to dive deep to find colder water.

Water temperatures generally do not go as low as air temperatures but the heat conductivity of water is many times greater than that of air. No cetaceans and few pinnipeds rely on fur for insulation, rather it is the subcutaneous blubber that forms a thick insulating envelope, usually several centimetres in depth, around a warm body core. The fur of seals has been thought to trap numerous small air bubbles to help insulation: indeed if a seal is shaved all over it shivers violently even in tepid water. More likely, the fur encloses a thin layer of water, warmed by the seal, which helps reduce the heat loss. Indeed this insulation is so effective that at times the problem is to dissipate heat rather than conserve it. The lack of epicrine sweat glands and inability to pant means we must look for less conventional means of heat loss.

Heat flux measurements of the skin of *Tursiops* have shown that in the resting metabolic state in 25°C water the insulation is only about 50 per cent as effective as that obtained from isolated pieces of blubber and skin. The difference is almost certainly due to the heat carried by the blood to the skin. Usually the skin is only a few degrees warmer than the water and a steep temperature gradient exists between the skin surface and the inner core. More consideration of this problem will be found in the section on the structure of the skin of cetaceans. What evidence there is indicates that the flukes, fins and flippers are the most important areas of heat dissi-

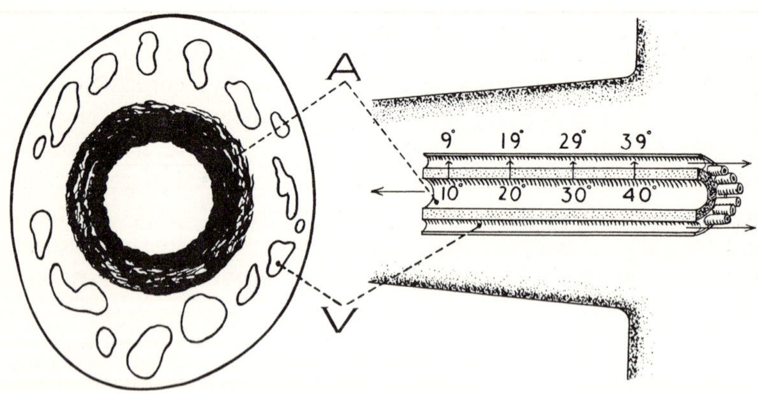

Figure 41. Diagram to show the venous plexus surrounding an artery to a flipper, illustrating the concept of counter-current heat exchange with the temperature gradient of arterial blood (A) and venous blood (V).

pation. In many marine mammals where these areas exist they are wholly or partially naked, capable of enhancing heat loss. One has only to feel the dorsal fin of any cetacean to be aware that its skin is often warmer than the air or water surrounding it. Some circumstances require a dramatic reduction in the heat lost from such areas. The blubber coat has to be equally effective over a wide range of temperature, at least in the deep diving and the migratory species. The coat must be functionally adequate when the animal is existing in polar waters as well as when vigorously swimming in tropical waters.

The flukes and fins of cetaceans appear to have a blood vessel pattern that seems admirably suited in this respect. Many of the arteries supplying the extremities are surrounded by a network of venous channels establishing what is believed to be a counter-current heat exchange system. Warm blood flowing to the peripheral part helps to raise the temperature of cold venous blood returning to the inner body core. Conversely heat is transferred to the returning blood avoiding loss to the exterior. Other vessels flow in and out of the fins and flukes in more conventional ways but there are numerous arteriovenous shunts to control blood flow and thus heat loss. These vascular arrangements acting in concert provide a considerable capacity for heat loss or conservation.

Other interesting problems must arise when a dolphin, its skin, flukes and flippers are continuously in seas at temperatures near 0°C. Could this have any effect on tissue metabolism and even nerve conduction? In Man there is numbness of peripheral parts, loss of sensation and even gangrene in fingers and toes, not to mention the acute discomfort when a numbed hand warms up. As far as we know marine mammals in the coldest seas do not experience these incapacitating discomforts.

Endocrine organs

The endocrine organs of cetaceans were much used in the early days of the relatively recent science of endocrinology. The organs of large whales were large themselves and were a source of crude extracts which could be refined for experimental and even therapeutic purposes. The celebrated anatomist John Hunter (1787) could not find the thyroid gland in the cetaceans he dissected but at least he did find the adrenal glands. In fact not very much is known about endocrine activity in marine mammals, and this is a field of investigation which could well be particularly productive. There are problems of growth, of the onset of maturity of both sexual and physical type, of metabolism and thermoregulation and many others where information of endocrine activity would be relevant.

Even the weights of endocrine organs, where available, give us some information:

	Pituitary		Thyroid		Adrenal	
	Weight (g)	g/kg Body weight	Weight (g)	g/kg Body weight	Weight (g)	g/kg Body weight
Large whale	74.5	0.0003	3450	0.06	1700	0.01
Dolphin	1.65	0.013	31.5	0.2	21.0	0.1
Seal	0.6	0.0003	11.7	0.07	10.9	0.1

The above figures, obtained from adult animals, indicate that the relative mass of both pituitary and thyroid compared with body weight is highest in dolphins. The amount of pituitary tissue per kilogram of body weight in a large whale is very small indeed when compared with that of land mammals (cat: 0.03, man: 0.01-0.14). The thyroid in marine mammals is significantly larger than in terrestrial mammals, but there is insufficient evidence to indicate whether thyroids from cetaceans of cold seas are relatively larger or more active than those from warmer seas. It must also be remembered that for several reasons there is much variation in weight of adult cetaceans during their life span. Of all cetaceans, the common porpoise *(Phocoena)* appears to have relatively the heaviest thyroid gland (0.7 g/kg in sexually active adults). At least in this species, already described as the smallest cetacean of the cold seas, the thyroid is of significantly larger size when compared with gland in other dolphins. Little is known about adrenals in dolphins except that adrenal weights relative to body weights in smaller cetaceans are greater than in terrestrial mammals. The highest ratio is found in *Phocoena* at 0.3-0.9 g/kg, whereas in a cow the ratio is 0.6 g/kg. The cetacean pituitary and thyroid will now be considered in more detail.

The pituitary

The earliest description, in 1885, of the cetacean pituitary noted that the only connection between the adenohypophysis and the neurohypophysis was loose connective tissue and a fold of dura reflected into the posterior sulcus. Later workers established that the pars intermedia and hypophyseal cleft were completely absent in *Tursiops truncatus*. Such findings seem to be characteristic of all species of Cetacea.

The pars distalis consists of the usual mammalian cords of acidophil, basiphil (mucoid) and chromophobe cells. The cords are separated by perivascular spaces containing capillaries and connective tissue elements. No pars intermedia is present but aggregations of colloid are conspicuous, notably in the region of the infundibular stem. On the basis of cell shape, cytoplasmic detail and granule size a variety of distinct cell types can be identified by electron microscopy in the dolphin pituitary: the maximum diameter of the granules is considered to be a better guide to their classification than the average diameter. When the cell types are compared with those present in the pituitary glands of other species, it is possible to identify somatotrophs, thyrotrophs, gonadotrophs and lactotrophs. Another cell type, known as a stellate cell, forms a marked feature of the delphinid pars distalis. Characterized by the absence of secretory granules, it has a stellate appearance with slender processes extending between and around adjacent secretory cells. Its cytoplasm contains mitochondria, scattered elements of ergastoplasm, free ribosomes and a small Golgi zone. Cytoplasmic filaments and lipid droplets are also seen in some cells. Marked and extensive perivascular channels are present, either between adjacent stellate cells or between a neighbouring secretory cell and a stellate cell. The perivascular channels are limited by desmosomes, contain numerous microvilli and can often be seen to form a communication between the secretory cells and the perivascular space. Small masses of colloid are present, entirely contained by these cells. The cytoplasmic boundary is often very thin and microvilli protrude into the colloid material.

Earlier workers have suggested that stellate cells could be related to the production of corticotrophin (ACTH), even so the precise cellular origin of ACTH still remains unsolved. Others have suggested either a supporting function for these cells or else some participation in a transport system. It has already been discussed as to whether, accidentally or deliberately, dolphins drink seawater. In this respect, the large quantities of colloid and the marked perivascular channels of the delphinid pars distalis are of particular interest. The tendency to produce colloid is more pronounced in some species than others but is no longer thought to be pathological. In rats, hypertonic saline administration or giving concentrated saline solutions instead of drinking water may induce colloid formation, as well as degeneration of basiphils and distension of the hypophyseal cleft with fluid. Increased colloid

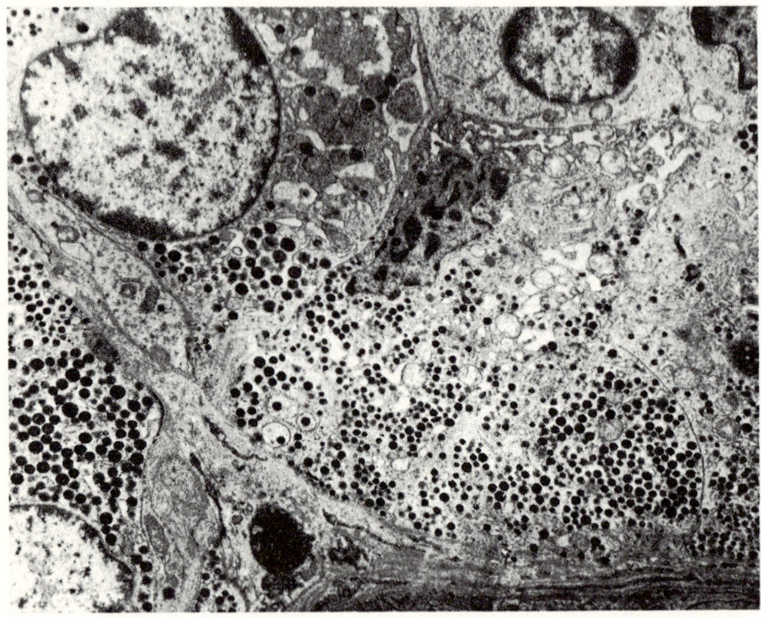

Figure 42. Electronmicrograph of the anterior lobe of the pituitary of a dolphin (Delphinus) *showing the granules in the various cell types.*

secretion in the cat has been obtained by perfusion with liquid of a low surface tension.

The potability of sea water is a problem of wide interest but whether excessive ingestion by dolphins could effect pituitary hormone production cannot be settled without experiment. However, the possibility that stress or captivity could have an effect on a dolphin's endocrine system should not be ignored. It has been assessed that the hormone content of the dolphin posterior lobe was lower in animals which had been in captivity for several days. One effect of stress is known to be a diminished secretion of gonadotrophins, somatotrophin and thyrotrophin and an increase in corticotrophin. A reduced output of gonadotrophins could be related to the apparent reduced reproductive activity and the infrequency of ovulation in certain captive dolphins, as judged by the number of corpora albicantia. Most of the dolphins examined by us had been held after capture for only a few days before death and were immature. There is no evidence for reduced synthetic activity of the gonadotrophs from their ultrastructural appearance, but it is impossible to obtain an overall assessment of hormone production from inevitably limited observations with an electron microscope.

Cetaceans lack a pars intermedia but they do not lack intermedin, which is present in the pars distalis. However, the cell type responsible for its secretion has not been identified. It is therefore of interest that a cell type identified in the dolphin pars distalis is similar in appearance to the pars intermedia cell of the rabbit pituitary.

The thyroid gland

The structure and function of the cells of the mammalian thyroid gland have been well described but little is known about the gland in mammals existing wholly or partially in a marine environment. The thyroid gland of *Delphinus* consists of two large lobes, one on each side of the upper part of the trachea, and joined by a narrow isthmus. The thyroid weight in this species varies from 7.8 to 17.0 g and the thyroid weight to body weight ratio varies from 0.15 to 0.31 (g/kg).

The gland is made up of small, often irregular follicles with a diameter between 35-125 μm. The follicles are composed of a layer of cuboidal cells, about 10 μm in height, surrounding a colloid-containing lumen and enclosed in a basement membrane. In addition to the follicular cells a second type of epithelial cell can be recognised, the light or parafollicular cells. Such cells are contained within the follicular basement membrane. They do not come into contact with colloid and they contain characteristic small, dense granules.

In the electron microscope the follicular cells have an apical surface with irregular arranged microvilli and occasional cilia which protrude into the colloid. Apical cytoplasmic outpushings, sometimes with colloid droplets, also occur, as well as "pinocytotic" invaginations of the apical plasma membrane. Beneath the junctional complex, the lateral plasma membranes separate to form well-marked intercellular channels which contain microvilli. The channels are limited by conspicuous desmosomes and are enlarged

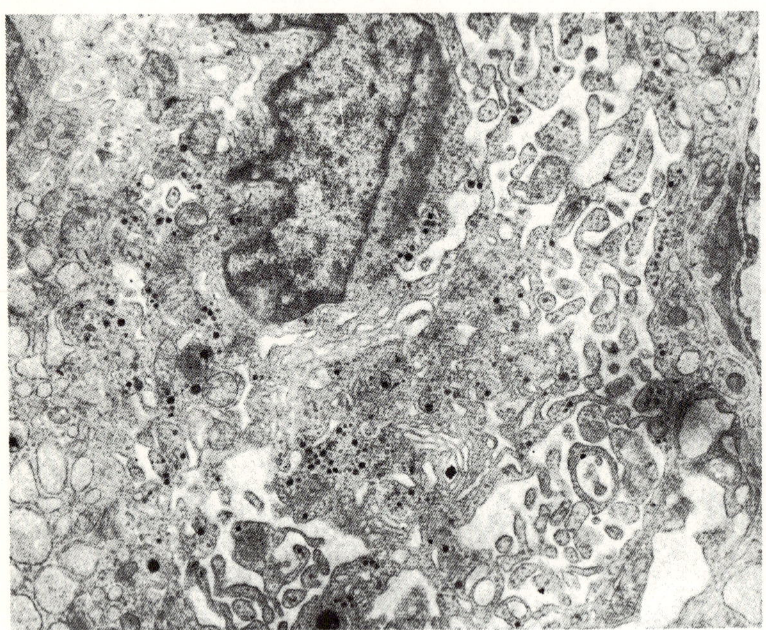

Figure 43. Electronmicrograph of the thyroid of a dolphin (Delphinus) *showing a light cell.*

towards the base of the cells. The base of many follicular cells is irregular and is covered by a basement membrane. The perifollicular capillary contains endothelial pores and is contained by a basement membrane. The endoplasmic reticulum, composed of cisternae and attached ribosomes, is widely distributed. Most cisternae are dilated, especially towards the base of the cell. Occasionally the membranes of the ER can be seen fusing with the lateral plasma membrane and opening into the intercellular channel. The Golgi zone is supranuclear in position and often appears horse-shoe shaped.

Various types of inclusions are present in the follicular cell cytoplasm. They can usually be separated into groups by their size and appearance. Near the apex of the cells are the apical vesicles which are membrane-bound, with finely granular contents and about 0.15 μm in diameter. Also seen in the apical region are small tubules between 500-700 Å in diameter: their full length cannot be estimated. Typical colloid droplets are present in the supranuclear region. They are about 1.5 μm in diameter, membrane-bound with a moderately electron-dense, finely granular content. Also in the region are found very electron-dense, membrane-bound granules about 0.5 μm in diameter. Another type of inclusion, found in large numbers in many follicular cells, is usually rounded with a pale homogeneous content, about 1 μm in diameter and surrounded by a halo of denser material. Many appear to coalesce forming large irregular bodies.

Following formalin fixation the light cells can be distinguished from the follicular cells by their small dense granules. They are oval cells, lying singly or in pairs inside the follicular cells, pushing into the intercellular space as far as the junctional complex. Intercellular channels, limited by desmosomes, lie between a light cell and adjacent follicular cells.

The cytoplasm contains the characteristic electron dense granules, contained by a smooth-membrane, and up to 0.15 μ in diameter. They are scattered throughout the cytoplasm but may be situated more towards the base of the cell. Some light cells may contain more granules than others: even so, the paucity of granules in some light cells is a striking feature. Rarely, a nerve process can be seen in relation to the base of a light cell.

Marine mammals live in an environment colder than is usual for most terrestrial mammals and consequently they exhibit adaptations of structure and function to help maintain body temperature. Even if the water has a temperature as high as 25°C, which can be regarded as in the upper range for most dolphins, there will be a considerable thermoregulatory problem, especially as water has considerably more (about twenty times) cooling effect than air. It has been suggested that one such adaptation resulted in an increased metabolic rate which enabled dolphins to keep in constant motion and therefore maintain body temperature. The fact that some marine mammals have particularly large thyroids and high thyroid-to-body weight ratio has been confirmed by all investigators. Although the ratios of thyroid weight to body weight in our dolphins show considerable variation (some animals were not healthy and exhibit the lowest ratios) they are consistently higher than those of terrestrial mammals.

It is apparent that dolphins have a high metabolic rate. The resting consumption of oxygen by *Phocoena* has been estimated at 400-500 cc/min, and in an inactive *Tursiops* it is 900 cc/min. This is about three times the resting rate for man on a surface area basis. Such an increased rate might be reflected in hypertrophy and increased weight of the thyroid gland. This could be due to a raised TSH production and thus result in a higher level of secretion of thyroid hormone. The picture of hypertrophy is possibly non-specific and might originate from a variety of causes such as diet, stress or disease. Certain diseases of dolphins, such as respiratory infections or parasitic infestations could cause stress by interfering with navigational or food-catching activities. The stress of confinement in captivity might be one factor causing increased TSH production. It is therefore interesting that all the dolphin thyroids that have been examined show signs of increased activity consistent with mild TSH stimulation. When seen with the light microscope the follicles are small with a high cuboidal epithelium. The follicles are often collapsed with infolding of the epithelium. Some thyroids have shown an exaggerated picture not unlike that of a toxic goitre. With the electron microscope the ER is found to be conspicuous, the Golgi zone large and numerous dense granules present in the apical cytoplasm. However, the microvilli are of normal size and not increased in numbers and colloid droplets are not abundant.

Although it is known that low temperature will also cause stimulation of the thyroid gland the long term effects of cold on the thyroid are not known. There is little evidence for the suggestion that the thyroid glands are bigger in animals living in cold seas than those in warmer seas such as off southern California. In any event dolphins are so well insulated and regulated against heat loss that it would probably need sudden falls of well over 10°C to bring about a response.

The occurrence of tubules in the apical cytoplasm of the dolphin thyroids has not been reported in other mammals. They are larger and different in general appearance from microtubules. They are similar in appearance to the tubules described in the apical region of the trophoblast cells in the marginal haematoma of a common seal placenta. They might aid in absorption or in transport of enzymes from the cell but their function in the dolphin thyroid has still to be explained.

The numerous large globules found in many follicular cells are similar to those seen in the thyroids of common seals. They are more abundant in pregnant and lactating seals. They consist of two components, lipid and traces of iron, and they might be related to ageing. Similar globules have also been described in the human thyroid, they contain lipid and their numbers decrease in diffuse toxic goitre. They might represent an energy store which the cells could use in increased need. In the young dolphins examined ageing is unlikely to be a factor. The large globules might, however, represent an energy store in the glands of animals with a high metabolic rate.

The intercellular channels are conspicuous in dolphin thyroids as are the foldings at the base of the cell. During chronic stimulation of the thyroid the basal processes of the follicular cells acquire greater complexity. Although synthesis, storage and secretion of the thyroid hormone have been fully investigated, little is known about how it actually leaves the follicular cells. It is usually assumed that the hormone leaves the base of the cell, perhaps by contact of vesicular and plasma membrane. The contact between the cisternae of the endoplasmic reticulum and the lateral plasma membrane in dolphin thyroids is therefore of considerable interest. In certain circumstances the hormone might leave the cell in a manner resembling the absorption of fat in the columnar cells in the small intestine.

The light cells described in the thyroid glands of dolphins exhibit, with minor differences, the general mammalian structural pattern. The presumed secretory product of these cells has been described variously as vesicles or as granules, possibly depending upon the species examined but more likely upon the method of fixation.

Desmosomes have been seen between follicular cells and light cells and between light cells. In dolphin thyroids, intercellular channels limited by desmosomes are seen between light cells and follicular cells. The presence of desmosomes between these cells, which may be of different embryological origin, may be accounted for on a functional basis as taking part in the formation of the intercellular channels.

It is likely that the morphology of light cells reflects their activity and in dolphins they are certainly in an active phase. They contain large amounts of ergastoplasm, with dilated cisternae and also large Golgi zones. The granular content of the cytoplasm varies, some cells even seem to be degranulated. The bones of almost all marine mammals show several specializations related to their marine existence. Sirenians have very dense bones whereas cetaceans have lighter, spongy bones which lack a marrow cavity. The whole sequence of ossification is delayed in cetaceans and a considerable quantity of bone must be laid down before adult life is attained. The activity of the light cells, therefore, could be related to an increased production of calcitonin, the action of which is thought to be related to the inhibition of bone resorption.

The thyroid gland is known to have an autonomic nerve supply to its blood vessels. In animals in which the ultimobranchial body remains as a separate endocrine gland (frog and chicken for example) nerve fibres have been described in relation to the secretory cells. It has been suggested that some secretory activity might be under sympathetic nervous control in these forms. Nerve processes have been found in relation to the light cells in the thyroid gland of dolphins, which may well be of significance in connection with the observations outlined above.

Reproductive patterns and organs

Marine mammals exhibit several interesting features in relation to reproductive activity. One of these, delayed implantation, occurs in most pinnipeds, and in all cetaceans so far examined there is a persistence of the corpus albicans, probably for life. Little is known about reproduction in quite a number of species mainly because of lack of specimens. On even relatively simple matters, such as age at reaching puberty or sexual maturity and the age of adults, information is scanty. Blue Whales are thought to be able to reproduce when five to six years old and similar figures have been given for other large whales. Dolphins reach sexual maturity between three and six years and porpoises at about three years, although there is now thought to be much variation and these figures may be too low for the larger dolphins.

There are several methods available for estimating the age of adults but none is known to be precise. Counting the persisting scars of corpora lutea, corpora albicantia, in the ovaries gives an indication of the number of ovulations during a whale's lifetime. The count could be of value if the pattern of reproductive events were known and if every ovulation were followed by pregnancy. Periodicity in the formation of baleen in mysticetes and in the deposition of layers of dentine in odontocetes has helped to assess age. Sectioning of the 'wax plug' found in the deeper part of the external auditory meatus of mysticetes has revealed incremental rings which appear to be related to the increase in size of the skull and thus of the meatus. All these periodic markings and rings are probably affected by metabolic and other activities. Not surprisingly there is often little correlation between the methods. By their application it has been estimated that the largest mysticetes and odontocetes have a life span of over 30 years, that Killer and Pilot Whales live for about 40 years and that small dolphins and porpoises survive for over 15 years. Unfortunately dolphins have not yet been kept in captivity long enough to provide much information on longevity.

More is known about reproduction in those cetaceans hunted for commercial purposes. Blue Whales mate in the winter months in the southern hemisphere; the gestation period is about 11 months and lactation lasts about 6 months. There is a probable interval between successive births of two years and a female could presumably have given birth to twelve offspring during her lifetime. Twinning is rare in cetaceans, though multiple births have been recorded. A newborn Blue Whale calf is about 5 m long, weighs two tons and can swim actively from birth. Sperm Whales mate from September to December in the southern hemisphere and give birth to a single young 15-16 months later. Lactation lasts about a year. The shortest gestation period, of ten to eleven months, occurs in small dolphins and porpoises with a lactation period of six to nine months. Many cetaceans have a limited mating season. Narwhals, Pilot and Killer Whales can breed throughout the year but there are periods extending over several months when the mating intensity is maximal. Some dolphins come on heat, and the males exhibit testicular activity in both spring and autumn in any year.

Ovulation was thought to occur once during the mating season, but evidence is accumulating that several types of cetacean are polyoestrous and that they may ovulate several times before fertilization occurs. It is not easy to distinguish corpora albicantia of oestrous cycles from those of pregnancy. Thus, even if the corpora do persist for the lifetime of the animal, there is no certain method of identifying the nature of each corpus and thus reconstructing an animal's reproductive history.

Mating in all cetaceans occurs in the water and although a preliminary period of sexual play may be prolonged for hours, the true copulatory act is thought to be of short duration. An account of mating in large whales goes as follows: "After some introductory love play, whales are said to dive, to swim towards each other at great speed, then to surface vertically and to copulate belly to belly. In so doing, their entire thorax and often part of their abdomen, as well, are said to protrude out of the water. They then drop back into the sea, with a resounding slap, that can often be heard far away". Another account in the Common dolphin: "The male persistently followed the female everywhere, rubbed against her and stroked his flippers against her. The female reciprocated his advances. He then would flip over on his left side and the female would assume a like position on her right side, and both animals brought their genital regions in a close alignment". More frequently the male turned on his back and swam directly under the female. There are also some accounts, in captivity, of inter-specific and inter-generic sexual behaviour but none of successful fertilization in such circumstances.

No photographs or detailed accounts of parturition in mysticetes are available, but recently there have been numerous births of

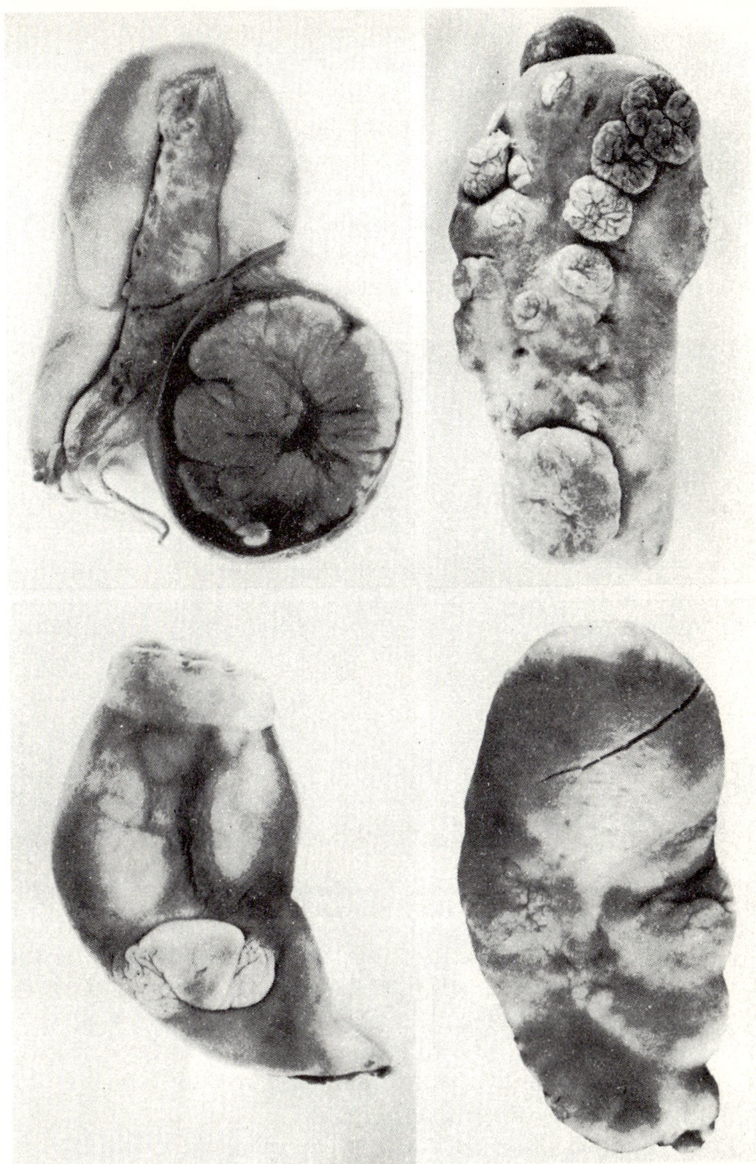

Figure 44. Ovaries of some odontocetes: top left, Tursiops; *top right,* Lagenorhynchus; *bottom left,* Stenella; *bottom right* Delphinus.

delphinids in oceanaria, and several have been recorded and filmed. The shape of the bicornuate cetacean uterus encourages a foetal lie with the flukes presenting and most births in odontocetes have presented in this way. The girth of the foetal dolphin is greatest in front of the flippers and thus passage of the foetus is easiest with the flukes born first. Nevertheless many fluke-presentations in dolphins in captivity have resulted in stillborn young. The stimulus to breathe in newborn dolphins is not known: but if the neonate survives birth it swims unaided or is helped to the surface

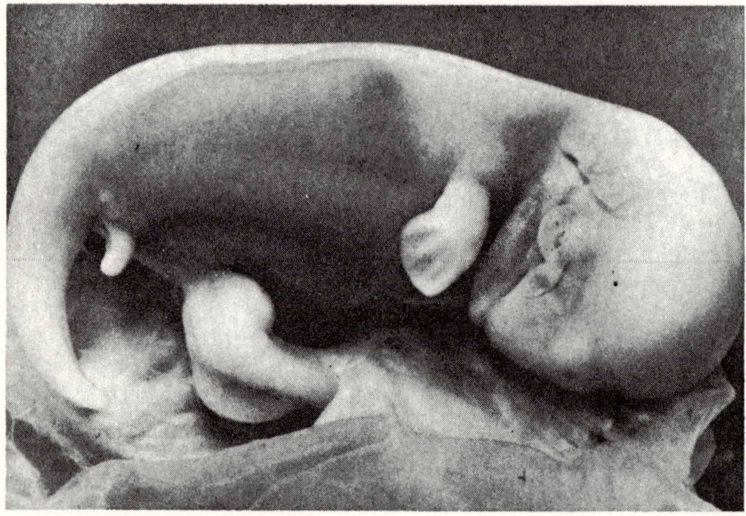

*Figure 45. Foetus of a white whale (*Delphinapterus*); note the digital rays in the fore flipper.*

for its first breath. The umbilical cord is relatively short when compared with that of other mammals and it is stretched taut when the neonate emerges. The cord breaks apart close to the umbilicus of the newborn immediately after birth and has not been seen to be bitten through by the cow.

Little is known about the details of placentation in dolphins because of lack of well preserved specimens. The cetacean placenta is diffuse and epitheliochorial and there are no features that can be especially associated with an aquatic existence or with diving. It is thus quite different from the pinniped placenta and exhibits many features more like those of ungulates. The chorion is thin and lined by the allantois: it is much pleated and is covered with villous processes. The pleats and villi fit into corresponding depressions on the folded uterine mucosa: the complexity of the folds varies throughout the uterus. Some parts display marked folding of the mucosa to give a sponge-like appearance which has been likened to the ungulate placentome, in which digitiform chorionic villi fit into crypts in the maternal caruncle.

Pinnipeds display many differences in their reproductive patterns. They give birth on land and mate on land or in shallow water. Many forms establish territories and collect harems of adult females from ten to about a hundred in number: other forms are monogamous. Mating takes place within days or a few weeks of birth. All pinni-

Figure 46. Bottlenosed dolphin and calf. Photo: Marineland of the Pacific.

peds, except walruses and perhaps the Crabeater seal, are believed to exhibit the reproductive phenomenon of delayed implantation: no cetacean has been shown to display this pause in development. The egg is fertilized and cleavage progresses until the formation of a blastocyst. This arrives in the uterine horn in the expected mammalian manner but no intimate contact is made with the uterine tissues. The zona pellucida persists and the cells of the blastocyst increase only slowly and slightly in number.

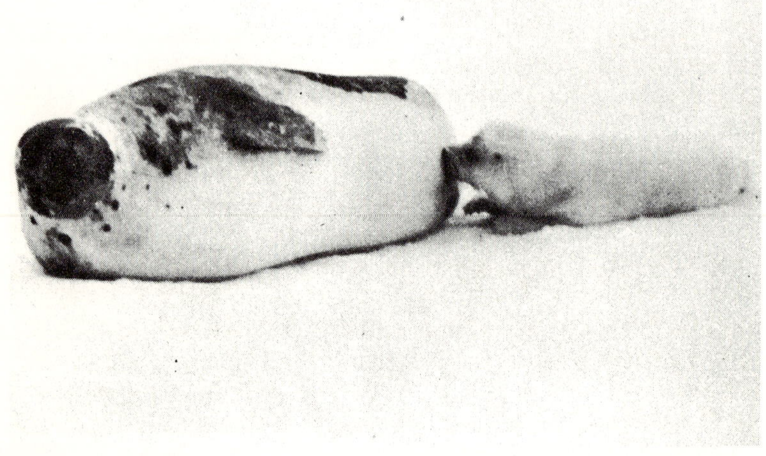

*Figure 47. Harp seal pup (*Pagophilus*) being suckled. Photo: Professor K. Ronald.*

The period of delay, strictly the interval between the time of arrival of the blastocyst in the uterine horn and that of its implantation, varies from about two to about four months in pinnipeds. There is as yet no evidence that it can last longer, even fourteen to sixteen months as it is now known to do in some badgers. Not all individual cows necessarily exhibit delay every year and the length of delay may vary amongst a herd. The factors influencing whether or not delay occurs may include the past reproductive history of the cow, its age, and whether the species exhibits out-of-season mating and ovulation. Delay in implantation in pinnipeds does not seem to be an effect secondary or related to other reproductive factors, such as moulting in the cow, migration, lengthening day or increasing sea temperature.

The cause of delayed implantation has yet to be found although there is an undoubted hormonal background. Whether there be suppression of progesterone production, or retarded liberation of luteotrophin, there is sufficient activity in the uterine tissues to supply nutrition to the blastocysts. Experimental work indicates that implantation does not occur until both the blastocyst and the uterine mucosa are brought to critical phases in their development. Various factors appear to be involved in bringing about this critical inter-relationship in different species. Light, climate and activity may affect the situation and possibly exterocrinological influences related to gregariousness. The Crabeater seal, said to be a solitary seal, is not known to delay implantation, the Southern Elephant seal has a complex harem system and has probably the longest period of delay in pinnipeds.

The chorio-allantoic placenta of seals is large and zonary, like a muff surrounding the middle of the foetus. Beyond the annular band the membranous chorion projects caudally and cranially to enclose the foetal head and tail. The placental band is labyrinthine with large, dilated maternal sinusoids lined by an irregularly thickened endothelium. The trophoblast in the labyrinth is separated from the maternal endothelium by a thick layer of perivascular substance that is of particular interest. It is no inert structure and it must play an important part in certain aspects of placental transfer. The relationship of foetal and maternal tissues is endotheliochorial and the barrier between the foetal and maternal circulations is strikingly thin and less than 1.0 μ thick in many places. Much of its thickness is provided by the perivascular substance. The thinness of the intervening barrier could be an adaptation related to facilitating gaseous exchange to a foetus that is in danger of becoming anoxic if the mother dives with a slowed heart rate. It is not known whether the foetal heart rate also slows during a dive, but a near-term foetus is able to display the diving bradycardia.

The marginal region of the placental band is bright pink or a dark red-brown. This zone has been called a 'marginal haematoma' in carnivore placentae. Similar regions are found in other places on the placental band in seals and show variations in complexity of structure and in size. In both fissiped and pinniped carnivores these regions are important sites of iron transfer to the foetus, as a result of breaking down of maternal red cells. They are particularly important where the foetus has a high blood volume and high red cell count and presumably a high demand for iron, as is so in seals.

The Brain and Central Nervous System

The brain of the largest cetaceans is heavier than that of any other mammal. The heaviest weight recorded is for the brain of a Sperm Whale, 9,200 g (about 0.03% of the body weight). A human brain weighs about 1,430 g (1.93% of the body weight). Only an adult elephant, amongst terrestrial mammals, has a brain heavier than man's at about 5,000 g. A Bottlenose Dolphin's brain weighs about 1500 g (1.2% of the body weight) and that of a porpoise is about 500 g (over 0.85% of the body weight). Brain weights in pinnipeds vary from 250 g in Common seals, 375 g in sea lions and up to 1,000 g in a walrus. An adult manatee has a brain weighing about 300 g. The weight of a brain is not in itself an indication of any particular degree of intelligence, the increase in size is mainly related to the surface area in mammals. The ratio of brain weight to body weight is also of little value when there is, as in cetaceans, so much variation in body weight due to factors involving seasonal changes in food, migration, lactation and disease. Mammals of equal body weight, though quite unrelated, which show increased brain weight and some indication of higher organisation of cerebral centres have been described as being highly encephalized. Efforts to devise coefficients of encephalization have shown to some extent that odontocetes have well developed brains, some not far below that of man. This does not necessarily mean that mysticetes are much less intelligent. Their body weight is made up of a high percentage of blubber which fluctuates markedly. It is quite possible that ratios of various parts of the brain to each other and to the total brain weight could indicate degrees of brain development with greater meaning.

Brain weights in some adult cetaceans in g

Amazonian dolphin *(Inia geoffrensis)*	400-500
Common porpoise *(Phocoena phocoena)*	470-550
Common dolphin *(Delphinus delphis)*	750-870
White-sided Pacific dolphin *(Lagenorhynchus obliquidens)*	1050-1300
Bottlenose dolphin *(Tursiops truncatus)*	1250-1850
Pilot whale *(Globicephala melaena)*	1850-2500
Sperm whale *(Physeter catodon)*	6000-9000
Fin whale *(Balaenoptera physalus)*	6000-7800
Man	1300-1430

The cetacean brain and especially the cerebral hemispheres give the appearance of being much foreshortened and widened transversely. There is some evidence, but no definite proof, that this shortening and brachencephaly might have evolved before and perhaps independently of the telescoping of the skull. The relation between the form of the skull and that of the brain could be considered as one of mutual adaptation occurring over a long period, in which any direct stress on the brain was minimal. There is for example a possible effect of intracranial tissues such as retial masses on brain shape which is virtually impossible to assess. There may be an asymmetry in brain shape, especially in odontocetes which have an asymmetrical skull, but the large size of the brain in mysticetes makes proper fixation and thus assessment of its real shape very difficult.

The dolphin brain is characterized by a remarkable development of the telencephalon with striking convolution of the cerebral cortex. The degree of complexity of the convolutions and even their depth approach a human brain. The sulcal pattern, however, is unlike that of man and bears more resemblance to that of carnivores and

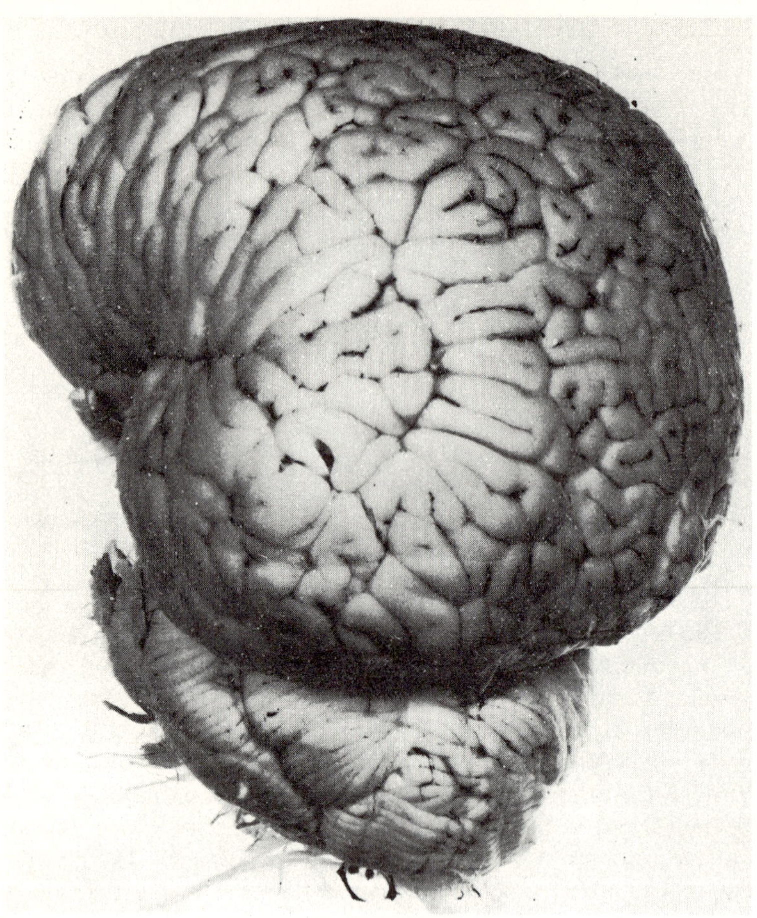

Figure 48. Photograph of the lateral aspect of a brain of Lagenorhynchus obliquidens.

ungulates. Investigation of the cellular arrangements suggests that although five or more layers can be discerned in the cerebral cortex (isocortex) of some dolphins, neurone density is variable but relatively low and the basic cytoarchitectonic structure is said to be somewhat different from that in other mammals. The thickness of the cortex is least in some river dolphins. Some early writers have suggested that the cetacean cortex has a very primitive structure but recent work has revealed many different types of neurones in the various cortical layers of dolphin brains believed to indicate a more differentiated cortex. None of these findings necessarily indicate a lowered mental capacity; for example in mammals generally cortical neurone density appears to decrease with increasing brain weight. The allocortex, the unlayered and more primitive

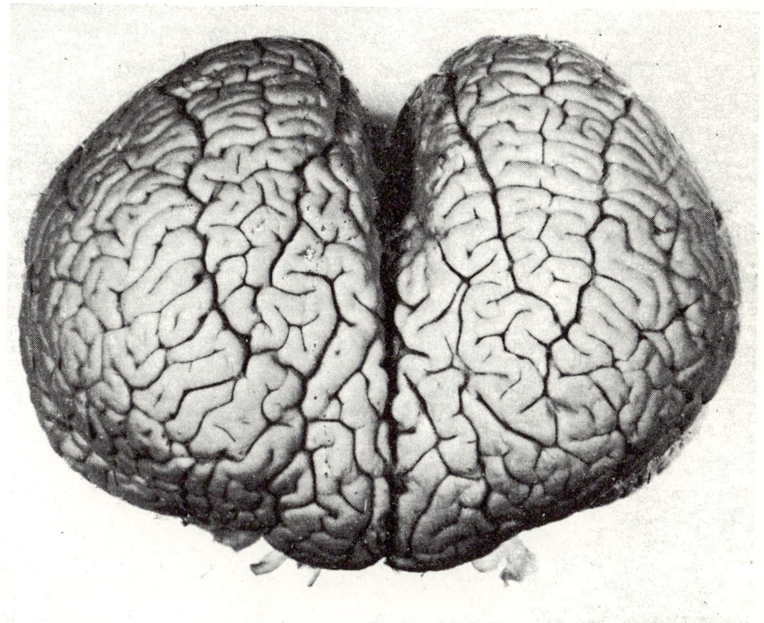

Figure 49. Photograph of the dorsal aspect of a brain of Tursiops truncatus.

cortex, has recently been shown to display a degree of differentiation greater than previously recognized. The great development of the cetacean forebrain has been correlated with the vast enlargement of muscle masses and the need for the more highly organized nervous system that complicated swimming movements would demand. Certain parts of the brain are indeed developed in this connection but it seems unlikely that an animal the size of a dolphin which uses its tail primarily for propulsion would have so large a brain both absolutely and relatively if other faculties were not involved.

Other parts of the cetacean central nervous system are also interestingly developed. Use of embryological material has shown that although arrangements in the cerebellum conform to the general mammalian pattern there are a number of characteristic features. There is an enlargement of the paraflocculus; this is probably associated with the numerous afferent fibres arising in the cerebral hemispheres and upper brain stem. There is also a large lobulus simplex and paramedian lobule. The lobulus simplex probably receives tactile sensation from the head region and the paraflocculus is associated with sensory afferents from the trunk, tail and flukes. Other parts of the cerebellum, such as the anterior hemispheres, the ansiform lobule and flocculo-nodular lobe appear to be much reduced in size. This could be associated with the relative lack of

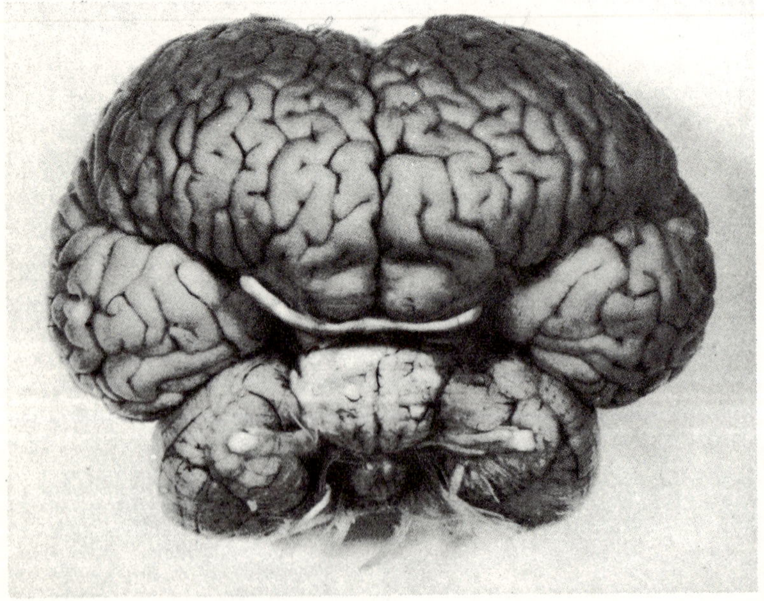

Figure 50. Photograph of the base of a brain of Phocoena phocoena.

demand for equilibration in a medium such as water in which a dolphin floats without difficulty the right way up and where it is not vital if it swims along upside down, as opposed to the much more complex situation facing a terrestrial mammal. In fact there is not much enlargement of the vestibular apparatus and the flocculo-nodular lobe, which receives vestibular sensory afferents, is reduced in size. Vestibular activity does not seem to be a factor increasing cerebellar size, but even if they do not have an increased sense of equilibration, they do display considerable powers of muscular co-ordination.

The cerebellum, the basal ganglia, the red nucleus and various other parts of the nervous system have been implicated in obtaining the control needed for swimming. The precise manner in which they work is quite unknown and such functions must depend on the integrated activity of many parts of the nervous system. Any attempt to associate specializations in the muscular arrangement, or activity of dolphins with particular characteristics of the brain, seems likely to be rather pointless.

The importance of auditory mechanisms in cetaceans is considerable in view of their ability to hear under water and not surprisingly the cochlear nucleus in the brain stem, the trapezoid body, the lateral lemniscus and the inferior colliculi are all enlarged, more so in toothed whales. All these findings emphasize the importance of the auditory pathways. Little is known about the auditory projections to the cortex, though a large area of the cortex must be involved; the large size of the temporal lobe is also of significance.

The auditory apparatus is also well developed in the small porpoise *Phocoena*; the cochlear nerve contains many fibres (100,000) and there are several times the number of cells in the cochlear nuclei as in man. This has been associated, as in other cetaceans, with the possession of an efficient system of underwater echolocation by *Phocoena*.

Only a subsidiary role has been attributed to visual, trigeminal and general sensory impressions as regards differentiation of the cetacean cortex. Disagreement exists as to the degree of visual acuity dolphins possess and whether they can see equally well in air as in water. It has been said that sight in air is poor because of short sightedness in this medium and that in water perception is limited to small moving objects or large stationary ones. The structure of the retina exhibits sparsely distributed ganglion cells each fed from many rods, which must reduce discrimination. Cones are thought to be absent, although a few workers believe they are present in some cetaceans. Rods are most densely packed in the small *Phocoena,* even so the number of fibres in the relatively small optic nerve is many fewer in *Phocoena* (80,000) than in large whales

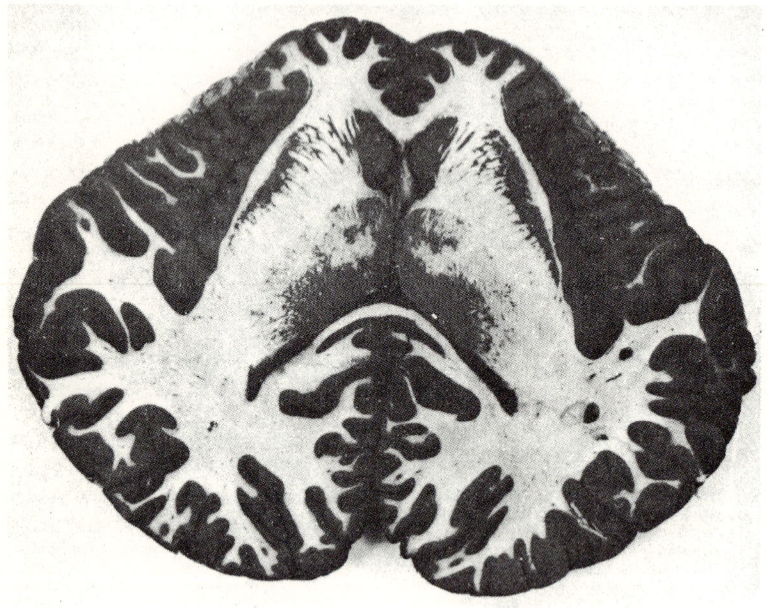

Figure 51. Horizontal section through a brain of Delphinus delphis.

(300,000). In the Gangetic Dolphin the optic nerve is reduced to a thin thread. Some cetaceans may be amphiophthalmic in that images of objects in different media pass to different segments of the retina with axes of unequal length or different orientation. Recent observations suggest that the sight of dolphins in air is as good as that of most mammals: whether or not they can echo-locate in air still has to be verified.

The lateral geniculate body, which can be a guide to the degree of visual acuity by virtue of its construction, is large in *Tursiops* but lacks any lamination. In cetaceans generally it is small and circumscribed, and the superior colliculus is quite distinct. Little is known about the visual cortex: some maintain it is absent but others maintain it is well developed. Bottlenose Dolphins can certainly see out of water and they look round with interest at their surroundings.

The trigeminal is the largest cranial nerve in many cetaceans and the oral region is very sensitive. The hairs, bristles and tubercles found in the snout region of cetaceans are thought to be tactile structures. The head region of dolphins is obviously an area of primary importance and is probably the region from which most sensory information arises. It has been argued from the structure of peripheral nerves that cetaceans have their cutaneous and proprioceptive sensibilities poorly developed in the rest of the body. It has been claimed that dolphins are insensitive to pain, but in certain regions, such as the urogenital area, odontocetes have an abundant cutaneous innervation.

Odontocetes possess an adipose cushion in the snout region above the upper jaw called the melon. It is profusely supplied by branches of the trigeminal nerve. The function of this region is not understood but it has been suggested that there are tactile endings which respond to changes in water pressure. It is also possible that the rich innervation of this region plays a part in controlling the amount of air retained in the respiratory pathway. The muscles of the region, supplied by the facial nerve which can be up to twenty-five times its size in man, are well developed. One function of the muscles might be to force air back into the trachea so that it could be used again for sound production. The muscles of the blowhole and melon region are enormously developed in the Bottlenosed whale which is believed to be a very deep diver.

Olfactory nerves and bulbs are absent in all toothed whales. They have been found in foetal mysticetes, but there is doubt whether they are present in adults. They may have been missed when the brain was removed. The absence or marked reduction of the olfactory nerves, bulbs and secondary olfactory connexions is not apparently related to the changed configuration of the cetacean brain, nor do they seem to have become reduced as a result of the more important part played by auditory centres. Investigation of the tertiary olfactory centres has led to ambiguous conclusions, mainly resulting from our lack of knowledge about those parts of the brain alleged to be concerned with a sense of smell. The characteristics of the cetacean hippocampus could be used to support either an olfactory or a non-olfactory argument as to its function. Cetaceans have no or little sense of smell in water: indeed the sea would not appear to make any contact with the regions innervated by olfactory fibres. Studies of the cetacean rhinencephalon and its related structures have helped little towards localizing olfactory functions. Investigation of the brain stem and spinal cord have supported the alleged lack of cutaneous sensibility in cetaceans. The smallness of the number of sensory fibres in the dorsal roots of spinal nerves, the small size of the posterior horns and of the sensory fibre tracts and the relatively small nucleus gracilis and nucleus cuneatus all seem to indicate a reduced sensory input. Part of the skin surface seems to be sensitive to air as the only stimulus to a dolphin to breathe would appear to be the feeling of air in contact with the skin about the blowhole when the animal breaks the surface. Dolphin skin may be sensitive to the accretion of dead skin and debris though possibly not to temperature changes. There is a rich cutaneous sensory innervation about the urogenital opening in both sexes; the penis is often used as a tactile organ for exploratory purposes. Dolphins in captivity enjoy the feeling of air bubbles against the skin and appear to gain pleasure from scrubbing or rubbing the skin.

The spinal cord exhibits certain features related to the modification of the fore-limb as a flipper, the absence of hind-limbs, the absence of hair and the increased importance of the tail. The spinal cord displays a marked cervical enlargement but there is only a slight lumbar enlargement. The groups of neurones in the lumbar region which normally innervate the hind-limb are absent. Those in the cervical region that would supply muscles of the hand are also lacking. The motor neurone columns in the thoracic and lumbar regions show much complexity in their arrangement, probably a direct response to the muscular changes associated with development and movement of the fluke.

Disease

More and more evidence is accumulating to indicate that dolphins are subject to a variety of pathological conditions both in the wild and in captivity. The situation seems to be somewhat different from that in large cetaceans in which a very low incidence of pathological lesions has been reported. We know virtually nothing about the incidence of various diseases in wild dolphins and as has been indicated earlier there is little accurate knowledge about causes of death and longevity. Captive dolphins have lived for a few days to periods of over ten years but most oceanaria lose a percentage of their dolphins every so often, sometimes for unexpected reasons. Some oceanaria have lost 80% of their stock over periods of 5 to 7 years, others have been more successful but only by rigorous control of water conditions in the pools or tanks, correct choice of food and strict attention to health of the dolphins.

Clean, uncontaminated water is essential for any dolphin pool and pools should not be overcrowded. Circular pools are best as they allow water to circulate and eliminate the build up of stagnant contaminated backwater in corners or recesses. Obviously natural seawater is best but if it is not replaced rapidly, it must be effectively purified through appropriate filters. A turnover rate of every 2 to 4 hours is recommended depending on the number of dolphins. Use has been made in inland pools of artificial seawater: the salinity should not be allowed to fall below 1% and preferably should be that of seawater, 2.7 to 3.5%. Dolphins can tolerate freshwater, indeed River dolphins spend their lives in it and marine forms will travel far up estuaries and rivers. The marine forms seem less buoyant in freshwater and it also affects their skin causing extensive peeling and increased liability to infection. The pH of the pool water should be between 7.2 and 8.0. Chlorine is usually added to control growth of microorganisms but the level should not rise above 0.4 parts per million otherwise there will be effects on the dolphins' eyes. Water temperature is also important, as well as the air temperature above the water, depending to some degree on the species of dolphin in a pool. Draughts directly across a pool should be eliminated. Pipes leading to the pool to or from pumps should be damped to avoid transmission of noise to the pool water. There are many other minor considerations for improving the pool environment; many dolphins like to scrub their skin on underwater brushes and all appear to thrive on exercise and being given something to do—a kind of occupational therapy perhaps.

One curious and so far inexplicable phenomenon does seem to beset wild dolphins and even larger cetacea. This is known as stranding. For an unknown reason, or reasons, one animal or even a large school swim ashore on sandy or shingle beaches and become stranded there to die. Death occurs sooner in larger animals because being out of water there is greater difficulty in raising the body to ventilate the lungs. The weight of the body can even cause the ribs to fracture and sharp spicules of rib to puncture a lung causing it to collapse. If the animal is pushed back to sea, or even taken out to sea and let free, it almost always goes straight back and strands again. Much has been written about stranding, even by Aristotle, and in Britain reports of stranded cetaceans have been published regularly by the British Museum (Natural History) since 1913. Between 1913 and 1966 a total of 1,547 identified cetaceans have been stranded on British coasts. Many explanations have been put forward to account for this self-destructive, almost lemming-like behaviour. The ancient writers suggested that fright, panic and mass hysteria, generated by storms, thunder or evil agencies caused cetaceans to strand. It was well known that dolphins could be driven into bays and on to the shore by men in a number of boats beating the surface of the sea with oars. Such practices have been carried out in Iceland, Faroe Islands and in Denmark for centuries. It seemed obvious to conclude that some similar disturbance caused stranding of wild animals.

More recently it has been maintained that disease could be an important factor in causing stranding. Virus infections could affect the brain of several animals simultaneously, as could parasitic infestation if all members of a school fed on the same contaminated source. Parasites are also known to be endemic in many cetaceans and could interfere with echolocation and navigational abilities. Disease could affect a dolphin so that it ceases to be able to take food, becomes emaciated and weak, unable to keep up with a school, and in such difficulties that currents and tides sweep it ashore. Loss of blubber from starvation, inefficient ventilation of the lungs and muscular weakness could force a dolphin deliberately to look for the support offered by a beach. It has also been contended that the efficiency of a dolphin's echolocation is at a low level when concerned with detection of soft sand. If true, how much worse would be the situation in an ill dolphin. It is not easy, however, to understand why one sick dolphin would lead its unaffected comrades to their doom. Perhaps it is only the diseased animals which strand and the remainder of the school stay out at sea. Certainly it seems that many stranded cetaceans are diseased, we have found several with gastric ulceration and a high proportion have parasitic infestations. If a stranded dolphin is recovered quickly from the beach, efforts to rehabilitate it in a pool are not often successful. Despite all such comments, it is still a mystery why strandings occur.

Of all the diseases so far known to affect captive dolphins, the most common are respiratory infections. Various organisms have been implicated and localized lung abscesses are often encountered: bacterial infections may be superimposed on parasitic lesions. Another frequent cause of death of captive dolphins is *Erysipelothrix insidiosa* which can cause an acute septicaemic condition and a more prolonged cutaneous phase with large rounded raised plaques on the skin. This pathogen can be transferred to men who handle dolphins and seals, giving rise to conditions such as speck finger, sealer's finger and blubber finger. Several examples of tuberculous sea lions and seals have been recorded but the species of mycobacterium was unknown. At least one man bitten by a tuberculous seal contracted pulmonary tuberculosis. Another, bitten by a dolphin, developed ulcers on the hand due to *Mycobacterium marinum,* an occasional cause of so-called 'swimming bath infection'.

Dolphins also develop ulceration in the lining of the forestomach. Large gastric ulcers have been found in stranded animals and also in those dying in captivity. The cause has been ascribed to dietary histamine, which is produced when certain fishes, such as Pacific mackerel, are thawed out and left at room temperature for a few hours. Parasitic infestation has also been considered a cause but

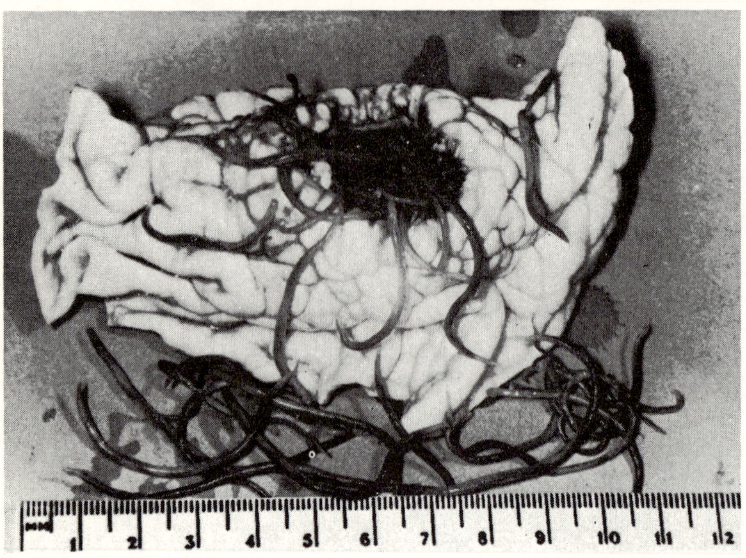

Figure 52. Photograph of an ulcer in the forestomach of Phocoena *infested with nematodes.*

might develop secondarily to the ulceration. It is possible that the stress of captivity may also be a causative factor, as well as the foreign bodies which are so often swallowed when opportunity allows. The list of assorted articles which have been ingested include coins, toys, rubber balls, tin cans, cups, plastic bags, bottles and a screwdriver. Many have been recovered through normal routes, others have needed active intervention to remove them, and a few have caused death by penetrating the wall of the alimentary canal, usually oesophagus, with ensuing infection. Several suggestions have been made to account for this odd behaviour. Dolphins are known to ingest stones and these may aid in breaking up food in the forestomach: hard foreign objects could subserve a similar function but it would not explain the swallowing of such a wide range of soft objects. Curiosity, the conditioning to eat dead fish resulting in the ingestion of anything thrown into the water, and a reduced gustatory sense have all been implicated, as well as hunger, gastrointestinal disturbances, parasites and nutritional deficiency.

There is a considerable literature on the parasites of marine mammals, and although a certain level of infestation appears tolerable to many forms it is a frequent cause of death. As well as external parasites such as whale lice, marine lice, barnacles and copepods which can cause severe skin lesions there are numerous internal parasites which invade the lungs, air passages, alimentary canal, mammary glands, liver, brain and other regions. Nematodes, tapeworms and flukes of several species are frequently found in dolphins and the cysts of *Phyllobothrium delphini* are encountered in the fat, muscle and peritoneum. In many delphinids the infestation has been heavy enough to have interfered with adequate functioning of the respiratory and alimentary systems.

In conclusion, it will be obvious that marine mammals possess many structural and functional attributes that enable them to exist successfully in their environment. All these attributes are worthy of our attention, not only because they provide examples of the adaptive process, but also in that they draw our notice, in their extreme form or manifestation, to the more usual situation in terrestrial mammals and make us re-think some of our ideas and concepts. There are still large gaps in our knowledge, for example why has a dolphin such a relatively large brain and for what purposes is it used? Man has still much to develop in the sea and much to learn about the deep ocean environment. Marine mammals have explored these regions for millions of years: surely there is much they can tell man on how to do it and how to exploit the seas.

FURTHER READING

Andersen, H. T. *The Biology of Marine Mammals* (New York: Academic Press, 1969).

Elsner, R., Franklin, D. L., van Citters, R. L., and Kenney, D. W. "Cardiovascular defense against asphyxia." *Science N.Y.*, 1966, *153*, 941–9.

Harrison, R. J. "The life of the Common Porpoise" (*Phocoena phocoena*). *Proc Roy. Instn. Gt. Br.*, 1971, *44*, 113–133.

Harrison, R. J. *Functional Anatomy of Marine Mammals* (London: Academic Press, London, 1972).

Harrison, R. J. and Kooyman, G. L. *Diving in Marine Mammals.* Oxford Biology Reader No. 6 (Oxford University Press, 1971).

Harrison, R. J., Hubbard, R. C., Peterson, R. S., Rice, C. E. and Schusterman, R. J. *The Behavior and Physiology of Pinnipeds* (New York: Appleton-Century-Crofts, 1968).

King, J. E. and Harrison, R. J. *Marine Mammals* (London: Hutchinson, 1965).

Norris, K. S. *Whales, Dolphins and Porpoises* (University of California Press, 1966).

Ridgway, S. H. *Mammals of the Sea* (Springfield, Illinois: Charles C. Thomas, 1972).

Immunity in a Modern Setting

by

G. J. V. Nossal

CHAPTER ONE

Antibodies and Lymphocytes

In the past hundred years medical scientists have made a monumental effort to place the important and ancient art of medicine on to a sound scientific footing. Amongst the first major triumphs of this recent scientific era in medicine was the discovery of immunization—a procedure by which people can be protected against infectious diseases through essentially harmless injections. The last major conceptual breakthrough in the field of practical immunization was the development of anti-poliomyelitis vaccines, made possible through the growth of polio viruses in cultures of monkey kidney cells growing in glass dishes. As this discovery dates back to the middle 1950's, you might be excused for thinking that immunology, the study of the cellular and biochemical basis of immunization, was a rather "dead" subject. Interestingly, this is not the case. In fact, immunology today is widely regarded as the "glamor" science in medical research, and it has been estimated that about 15% of the total world wide effort in medical research is going into immunological and related projects. What are the reasons for the "boom" in immunology research?

There are four chief reasons for the wide spread interest in immunology today. Three of them are practical, and one theoretical. First, it has been realised that a large variety of common and serious human disorders, many of them having nothing to do with ordinary infectious diseases, are due to the immune system going wrong in some way. If you like, this can be thought of as the body making "bad antibodies", with activity harmful to components of the body itself. Amongst the diseases falling into this "autoimmune" camp we can include disorders such as arthritis, many chronic liver and kidney diseases, many anemias, chronic degenerations of the endocrine glands and particularly of the thyroid and adrenal, and probably a variety of the chronic degenerative conditions of the nervous system such as multiple sclerosis. Secondly, it has been realised that the immune system, so powerful in combating invasion of the body by bacteria and viruses, can also play a role in helping to defend the body against the growth of cancer cells. Thirdly, immunology is very involved in the controversial field of organ transplantation. Surgeons have learnt to perform a variety of technical miracles in the grafting of organs from either living or dead donors into the bodies of recipients whose own organs have been destroyed by disease. However, the immune system of the recipient recognises the foreignness of the grafted tissue, and promptly mounts a severe immune attack. Thus the eventual fate of that organ transplant depends on the skill of the immunologist in combating this immune rejection phenomenon even more than it does on the skill of the surgeon.

These practical realisations have provided almost half the impetus for the development of immunological science in the last decade. However, an equally powerful impetus has been derived from basic science. Unquestionably the signal achievement of the biosciences in the last twenty years has been the cracking of the genetic code. From Professor Phillips' lectures you will realize that we now know a very great deal about the chemical nature of heredity, and about the way the message of the genes is encoded and later translated into the synthesis of the various protein enzymes that a cell needs to perform its various tasks. However, nearly all the information that we have on this point has been derived from very simple life forms such as viruses and bacteria. Before this knowledge can be translated into something of practical use, and even before it can achieve its full flowering as a piece of pure science, it is necessary that we understand the nature of gene regulation in higher life forms, and preferably in mammals. A very great deal of progress has been made in the understanding of the genetic basis of antibody formation, and it is hoped that this will be the model that brings the Watson-Crick revolution fairly and squarely into mammalian physiology.

History of Immunology. Modern immunology is really only one hundred years old. It dates from the discovery by the French scientist, Louis Pasteur, of the microbial origin of infectious disease, published in 1865. Admittedly, a major empiric contribution had come from Edward Jenner in 1798, when he discovered that milkmaids who had caught cow pox (vaccinia) from a member of her herd never subsequently came down with the dreaded smallpox (variola). Jenner inoculated human beings with fluid from a cow pox sore, which resulted in a mild inflammation only. This treatment conferred protection from the infinitely more serious related disorder of smallpox. However, although the studies of Jenner were well documented and controlled, they were a real shot in the dark. Jenner lived in an age when no one knew about viruses or the true nature of infections. His study gained immensely beneficial practical results for the one disease in which he was interested, but research on immunity really lay static until Pasteur found out that microbes caused disease. After Pasteur's discovery, things moved rapidly. Pasteur himself developed a vaccine against rabies, and in 1890 one of his own collaborators, Doctor Roux, and independently the German investigator Von Behring discovered antibodies. These are protein molecules circulating in the blood stream and have affinity for a specific virus or bacterium, or a toxin produced by a micro-organism. The antibodies thus neutralised the ill effects of infection, and in 1891 Von Behring found that antibodies produced against diphtheria in a horse could help to cure the disease when it arose in man.

There followed a period of about forty years during which immunology moved in essentially two directions. On the one hand, immunology became an important subdiscipline of microbiology. The causative organisms of a very large number of diseases were discovered one by one, and as the organisms were cultured and purified, so it became possible to make vaccines against them. Immunization against a large variety of infections, such as smallpox, yellow fever, diphtheria, tetanus, whooping cough, and (where necessary) typhoid, cholera, plague and anthrax gained wide community acceptance. The second direction in which immunology moved was as a branch of biochemistry. Scientists knew that antibodies were present in the blood, but the job of purifying them and analysing them turned out to be a long and difficult process. It was in this area that the German Paul Ehrlich and the Austrian Karl Landsteiner made their great contributions. Both of these mainstream traditions in immunology have their modern counterparts. As regards infectious disease control, the technology of growing viruses in the test tube turned out to be much more difficult than that of growing bacteria, so that anti-viral vaccines of most sorts were developed only relatively recently. Amongst these we can include vaccines for poliomyelitis, measles, mumps, and German measles. For Western communities, this really leaves infectious hepatitis as the only major viral condition for which we do not yet have a vaccine, and this is awaiting better means of growing the hepatitis (jaundice) virus in the test tube. For the chemical tradition, it is pleasing to report that many antibodies are now able to be analysed by the most sophisticated techniques of modern protein chemistry, as we shall see below.

However, towards the middle 1950's a third tradition made its appearance, and it is possibly the most important and rapidly moving one at the present time. This is one we might term the cellular tradition—a new and detailed look at the family of white blood cells which we term lymphocytes, and which are the factories for the synthesis of antibodies and the starting point for all immune responses. The nature of these lymphocytes, the rules governing their birth, activation and proliferation, their role in autoimmune diseases, in cancer, and in transplant rejection—all these are phenomena which have come into clear focus only over the last fifteen years.

Antigens

One of the curious things about immunology is that it is virtually impossible to define antigens without defining antibodies, and conversely impossible to define antibodies without defining antigens. An antigen is simply any material which, on injection into the body, causes the production of an antibody complementary to it.

Living cells are made up of many different compounds, including water, minerals, small organic molecules such as hormones and vitamins, and a host of other materials. However, the most characteristic molecules of living forms are the four great families of macromolecules: proteins, carbohydrates, lipids, and nucleic acids. These are big molecules, composed of many small parts; in other words, they are polymers. Proteins are made up of amino acids, carbohydrates of sugars, and lipids or fats of glycerol and fatty acids. Nucleic acids are more complex and consist of units called nucleotides strung together; each nucleotide consists of a phosphate group, a sugar, and a base. We can symbolically show the formulas of these molecules by drawing them on a piece of paper. However, we must remember that in reality they are three-dimensional, and if we think, in three-dimensional terms, of a portion of a macromolecule consisting of three to ten average-sized building blocks such as sugars or amino acids, we obtain a picture of one unit of antigen, or in technical terms, one antigenic determinant. It is an area or contour of this size which locks with one molecule of antibody. Virtually no work of an immunological nature has been done with fats, but each of the other three classes of macromolecules can act as antigens. Now as one molecule of protein, for example, can contain hundreds or even thousands of component amino-acid units, clearly a single molecule of antigen can, and usually does, contain a number of antigenic determinants. We are now in a position to define our terms: An antigenic determinant is a molecule or portion of a molecule capable of binding firmly to the combining site of an antibody; an antigen is a molecule containing one or more antigenic determinants.

Though most of the antigens in nature are proteins or carbohydrates, it would be misleading to suggest that no other chemical configurations are antigenic. In fact, chemists can hook quite small molecules, termed haptenes, to a protein "carrier". When this haptene-protein conjugate is injected into an animal, a variety of antibodies are formed. Some of these recognize and unite with the "carrier" portion, others with the haptene. Almost any type of chemical can act as a haptene, or in other words it is possible to make antibodies against almost anything. The exception is that animals cannot make antibodies against certain of their own bodily constituents, and we will have occasion to return to this point a number of times.

Antibodies

What, then, are antibodies? They are protein molecules present in the serum which have the capacity to unite with, and bind firmly to, an antigenic determinant on an antigen molecule. Moreover, they are highly specific molecules. For each antigen, there is a corresponding different antibody. As with locks and keys, only certain pairs fit. This is illustrated in Fig. 1-1. Although the union between antigen and antibody is a firm one, it can be reversed by chemical procedures. In fact, our purest preparations of antibodies are made by dissociation of antigen-antibody complexes.

As a rule antibodies are made only in response to the entry into the body of some antigen. In nature this will usually take place after infection with some foreign invader. In the laboratory and

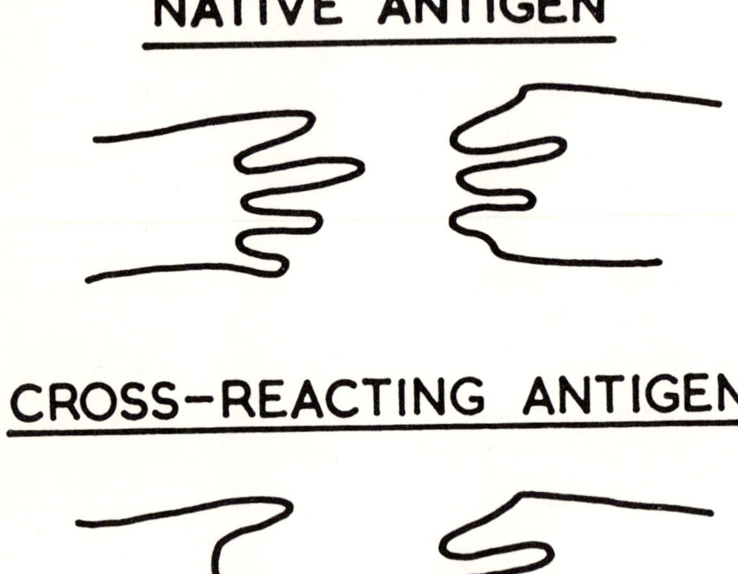

Figure 1-1. Schematic view of antigen-antibody union. In the top half, an antigenic determinant is about to unite with highly specific antibody. Note the closeness of fit of the antigenic determinant and the antibody combining site. In the lower half, the same antibody (right) can also combine with a related, "cross-reacting" antigen, but not as well. In fact, the antibody system is both degenerate and redundant.

clinic it usually occurs after an injection of an antigen under the skin or into a muscle. Although food contains many antigens, these usually cause no antibody production because they are broken down into much smaller fragments by the digestive enzymes before being absorbed into the bloodstream.

Using the electron microscope, one can actually see antibody molecules. However, usually we detect their presence and measure their concentration in the serum by much simpler means. These

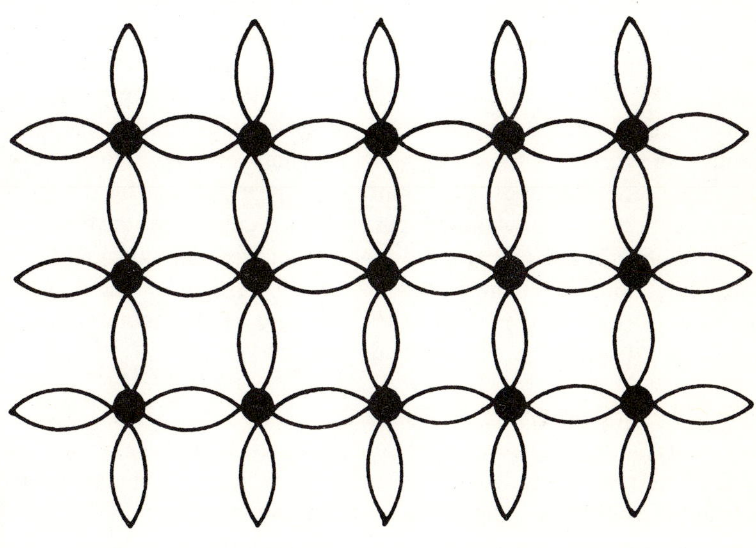

Figure 1-2. Symbolic representation of the formation of an antigen-antibody lattice, which can cause precipitation or agglutination of the antigen. The antibody molecules have two combining sites (always identical). The antigen is depicted as having four antigenic determinants. Though the model does not show this, these determinants can be either identical repeating units, or different from each other. In the latter case, the antibody molecules shown would be a heterogeneous collection, but a lattice would still form.

antibody titration methods depend on some observable interaction between antibody and antigen. For example, we can measure the capacity of a serum containing antibodies (antiserum) to form a precipitate with a solution of the antigen. The precipitation occurs because antibody molecules have two combining sites and antigens usually have more than one antigenic determinant. Thus a lattice structure is built up, as shown in Fig. 1-2.

If the antigen is a preparation of whole bacteria, the lattice formed results in very large particles readily seen with the naked eye. Such lattices are called agglutinates. Similarly, antibodies made against foreign cells such as red blood corpuscles can cause visible agglutination of the cell antigen. These reactions are widely used by immunologists because one can measure the concentration of antibodies in a serum very quickly and without expensive equipment. Other antibody measurement procedures depend on the antibody interfering with some vital function of the antigen. For example, one can assess how many virus particles can be rendered noninfectious, or how many enzyme molecules nonactive, by a sample of serum. There are also many other ways in which antibody molecules manifest their presence to the investigator, but they all depend on the same property—the capacity of the combining site to form a firm union with an antigenic determinant.

The Chemical Structure of Antibody

For reasons that are not yet completely clear, but which will be discussed, nature has decided to make five chemically different kinds of antibodies. Most antigens cause the production of all five types. A long time ago that portion of the serum proteins to which antibodies belong was termed the globulin fraction, and antibodies in general were termed immunoglobulins, or Ig for short. The main chemical classes of Ig are known as IgG, IgA, IgM, IgD, and IgE. The IgG is the most common type of antibody in human beings and other mammals. It used to be known as "gamma globulin", and though this word has become too imprecise for scientific purposes, it is still used by many practicing doctors to describe certain passive antibody shots given to prevent diseases. IgG has a molecular weight of 155,000 and is made up of about 1,330 amino acids. We will discuss its detailed structure below. Most IgA molecules are much the same size as IgG, but some are bigger. IgA has the special property of crossing cell barriers readily. For this reason secretions such as tears, saliva and mucus in the gut are rich in this particular type of antibody. It serves a valuable role in keeping inner body linings free of disease. In fact, it can be thought of as a kind of protective antiseptic paint guarding the various potential portals of entry of infection. IgM is a much bigger and more complex molecule than the others. It has a molecular weight of over 900,000 and contains a significant amount of carbohydrate. It appears to be the most primitive form of antibody, being that which developed already in the primitive cartilaginous fishes. Also it is the first to appear in the serum following an antigen injection.

Detailed Structure of IgG

Like many proteins, the IgG antibody molecule is really a hybrid molecule, made up of four submolecules or *chains*. These are hooked together both by covalent bonds in the form of disulphide bridges, and by non-covalent forces. A schematic view of an IgG molecule is shown in Figure 1-3. The molecule shows a high degree of symmetry. It will be noticed that there are two patches of the molecule that are labelled combining sites. These are the regions of the molecule actually involved in union with the antigenic determinant. Every IgG molecule is double-headed, with two combining sites, and the two sites are always identical to each other. It can also be seen that the combining site is formed by both heavy and light chains. The combining site is really the "business" portion of the molecule. It can be pictured as a shallow crater. When antigen and antibody unite, the antigenic determinant pokes into the crater. The better the fit between antigenic determinant and combining site, the tighter the unions, the more effective the antibody.

Animals and people can manufacture antibodies to a vast variety of different antigens, and thus there is a very large number of possible different antibodies. Interestingly, no one knows exactly how many different antibodies a given animal can make. It is likely that the number is larger than a million, but just how much larger no one can say. We do know, however, that the exact shape and structure of the combining sites are dependent upon the sequence

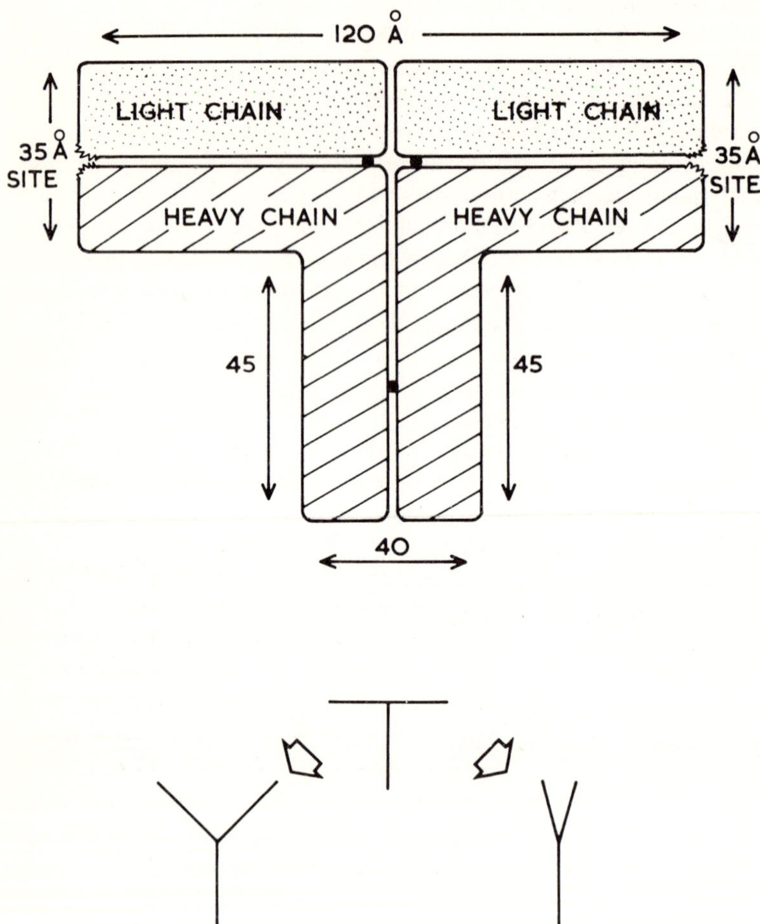

Figure 1-3. Diagrammatic view of an IgG antibody molecule. Note two heavy chains and two light chains, and also the symmetry of the molecule. Imagine an axis of rotation near where the four chains meet allowing flexibility of the molecule. Note that the combining site is a region to which both heavy and light chains contribute. The chains are held together by disulfide bridges (black dots) and also by non-covalent forces. The diagram is derived from work by the late Dr. Robin Valentine, who studied the IgG molecule in the electron microscope, magnifying it about half a million fold.

of amino acid building blocks in the heavy and light immunoglobulin chains that constitute the antibody.

If we look a little closer at different IgG immunoglobulins from the point of view of the amino acid sequence of heavy and light chains, we see a situation as depicted in Figure 1-4. Lining up the different antibodies, we see in fact that one half of the light chain and three quarters of the heavy chain are absolutely identical between the different antibodies. The variability, that which gives the unique combining properties to a particular antibody, is confined to a stretch of about one hundred and ten amino acids, and it is on the basis of amino acid variation in this section that the whole heterogeneity of different antibodies rests.

Two features of this unusual structural arrangement warrant comment. First there is no other known example of a bodily protein where there is this pattern of a "changeable" portion followed

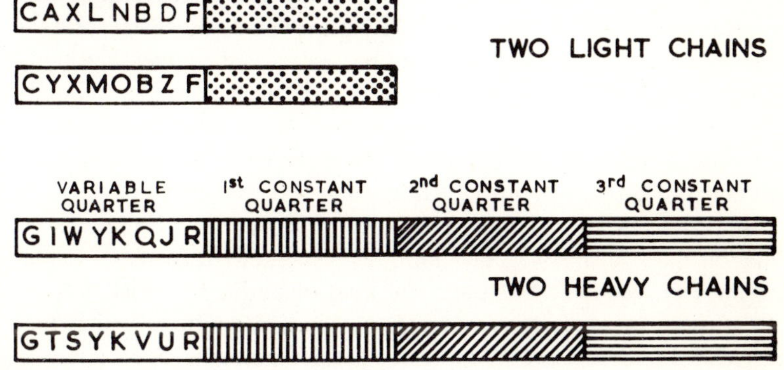

Figure 1-4. Schematic representation of two light chains and two heavy chains from two different antibodies of IgG class. The letters of the alphabet symbolize amino acids. Only eight are shown, but each stretch really contains c. 110 amino acids. Note that for the variable region some amino acids are constant between different antibody chains, and some are variable from antibody to antibody. One half of the light chain and three quarters of the heavy chain are constant from antibody to antibody. There is sequence homology between the different constant regions of the heavy chain, and between these and the constant portion of the light chain. It is thought that in evolution, there was repeated duplication of a primordial gene coding for a protein of half-light chain size, followed by mutation.

by a constant stretch of amino acids. Secondly, as far as we know, the sequence of amino acids in a protein reflects accurately the coded sequence of nucleotides in the DNA of the structural gene for that protein. In other words, for each of the vast array of globulin genes able to be assembled by a living animal, there must be a corresponding gene in the cell manufacturing the chain. Of course, as any heavy chain can combine with any light chain, 10,000 different light chain genes and 10,000 different heavy chain genes could lead to the production of one hundred million different antibodies.

Design of the cellular system responsible for antibody production

The reader may have already guessed one of the key dilemmas in understanding the regulation of antibody production. On the one hand, the structure of antibodies appears to be dictated by genes in cells, as indeed is the structure of every other protein manufactured in the body. On the other hand, the response system seems to be able to generate an enormous variety of different antibodies, on antigenic demand. The way that nature has designed the cellular system which allows this goal to be accomplished is remarkably cunning. The cells which respond to antigen belong to a family of white blood cells called lymphocytes. In an adult mouse there are about one thousand million lymphocytes, widely deployed around the body, and in an adult man there would be some hundreds of times more lymphocytes still. Each of these lymphocytes has on its surface certain antibody receptor molecules. In all probability, each cell has only one kind of antibody on its surface, and this is an accurate sample of the kind of antibody which that cell will make and export in vastly greater amounts upon stimulation by antigen. Therefore, when a particular antigen is injected into the body, its molecules diffuse through the body fluids and encounter various lymphocytes. 99.99% of the lymphocytes will have receptors predestined to react to some other antigen, and therefore will display no interest in the antigen. One lymphocyte in 10,000 or even 100,000 will be predestined to react to that particular antigen, and a firm antigen-antibody union at the surface of the lymphocyte will result. This is the process which Burnet has termed "clonal selection". When a lymphocyte meets an antigen that has affinity for the receptor on its surface, the resulting antigen-antibody union at the surface of the cell triggers a complex series of metabolic changes in the cell. The chief results are a rapidly increased rate of cell division, and growth in the cytoplasm of the cell of an extensive and efficient protein manufacture, assembly and export plant. Four or five days after the initial contact with antigen, the single lymphocyte initially triggered will have generated a population of some hundreds of progeny cells, all synthesizing the specific antibody at maximal rate. The clonal selection theory is depicted in schematic form in Figure 1-5.

One of the great puzzles of modern immunology is just how all the lymphocytes of an animal came to be so different from each other, each having different receptors on the surface. The most widely held view on this point is that lymphocytes undergo constant mutations during the growth of the animal in the genes controlling antibody synthesis, and that this process of "somatic mutation" is actually the generator of diversity amongst lymphocytes.

As well as making progeny cells capable of secreting antibody at a high rate, a lymphocyte triggered into antibody production also makes more of itself. This means that at the end of an immune response, we may end up with ten times more of a particular kind of lymphocyte, with a receptor having affinity for the antigen concerned. Thus, if that antigen is injected a second time some weeks after the first injection, it encounters a higher number of responsive lymphocytes, and a more rapid and vigorous immune response is the result. This is a condition that we term immunological memory.

Many antigens on entry into the body are of too large a size to interact effectively with receptors on lymphocytes. This is certainly the case for many of the bacteria and viruses entering the body during a natural infection. There exists a system of scavenger cells called macrophages which have the specific task of breaking such large antigens down into smaller, lower molecular weight forms. Thus, for many antigens, a sort of "processing" by macrophages is a necessary prelude to an effective immune response.

Regulation of immune responses

Another extremely tricky prerequisite for the immune system is

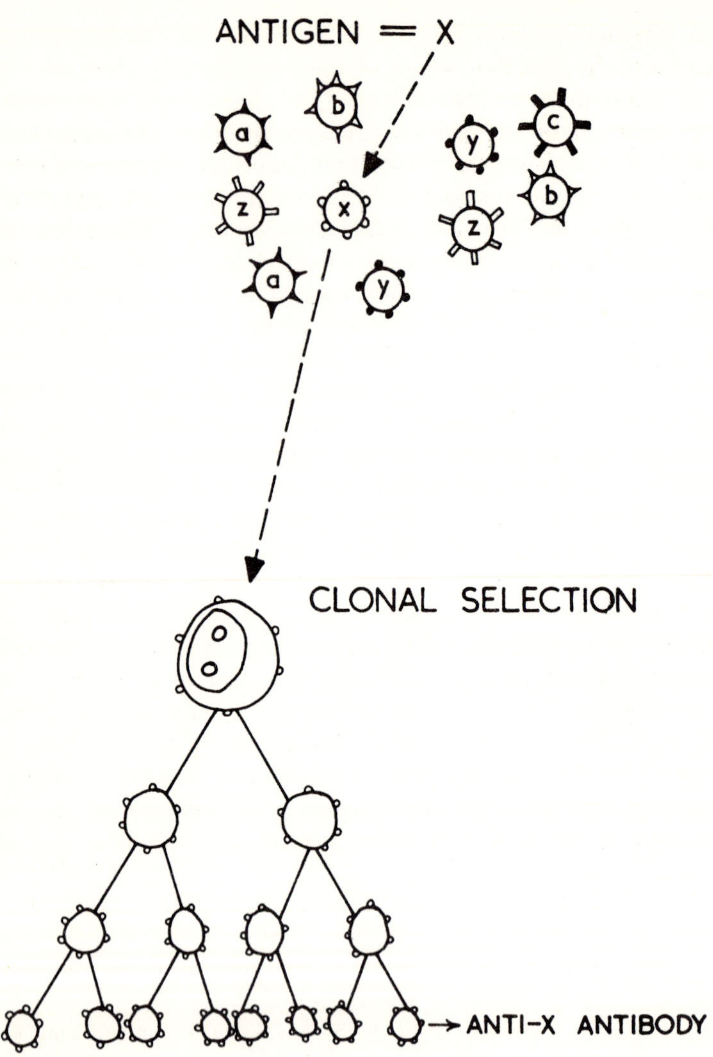

Figure 1-5. Burnet's clonal selection theory of antibody formation, formulated in 1957 and proven by 10 years work at the Hall Institute and elsewhere. The antigen x, injected for the first time into an animal, encounters a lymphocyte with anti-x on its surface. It selects this cell from all the others, stimulating it to divide and differentiate. Lymphocyte x may be one in a million. It is probable that this fantastic diversity amongst lymphocytes is generated by somatic mutation amongst lymphocytes.

that lymphocytes need to be able to discriminate between "self" and "not-self". In other words, lymphocytes must be ready to react to invasion from the outside world, but must not be allowed to form antibody to normal, healthy components of the body itself. Indeed, if animals started forming antibody against their own serum proteins, enzymes, hormones or cell constituents generally, then a form of civil warfare would have broken out in the body, and disease and possibly death would result. Therefore it is clear that lymphocytes with potential reactivity against self components must somehow be inactivated or eliminated. This is the process that we call immunological tolerance. It is now possible to induce experimentally a state of immunological tolerance even to foreign antigens, and the way this is usually done is by overloading the immune system by large quantities of antigen over long periods of time. The detailed cellular mechanisms whereby an antigen induces tolerance are still poorly understood, but it is clear that in principle an encounter between lymphocyte and antigen with affinity for its surface receptor can lead to either the events of immunity (as described above) or to an inactivation of the cell that we call tolerance. The induction of tolerance is one of our goals in organ transplant situations.

Antigen is thus the key driving force, the key regulator in immune responses. There are other subtle controls, however. For example, the end product, antibody itself, plays an important feedback regularity role. The IgM antibody, produced early in an immune response, in fact acts as a positive feedback force; and the IgG antibody, synthesized later, exerts a negative feedback action. Thus the system seems to be designed to get rolling in an accelerated way soon after infection strikes, but to avoid the continued synthesis at high levels of excessive amounts of antibody after peak levels of antibody are achieved in the serum.

Cellular versus humoral immunity

So far we have been talking as if antibody secretion by cells was the only important immunological mechanism. However, this is not true. What we have been describing so far is called the "humoral" immune response, mediated by a special kind of lymphocyte that we call a B cell (for bone marrow-derived lymphocyte). There is as well another major immunological system dependent on "T" lymphocytes, cells derived from an organ in the chest known as the thymus. These thymus-derived lymphocytes never secrete antibody in large amounts. Rather, they have antibody-like receptors on their surface and, on contact with antigen, cause localized toxic damage. This branch of immunity is termed cellular immunity, and it is cellular immunity that plays the key role in the rejection of organ transplants and in defences against cancer. In fact, one of the most rapidly moving areas of modern immunology is research trying to gain insights to the differences between the T and the B cell systems, and trying to understand just what T cells do when they meet their antigen. Most antigens on entry into the body activate both T and B lymphocytes, but to very differing degrees. If we could learn how to manipulate these differential responses, we could not only improve vaccination procedures, but might have powerful weapons of importance in organ transplantation and cancer, as we shall see in later chapters.

The immune response as a paradigm for other biological learning systems?

One of the interesting things about the immune system that emerges from a consideration of clonal selection is that it is a "learning" system in the sense that antibodies are formed and poured out into the serum only when antigen instructs the system to do so; yet it is also a "selective" system, in which normal Darwinian rules apply. In other words, the system learns to form antibody through accelerating enormously and amplifying a synthetic potential which already exists in the animal before the antigen ever came along.

The design of this system has intrigued workers concerned with trying to understand how other biological learning systems function. The central dogma of molecular biology is that information flows from DNA, through messenger RNA finally to protein, and not in the reverse direction. Although some recent evidence has emerged which suggests that RNA can sometimes reverse this process, causing a corresponding DNA to be formed, there is as yet no example of protein being able to break the dogma. Examples of other learning systems are brain functioning and, in a more restricted sense, functioning of the liver (which never knows beforehand what kind of toxic product it must deal with, but which can activate many different systems of detoxifying enzymes). It is intriguing to speculate whether the tricks of amino acid variation and somatic mutation that appear to be at the basis of antibody formation have also been seized upon by nature for some of these other learning systems. In fact, we know virtually nothing of the biochemical basis of learning in the brain, but as our knowledge of the immune system becomes more and more detailed, it is possible that concrete models might emerge which could then be experimentally tested.

CHAPTER TWO

The Immune System and Cancer

Cancer is widely regarded as one of the three most challenging problems left for medical science to tackle. The other two are heart and arterial diseases, and mental disease. Many people have a morbid fear of cancer, a fear which is not helped by the widespread ignorance about the central facts. Cancer is indeed a killer—it ranks second only to heart and arterial diseases in mortality statistics for most Western communities. In fact, about one person in five will eventually die of cancer. It is disturbing to find how often even quite intelligent and educated people will stop you in the street and ask "Have you a ray of hope in the cancer field yet? How long will it be till we have a cure?" Of course the fact is that some cancers can be prevented and many can already be cured. Our knowledge of the cancer process and our ability to prolong lives is increasing constantly. As cancer undoubtedly represents a wide spectrum of different diseases, it is unlikely that a sudden breakthrough or panacea cure will suddenly emerge, and certainly no responsible investigator would claim that an immunological approach will soon bring all the answers. However, multiple useful links between immunology and cancer research are developing, and the exploration is proving helpful to both disciplines. Therefore, in this chapter, we shall examine what impact immunology has had on cancer research so far.

What is cancer? It is a form of overgrowth of tissue that we term a malignant tumor. Tumors or neoplasms are the names that we give to any growth of tissue that is beyond the limits of health or normalcy. A gardener will get callouses on his hands, and a weightlifter will make certain muscles grow in size. These are normal responses of body cells to extensive usage. Tumors are growths which fall outside this range. They can be broadly divided into *benign* or *malignant* tumors.

Benign tumors, as the name implies, grow relatively slowly. They do not grossly invade the surrounding tissues or diffuse widely through the body. A good example of a benign tumor is an ordinary mole; another, the round, fatty lumps that quite a few people have on their backs or elsewhere, which are called lipomas. If a benign tumor is externally situated, it has at worst a cosmetically harmful effect. However, benign tumors in internal organs can be quite serious and even fatal. For example, a benign tumor in the brain can grow ever so slowly, causing no symptoms at all until it begins to press on vital nerve centers. If surgery is then undertaken, a complete cure can result, but if the case is left untreated, the patient may eventually die.

A malignant tumor is characterized by more rapid growth, invasion of surrounding tissue and tendency to spread through the lymphocytes or via the bloodstream to give satellite areas of tumor growth, usually called *metastases* or secondaries, in organs remote from the original tumor site. Unfortunately, malignant tumors are quite common in both sexes, typical examples being cancer of the lungs, intestine or stomach in men, and of the breast or uterus in women. If a cancer is treated by surgery (or in certain cases by high doses of irradiation) before spread from the original growth site has occurred, then a total cure will usually result. If spread is confined to the local draining lymph nodes a careful dissection of these and total removal can succeed in achieving a cure, but this is less common. If widespread metastases are present, total cure is highly unlikely but a variety of treatments exist which can give the patient years of extra useful life free of pain. Even in this area, the prospects are not quite as gloomy as they were.

It is rather doubtful whether it is proper to refer to cancer as a single disease entity. In all probability, cancer is a blanket term which describes a number of totally different and possibly causally unrelated conditions. In the last century an enormous amount of observational knowledge has been collected about human cancer. We can give quite elaborate and detailed descriptions of the microscopic appearance and growth behavior of cancers of each organ. We have many helpful diagnostic tests to allow early detection.

We can give accurate statistical estimates of the likely responses of the different forms of cancer to treatment. In a few instances such as lung cancer, or skin cancers of white Australians living in the northern states, we can describe the most probable causative factor—cigarette smoking, on the one hand, and excessive exposure of fair-skinned people to the sun, on the other. However, all work on human cancer is bedeviled by the fact that opportunities for experimentation are of necessity very limited. Because of this, we have only primitive insight into the origins, causes, and predisposing factors in the various malignancies. For this reason, many researchers in the cancer field have devoted their lives to a study of cancer in experimental animals, predominantly mice. This is motivated by the hope that a thorough understanding of the causes and effects of cancer in the mouse will illuminate the darker aspects of our knowledge of cancer in man. Virtually all our knowledge of the immunology of cancer comes from work on inbred mice, and we have already realized that humans are far from being a genetically uniform population. It may therefore be worthwhile to issue a chastening reminder that the direct relevance of much recent advance in cancer immunology to human malignancies remains to be proven.

Phases of the cell mitotic cycle and characteristics of self-renewing systems

We are now in the position to be a little more specific about the derangement in cell multiplication and differentiation implied in the word cancer. The different phases in the life cycle of a cell are shown in Figure 2-1. Immediately after a mitotic division, a cell enters a phase termed G_1, for first growth phase, during which it grows, and frequently synthesises both protein and ribonucleic acid, but does not synthesise DNA. The second phase, called S, involves an actual duplication of the genetic material DNA, and is relatively constant for most cells of a given species, the S phase for the mouse being about 6 hours. Then follows a short phase G_2, followed by the actual mitotic event itself, termed the M phase. This results in the creation of two progeny cells, which then re-enter G_1.

Of course, many cells in the body do not divide at all, or else

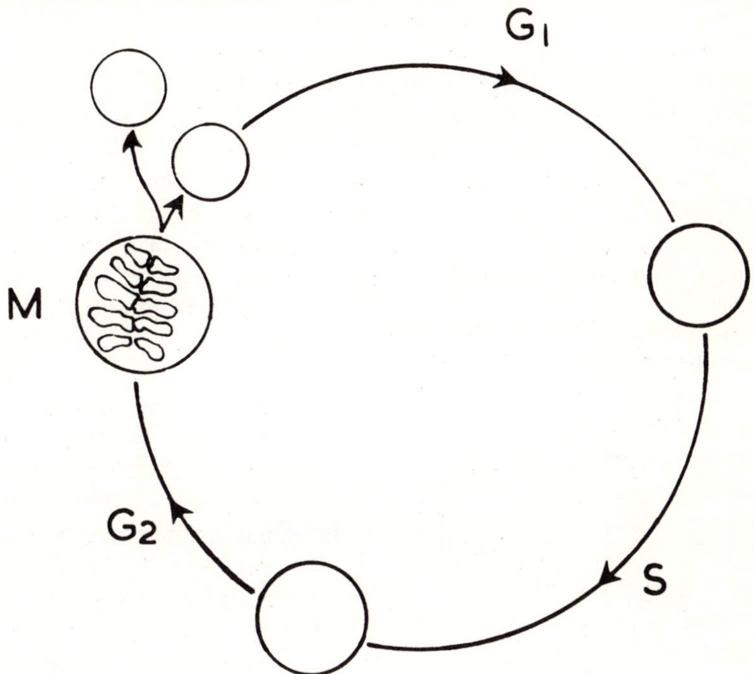

Figure 2-1. The four phases in the life cycle of a cell. Immediately after mitosis, the cell enters the G_1 or first growth phase. During this period it grows but no DNA is synthesized. Then follows the S phase, or DNA-synthesis phase, during which the amount of genetic material is exactly doubled. After this, there is the G_2 or second growth phase, followed by M, the period during which cell division or mitosis actually occurs.

only divide when some special stimulus comes along. Thus, the cells of the brain, or neurones, do not divide at all in man after the age of two years. The cells of the liver divide very infrequently, unless some damage to the liver makes it necessary for regeneration to occur. The antigen-reactive lymphocytes of the body only divide when stimulated by the effects of the immunising antigen. In all these cases, we can refer to the state of the cell as being G_0. When, for some reason, a cell in the G_0 phase is stimulated by an inducer of cell division, it enters G_1 and then subsequently the other phases of the mitotic cycle.

Having encountered the various phases of the mitotic cycle, we must now give some attention to the various stages of cell maturity involved in the concept of differentiation. Take the example of a cell lining the mucus membrane of the gut. The processes are schematised in Figure 2-2. The cells at the tip of the little process or "villus" depicted in the figure are nondividing. They are mature cells, and will in fact shortly be shed into the lumen or inside of the gut cavity. They are the differentiated offspring of more primitive or undifferentiated cells sitting in the basal portion of the villus. When one of these cells divides into two, the end result is the production of one primitive cell similar to the original cell, and one more differentiated daughter cell. In all probability, the actual division itself is symmetric, but one of the two progeny cells is pushed by mechanical forces into a different micro-environment, there coming under a different set of stimuli, and thus the functional end result of the process ends up an asymmetric division. There are many examples of self-renewing cell multiplication systems in the body, of which the best studied are lining cells such as skin or intestinal epithelium; and the production of circulating blood cells such as red cells or the various forms of white cells.

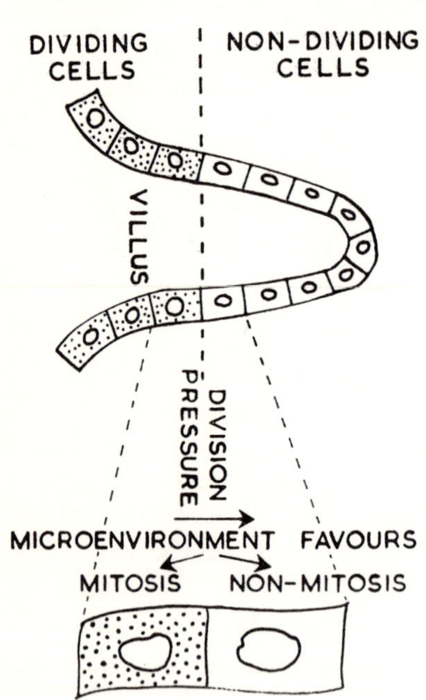

Figure 2-2. A portion of the intestinal lining showing division in cells at the base. Mechanical pressure forces cells upwards. Soon they reach an area where microenvironmental conditions do not favour cell division.

In all of the systems of cell birth and cell death, the cell is influenced by its immediate environment. This means it is subjected to a variety of molecules that impinge on its surface, and that act as either stimulators or repressors of cell division. Amongst the best studied examples of stimulators and repressors of cell division are the hormones, the products of the endocrine glands. However, there are many more, and some of these are still poorly understood. Nevertheless, we can enunciate in Figure 2-3 a general rule about cancer. It is simply that any physiological situation in which, for long periods of time, specific cells are subjected to an environment where there is a serious imbalance between stimulator and repressor influences, cancer is a likely outcome. At first, a cancer arising through such an imbalance of stimulators and repres-

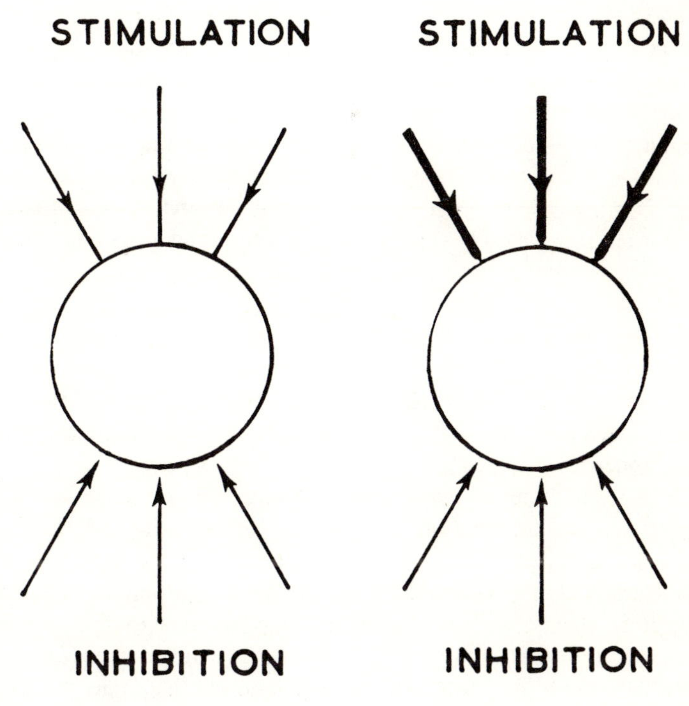

Balanced cell division.

Excess stimulation can lead to cancer.

Figure 2-3. Chronic and excessive stimulation of cell division, such as irritation of bronchial lining by cigarette smoke or of sensitive white skin by ultraviolet light, can lead to excess cell division and eventually to cancer.

sors is dependent, for its continued growth, on a continuation of the imbalance. This is frequently a phase of hormone dependency, and provides the rationale for the treatment of some cancers, such as certain forms of breast cancer and prostatic cancer, with large doses of hormones taken orally. However, eventually the dividing cancer cells assume a more or less total autonomy, at which time hormone treatment is not effective.

An appreciation of Figures 2-2 and 2-3 will soon reveal that an increased rate of cell division is not necessary for the causation of cancer. Rather, many cancers can be regarded as a form of maturation arrest. Regardless of the rate of cell division, a cell population will enlarge if the functional asymmetry seen in Figure 2-2 is abrogated. Were both daughter cells in the self-renewing system to remain primitive, and their progeny also, a continuation of this lack of differentiation would soon greatly enlarge the size of the cell population. This would be the case whether the cell divided every twelve hours, every twenty-four hours or even two days. In fact, it is now known that leukemic white blood cells actually divide rather more slowly than normal white blood cells, the fault being in the incorrect differentiation of the cell lineage.

Theories on the causes of cancer

There are basically three major theories on the way that cancer arises, and it is important to realise at once that these are not mutually exclusive. First, it is known that certain chemicals can lead to cancerous changes in tissues, particularly if these are applied in a repeated fashion over long periods. Good experimental examples are the application of coal tar to the skin (where the carcinogenic hydrocarbons lead to skin cancers) or the ingestion of certain dyes which predictably lead to the development of liver cancer. It is clear that lung cancer in man belongs to this group of chemically-induced malignancies. Bladder cancers frequent in workers in certain dye factories provide another example, as did the old fashioned carcinoma of the scrotum amongst chimney sweeps. In this latter example, and in many laboratory counterparts, mechanical irritation probably played a helping role in cancer induction.

The second major inductive force in cancer is virus infection. In experimental animals, it is clear that certain viruses can induce a variety of tumors, the case perhaps being best established for leukemia. However, we are by no means dealing with an ordinary infectious event as is the case for influenza or measles. In fact, most cancer viruses thought to be responsible for animal cancer are transmitted not by direct contact or air borne infection, but "vertically" from parent to offspring. In the case of many cancer viruses, infection is very widely spread through the body, even though only certain organs actually develop tumors. Therefore a pregnant mouse may pass the virus to its embryos at a very early stage of its development. In many cases, the actual development of cancer depends upon subtle interaction between the virus and the genetic constitution of the infected animal. Thus, leukemia can be produced in one strain of mice but not in another. The degree of genetic involvement is far greater than is the case for most normal infections. All of this adds up to the fact that virus infection may be only one of several factors leading to the eventual malignant process.

In the human, we cannot state with certainty whether any cancer is caused by a virus. Very active work on the isolation of a possible leukemia virus is going on, but it would be fair to say that the work so far has been less conclusive than that in experimental animals. There are two other malignancies in which strong hints of a viral association are present. The first is a tumor of the jaw occurring chiefly in Africans, and termed Burkitt's lymphoma; and the second is a cancer occurring in the tissue at the back of the nose and throat, chiefly in Chinese. In both cases, it is suspected that a virus similar to the agent which causes glandular fever is involved. Of course, the key puzzle is that vastly greater numbers of people become infected with the glandular fever virus than ever develop these relatively rare cancers. This is just another illustration of the confused state of cancer virology, and of the probably dominant importance of genetic considerations.

The third theory of cancer origin relates to the possibility that the cancer cell really represents a reversion to an embryonic pattern of cell division. It is a characteristic of embryonic cells to create an ever-enlarging cell population. The evidence for the embryonic reversion idea comes from immunology, as we shall see below. Two key examples of where embryonic reversion has been claimed for human cancers are cancer of the liver and cancer of the colon.

From a human point of view, the key thing to remember about cancer-promoting influences, be these viruses, chemicals, irritants or virtually unknown factors, is that not all people are at equal risk. It is sheer commonsense to avoid such obvious promoting factors as cigarette smoking, but there are still so many unknowns, particularly in relation to the importance of host genetic factors, that there is little virtue in brooding over possibilities.

Tumor-specific antigens

The key contribution that immunology has made so far to the cancer field is the realization that cancer cells are immunologically different from the normal cells of the patient or animal bearing the cancer. In other words, there are chemical groupings on the surface of a cancer cell that are sufficiently different from the normal components of membranes of normal cells for that difference to be recognised immunologically. It is possible to prepare antibodies against these so-called tumor-specific antigens. Most of the antigens concerned are rather weak ones, and so the amount of antibody made by an animal injected with cancer cells against tumor-specific antigens is small. Various specialised techniques had to be used to prepare tumor-specific antibodies. Once again, it appears that identifiable tumor-specific antigens fall into three groups: virally induced; chemically-induced; and embryonic. However, there are some important differences between the three varieties. In the case of virally-induced tumor-specific antigens, a given cancer virus will cause the appearance of the same antigen in each of twenty mice having the relevant cancer. In the case of chemically-induced malignancies, the same chemical can be used to cause cancer in twenty mice, but the detailed chemical properties of the tumor-specific antigen will be different in each of the twenty. This is because it is believed that the chemical really promotes the emergence of certain somatic mutations amongst cells, and as mutations are a random process, there is no reason why the mutation should take exactly the same form in the twenty cases. Thirdly, the embryonic type tumor-specific antigen is different from the normal adult pattern of cells, but there is similarity to antigens present on the surface of perfectly normal embryo cells.

There are practical consequences flowing from these differences in tumor-specific antigens. If one were ever to contemplate using a tumor cell as a kind of a vaccine for cancer treatment, it is clear that for those cancers that are of a virally induced nature, the cancer of person A would serve as a sort of universal vaccine, having antigens on its surface identical to that of patients B, C, D and E, provided always that we are dealing with the identical cancer. On the other hand, for chemically induced malignancies there is something unique about the tumor-specific antigens of patient A, and the only potential source for a vaccine would be the cancer cells of the actual patient himself. In the case of embryonic tumor-specific antigens, the situation would be similar to that pertaining in virally induced cancers. Different patients with the same form of cancer would exhibit the same embryonic antigen.

Possibilities of immuno therapy in cancer

At this stage, many readers will be wondering about the question of whether all of this work on tumor-specific antigens has any practical consequences. The answer is that while no thoroughly evaluated immunotherapeutic manoeuvre has yet gained full acceptance in the medical literature for any form of cancer, much human clinical trial work is proceeding, and it seems highly likely that, within five years at least, some forms of cancer will be being treated routinely by immunotherapy.

If cancer cells are antigenically different from normal cells, why does the body's immune system not mount an attack on the cancer rather similar to the attack that is mounted on a foreign organ transplant? There are three answers to this question. The first is that almost certainly many cancers are defeated by the vigilance of the immune system before they ever arise—nipped in the bud, as it were. In fact many authorities believe that the key reason why the cellular immune system really evolved is as a kind of surveillance mechanism against cancer. The second point is that tumor-specific antigens are only slightly different from the normal components of host cell membranes, and thus they are weak antigens. The immune response against them is feeble, and there may be instances where the unaided immune system of the patient cannot win out against the intense proliferative drive of the abnormal cancer cell. In this case we have a war waged between cancer and host defences, but being won by the cancer. The third explanation is more subtle, and involves a new concept which we term immunological enhancement. We have already mentioned that the immune system can be broken into two parts—a T lymphocyte system responsible for cellular immunity and a B lymphocyte system responsible for humoral immunity (antibody formation). It turns out that many antibodies are extremely inefficient at killing cancer cells, even when they have specificity against tumor-specific antigens on the surface of the cancer cell. The reasons for this are rather complex and technical, and have to do with the differing abilities of various antibodies to activate a complex system of killer enzymes known as the complement system. In many cases of cancer, both animal and human, antibodies against tumor-specific antigens are in fact made, but do not kill the cancer cells. In that case, the antibodies are not a help to the patient, and are not even a neutral entity. They can be a positive help to the tumor, a positive harm to the patient. This is because they may attach to the tumor cells, and cover up the tumor specific antigen, thus providing a kind of barrier cream which prevents T lymphocytes from reaching the tumor. The T lymphocytes, responsible for cell mediated immunity, are efficient killers of cancer cells. However, they can obviously not reach their target if it is covered up by a layer of antibody. Antibody acting in this way is known as enhancing antibody, and the descriptive term for the overall situation is immunological enhancement. Some years ago, it was thought that immunological enhancement was a kind of a laboratory curiosity. At the moment, we know that it is an important phenomenon both in respect of cancer, where it is an undesirable thing, and in respect of transplantation where, as we shall see in the next chapter, enhancement can be a desirable thing.

What might the goals of an immunological approach to cancer actually be? First and foremost, it is hoped that all of these branches of research will give us a fuller insight into the nature of the cancer process. In many respects, it is pointless to speculate too far further than this, particularly as the whole field is moving so rapidly at the present time. Amongst the readily foreseeable goals, however, are the following:

1. Identification of tumor-specific antigens, or antibodies against tumor-specific antigens, may be a considerable help in the early and correct diagnosis of cancer, particularly because of the well known sensitivity of immunological methods.

2. It might be possible to devise ways of immunizing patients with cancer against their own tumor-specific antigens present on the cancer cells, in such a way as to activate selectively the T lymphocyte system, thus providing a family of killer cells and avoiding the problem of immunological enhancement.

3. It may be possible to raise antibodies that are non-enhancing and effective activators of the complement system, thus providing killing of tumor cells by antibody.

4. It may be possible to chemically couple anticancer drugs or radioactive isotopes to anti-tumor antibodies, thus providing a kind of golden bullet that brings the destructive agent selectively to the site of the cancer.

5. In certain selected cases, it may be possible to identify cancer-promoting viruses and to vaccinate whole communities against these, although at the present time this goal seems further in the distance than some of the other ones.

Conventional treatment of cancer

Quite apart from these speculative matters just discussed, it must not be forgotten that there are a number of very effective methods of treating cancer available already in clinical medicine. Chief amongst these are surgery, radiation with deep X-rays or some other form of ionizing radiation, and the use of a variety of drugs which influence the rate of cell division, this last area being termed chemotherapy. There is thus no reason for the cancer patient to despair. Apart from the hope generated through immunology research, it is wise to stress that prevention through the avoidance of known carcinogenic influences and early diagnosis, together with early treatment, represent the corner-stones of any sensible approach to cancer.

CHAPTER THREE

Truth and Fiction about Organ Transplants

Few advances in medicine have been introduced amidst so much publicity, or have aroused so much controversy, as the field of organ transplantation. The reasons for this are not hard to find. There is something very spectacular and dramatic about replacing an organ that has been destroyed by disease, and whose failing function is threatening the patient's life, with a healthy organ from a living human volunteer or (better still) from a recently deceased person such as a victim of a car accident or a fatal brain hemorrhage. Moreover, when the organ concerned is the human heart, it is little wonder that the public interest is aroused.

Unfortunately, much of the publicity that has filtered down to the average layman has assumed a particularly lurid aspect, and the situation has not been helped by the rather unseemly scurry to get onto the bandwagon which followed the initial few cardiac transplants. Therefore it is important to outline the facts as they are, stressing both the value and the dangers in transplantation procedures, and giving some hints as to the direction that this field is likely to take.

In clinical medicine, the field of organ transplantation is usually thought of in terms of the saving of relatively young lives. Most deaths in people over say 65 or 70 years are due to diseases that are widely diffused through the body. Hardening of the arteries (arterio-sclerosis) and cancer are the two most common, but there are many other examples. In younger people, multiple organ diseases are also common, but there is still a large group of serious disorders where essentially only one organ has gone wrong. The commonest cause of death in middle aged men, for example, is a heart attack. This is due to the occlusion of a blood vessel bringing oxygen to the heart muscle. Sometimes, the disease in the blood vessel which leaves the blockage is widespread through the body, but often it is practically confined to the small arteries supplying the heart. The whole person dies because of malfunction of one working part. There are other serious, common disorders which do not kill quickly but impair health and shorten life. Among these is diabetes, which is due to poor function of certain cells in the pancreas. A reasonably common cause of death in younger life is chronic kidney disease. This may be the end result of repeated infections, or due to the progress of various types of auto immune disorders. Primarily, it is to all the disorders due to the failure of a particular organ that the field of transplantation is most relevant. The concept involves substitution of a good organ for the one destroyed by disease. If it were possible to use all the healthy organs of all people who die from whatever cause to replace effectively the diseased organs of sick people, the savings in human life and suffering would be incalculable. We are a very long way from that goal, but still closer to it than anyone would have dreamed possible as recently as a decade ago.

Early experiences in human organ transplantation

The tissue that was first extensively used in transplantation was skin. During World War II, much experimentation went on about the use of skin grafts, from both cadavers and volunteers, to cover large areas in victims suffering from burns. It was noticed that this physiological method of treating burns did indeed provide a very satisfactory first-aid dressing, helpful in the prevention of excessive fluid loss and in reducing infection. However, the skin never took permanently. Unlike skin taken from a healthy portion of the burnt patient's body, which healed and became fully integrated into its new locality, the skin transplanted from another individual, within a very few weeks, died and was sloughed off. We owe to the Scottish plastic surgeon Thomas Gibson and the British biologist Sir Peter Medawar the realization that this rejection process represented an immune response on the part of the lymphocytes of the burnt patient.

The techniques for skin transplantation were well established, as grafting of skin from one area of the body to another is part of the daily stock in trade of the plastic surgeon. However, with the kidney it was another matter. A complex organ such as the kidney is not as tough as skin. It will only survive in a transplantation situation if an adequate blood supply is established within an hour or so of its removal from the donor. This statement is subject to the reservation that techniques of organ preservation, of a rather complex and technical nature, and involving perfusion of the kidney with blood-like fluids, are making rapid strides. Therefore, clearly kidney transplantation could not be entertained until surgical techniques had been developed to enable the suturing of very fine vessels involving in this case the artery and vein of the kidney and also the tube which leads urine from the kidney to the bladder, called the ureter (Fig. 3-1). We owe much to an intrepid team of surgeons in Boston who pioneered the technique of kidney transplantation. Unfortunately, their early work ran into trouble, again because of host immunological defences, but it did show that kidney transplantation was technically feasible, and that the transplanted kidney functioned remarkably well until the processes of immunological rejection got under way.

The only exception to the rule of immune rejection is the case of identical twins. Here we are dealing with two individuals derived from a single fertilized ovum, and therefore having exactly the same complement of genetic material. Identical twins are alike in their blood groups and also in the genes (called histocompatibility genes) responsible for determining the chemical nature of the cell surface, and so a transplanted kidney will be identical in its chemistry to the kidney of the recipient. Thus, the cellular immune attack does not occur, and if the transplant has been satisfactorily performed from a surgical viewpoint, it will be a long-range success. The only danger is that the kidney disease which initially attacked the patient's own kidney may, some years later, recur in the transplanted kidney; and we are getting much better at preventing this from happening.

Immunosuppression in human organ transplantation

The cellular immune attack against a transplanted organ will vary in intensity depending on the detailed differences between the donor and host in their antigenic constitution. There are two broad

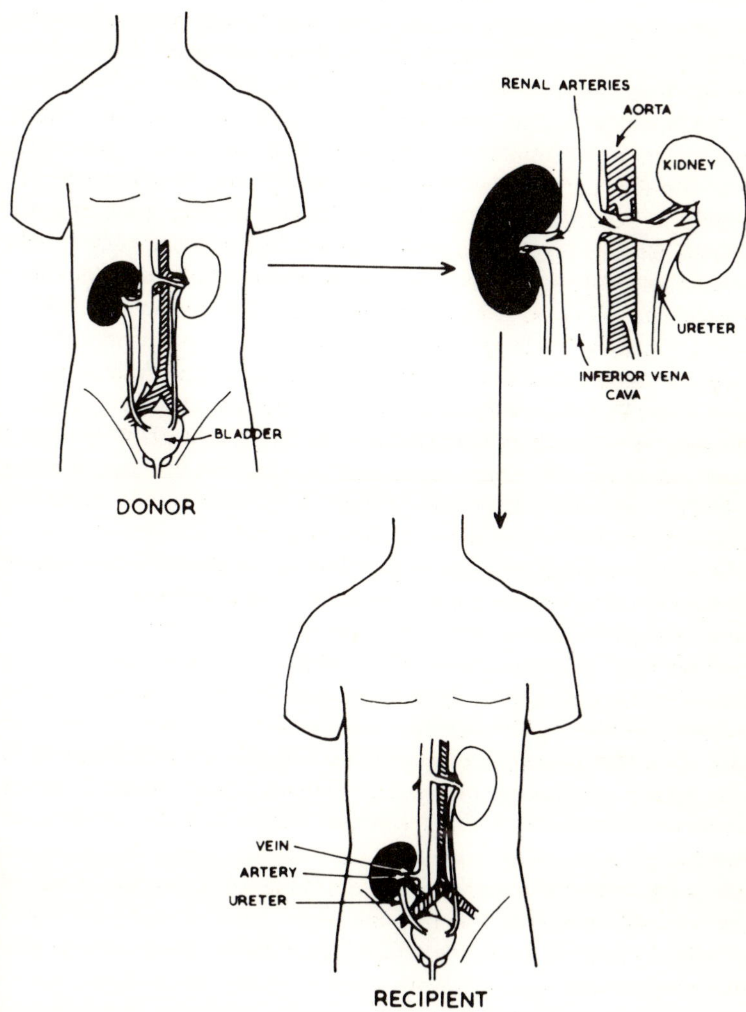

Figure 3-1. The technique of kidney transplantation in man. A new kidney is inserted into the lower right hand quadrant of the abdomen. Blood vessels and ureter are connected to blood supply and bladder respectively. The kidney frequently begins to secrete urine before the surgeon has closed the abdomen.

ways by which we can confine the intensity of immune rejection of grafts to tolerable limits. The first is an artificial suppression of the immune response by certain drugs and treatments; the second, an attempt to minimize histoincompatibility of donor and host by genetic matching procedures.

Treatment aimed at reducing the immune attack to an antigen is referred to as immunosuppression. The drugs and treatments used for this purpose fall into three groups; hormones which resemble in action a product of the adrenal gland called cortisone; poisons which prevent the proper division of the lymphocyte cells invading the organ transplant; and an agent called ALG or anti-lymphocyte globulin, which we shall describe shortly. Cortisone acts at various levels in damping down immune responses. First, excess amount of cortisone in the blood can be directly destructive to lymphocytes. Secondly, cortisone can interfere at a local level in the processes by which lymphocytes kill target cells. Thirdly, cortisone can lower the inflammation and destruction that arises when antibody molecules meet their target. These latter two modes of action, termed anti-inflammatory, are probably the most important. The cortisone group of drugs is certainly an invaluable sheet-anchor of human transplantation and, incidentally, also in the treatment of auto-immune diseases. However, like most powerful drugs, they are not free of side-effects. One effect of prolonged cortisone administration is universal: the person develops a characteristic puffiness of the cheeks and tissue over the jaws, giving the whole face a broader, rounder appearance. This moon-faced look is particularly distressing to women patients. Other side-effects of cortisone are fortunately less common. They include softening of the bones leading to backache and the development of peptic ulcers. For these reasons, it is desirable to keep the dose of cortisone-like drugs as low as possible.

While cortisone alone was used in human organ transplants, the success rate was relatively low. It improved greatly when a drug called azathioprine was developed; this was first used in human beings in 1962. It is typical of a group of drugs which was first developed for the treatment of cancer. The central purpose of these drugs is to interfere with processes essential for cell division. Naturally, if all cell division in the body were to cease, death would result. Many of the body's systems depend on constant cell renewal. The skin, the hair follicles, the mucous membranes, and all the cells of the blood depend on constant replenishment of cells through the process of division. How, then, can a drug like azathioprine be helpful? This depends on two facts. First, not all cells divide at the same rate, and those systems depending on rapidly dividing cells will be more severely affected by treatment with these drugs than tissues in which cell division is slower. This is the rationale of the use of such cell-division poisons in cancer treatment as, on the average, cancer cells divide more rapidly than normal cells. Of the body's different functions, none depends more intimately on rapid cell division than the immune response. Therefore, immune responses can be greatly reduced by doses of cell-division poisons considerably smaller than those which would impair the integrity of the skin or the intestinal lining. The second consideration is that not all dividing cells are influenced to exactly the same degree by a particular drug. Some drugs hit the hair follicles more than others; some have a special tendency to depress bone marrow function. Azathioprine has been chosen for immunosuppression because it tends to hit cells of the lymphoid system harder than it does any other type of dividing cell. In carefully chosen doses, it can depress immune responses while allowing adequate, though perhaps diminished, amounts of red cells and granulocytes to be made. Fortunately, it can block cellular immunity without disturbing hair growth or intestinal function.

The third agent, anti-lymphocyte globulin, has come into clinical usage much more recently, and we have had only two or three years to reach an assessment of its value in kidney transplantation. The principle on which ALG therapy is based is the following; if immune responses are based on lymphocytes, does it not make much better sense to inhibit the lymphocytes directly than to diminish cell division in general? ALG is made by taking human lymphocytes and injecting them into a horse or other suitable animal. The horse will make anti-human lymphocyte antibodies. Unfortunately, it makes antibodies against other human antigens present in trace amounts in the lymphocyte suspensions. These are largely removed by a process called absorption before ALG is used. If they are not eliminated completely, ALG can have some toxic side-effects. Its main action is on lymphocytes. If the dosage can be pushed sufficiently high, ALG can deal a death blow to the lymphocytes causing graft rejection, but toxicity frequently prevents full adequate doses being used.

The best results in clinical immunosuppression are achieved when these three drugs and agents are used in combination. As each has a different mode of action, the toxic effects are not additive, but the immunosuppressive effects are. The dosage of each drug must be carefully adjusted by the physician in charge for each particular case. It usually starts off quite high in the immediate post-operative period, and is then gradually reduced. At intervals, the lymphocytes seem to muster themselves from defeat and mount a sudden counterattack. This troop movement can be spotted by the alert physician. It is termed a "rejection crisis". The treatment is immediate administration of much larger doses of the immunosuppressive drugs. These are given for a few days in doses that would be quite toxic if maintained for long, but which are relatively safe over short periods under strict medical supervision in hospital. This often allows the rejection crisis to be rapidly overcome. Rejection crises are most common in the first six months after a transplant, and relatively rare after a year.

All of these methods of immunosuppression share a major disadvantage. They are not specific for just those lymphocytes that are attacking the graft. In fact, they are all active against lymphocytes generally. This frequently means that the transplantation is more susceptible than normal to infections, and moreover that this immunological surveillance system against cancer is depressed, rendering him statistically more likely to get cancer than another person. Fortunately, it is frequently possible to reduce the drug to non-dangerous levels after the first year or two after a transplant, the maintenance doses required not representing too much of a danger. We will encounter shortly the principle of immunological tolerance, which is a much more desirable one than immunosuppression from a theoretical viewpoint.

Genetic matching in human transplantation

While all grafts, with the exception of identical twins, suffer some degree of immune attack, the intensity of that attack will depend on the degree of genetic disparity between donor and host. First of all, the donor and host must be matched for the ordinary blood groups, and then it is desirable to achieve a close matching of the kidney cells themselves. Fortunately, most of the antigens present on kidney cells, or on the cells of other organs for that matter, are also present on blood white cells. Therefore, genetic matching can be achieved by taking a sample of blood from donor and host and subjecting both to a variety of serological tests. Unfortunately, the actual techniques of white cell matching are very much more complex than those of red cell matching, and moreover the degree of heterogeneity in a human community for major transplant groups is much larger than for the major blood groups. Thus satisfactory matching can only be achieved when the pool of potential recipients and donors is rather large, say in the hundreds. A number of interesting collaborative international projects are under way for the sharing of facilities between different centers, or even different countries. The best example is "Eurotransplant". In this program, specially preserved kidneys may be flown hundreds of miles to reach the patient most nearly matching that particular kidney.

Clinical results of transplantation for the various organs

By far the most progress has been made in the field of kidney transplantation. Over three thousand kidney transplants have now been performed in man. The success rate is encouragingly high. Results are near perfect for identical twins and the best centers are obtaining satisfactory results in over 80% of living volunteer transplants and about 60% in cadaveric transplants. One further gains the impression that results are improving all the time, as groups build up great experience. The reason that the kidney leads the field so clearly is not that there is something rather special about the histocompatibility antigens of kidney. Rather, it arises from a number of circumstances. First of all, renal transplantation is technically much easier for the surgeon than the transplantation of organs such as the liver or the lung. Secondly, and much more importantly, the renal transplant patient can be got into very satisfactory pre-operative condition by the use of the artificial kidney. This is a machine which removes waste matter from the blood that is shunted through it from the patient's circulation. In fact, if the transplant is a failure, the patient can go back onto the artificial kidney two or three times per week and await his chance of a further transplantation.

In general, the results of heart transplantation so far have been rather disappointing, but there is a handful of quite notable exceptions of patients who live for over a year with another heart. Factors which one might cite as creating special difficulties in heart transplantation include the very poor condition of patients when they present for the operation (obviously no one would contemplate replacing a person's heart unless the patient were on the very brink of death) and the absence of any support device such as the artificial kidney to help the patient over rough spots. Kidney patients can afford to have a poorly functioning organ for days or even weeks; if a heart transplant patient has a rejection crisis, it is a much more serious matter. The worst problem, however, relates to the development in the heart transplant recipients of a very severe vascular disease in the blood vessels going to the heart muscle itself. It appears that it is the lining of the blood vessels that bear the brunt of the immunological attack, and the severe vessel wall disease appears in transplant recipients despite best efforts at immunosuppression. This immunological coronary vessel disease is curiously like the spontaneous coronary artery disease which causes heart attacks. This has even led researchers to wonder whether some forms of spontaneous coronary artery disease might be immunological in origin.

Liver transplantation, though extraordinarily difficult technically, is making steady progress in three or four selected centers. Obviously, as for heart, one can think only of cadaver donors. A liver transplantation is a much more difficult operation than a heart transplantation, as the post operative technical surgical complications which ensue are, unfortunately, rather frequent and serious. Again, there is only a handful of long term successes to report in this field to date, but progress is being made.

Lung transplantation in the main has been extremely disappointing in the experimental animal, and, probably for this reason, relatively little work has been done in man. One or two apparently successful lung transplants have been reported. However, the lung is an organ with an extremely rich capillary blood supply, and as the lining of blood vessels is such a vulnerable target for immunological attack, the lung would seem to be an especially difficult organ to transplant from the immunological viewpoint. It seems likely that this operation will only become feasible when better methods of immunological control are available than today.

Most diabetics are treated adequately by injections of insulin, but some go on to develop severe kidney complications. There have been a series of such cases treated with combined transplantation of pancreas and kidney. Again we are dealing with a heroic surgical exercise, and so far survival has been limited to a few months. Similarly depressing comments can be made about the few attempts to transplant large segments of intestine.

As the non-renal field will probably look somewhat bleak to the reader, let us end on a more cheerful note. There are many less dramatic transplantation procedures that have a high success rate. The best known is the cornea of the eye, which normally has no blood supply and is relatively inaccessible to attacking lymphocytes. Also, many materials are of low inherent antigenicity, particularly after the cells in them have been destroyed by some preservative. These include tissues such as cartilage, tendons, bone chips, sheets of fascia and walls of large blood vessels such as the aorta (always provided the vulnerable inner lining is not present). Transplants of tissues such as these form an important part of eye surgery, plastic and reconstructive surgery, and orthopaedic surgery.

The reason for maintaining a certain guarded optimism about transplantation of major organs other than the kidney is that the methods of immunological control, and indeed the level of understanding of the problems involved, are increasing steadily. The biggest hope attaches to the concept of immunological tolerance, which we must now consider.

Immunological Tolerance

We have mentioned already that animals do not form antibodies against their own tissue components—they tolerate them, in fact. This capacity to fail to form antibodies is acquired during the embryonic life of the immune system. It is not an innate genetic characteristic, but rather something that lymphocytes must actively learn. The most dramatic way to illustrate this is through a very remarkable experiment that is being performed in many centers, following the pioneering work of Tarkovsky and Mintz. One can take two separate mouse embryos at an extremely early stage of their embryonic development, and by special tissue culture tricks, one can fuse the two embryos together. These can then be re-transplanted into the uterus of a mouse, and in a proportion of cases, a viable mouse emerges after the normal twenty day gestation period. This is not a monster with two heads, but rather a normal mouse with four parents! It is a mosaic of two different sorts of cells, one half derived from one set of parents, the other from the other set of parents. It has been shown that the immune system of such mice is also mixed, though not always on a fifty-fifty basis. Even when the two sets of parents differ from one another by a great distance from a genetic point of view, there is not civil warfare in the body, with the lymphocytes of one parental line attacking the lymphocytes of the other parental line. Rather, there is a state of immunological tolerance. Over embryonic development, the lymphocytes of one partner have become tolerant of the antigens of the other partner, and vice versa. In recent years, experimental immunologists have been learning some of the rules that lead a lymphocyte to respond to an antigen not by antibody production but by tolerance. In fact, in the experimental animal it is possible to induce tolerance rather readily, even in the adult. Most of the work has been done by using purified model antigens, such as pure proteins, rather than the chemically extremely complex and poorly understood histocompatibility antigens on the surfaces of kidney cells. Amongst the many factors which conspire to make an antigen tolerogenic rather than immunogenic are high dose, comparatively low molecular weight, and a specific configuration or clustering of antigen determinants when these attach to the lymphocyte surface. Operationally, tolerance induction in the experimental animal can frequently be aided by the use of immunosuppressive drugs, though again these have to be used in dosages that might be rather dangerously high in man.

The beauty of immunological tolerance is its specificity—rather than blanketing out the activity of the whole immune system, one simply deletes a specific word from the immunological dictionary. The tolerant animal is tolerant only of the antigen A which was used to render it tolerant. It can still respond quite adequately to B, C and D etc. As we learn more about the human histocompatibility antigens, and as the very vigorous current research aimed at their identification and purification proceeds, we can look forward to a time when these principles of immunological tolerance will be applied in the transplantation field.

Enhancement in organ transplantation

Once again, this paradoxical phenomenon of enhancement which we encountered in the last chapter has turned out to be of importance in organ transplantation. It is believed that many of the patients that have come into a reasonably harmonious relationship with a donated kidney are being helped by enhancing antibodies that are protecting the kidney from killer cell activity. The importance of enhancement has been shown extensively in experimental animals, but once again the principle is considered rather dangerous for extensive clinical trial at the moment. To my knowledge, there has only been one attempt to inject a human kidney graft recipient with enhancing antibodies against the kidney graft. At the time of writing, this appears to have been a reasonably successful experiment. It would appear that for the long term transplant recipients that are now requiring only quite minimal dosages of immunosuppressive drugs, their happy situation arises from a combination of four factors. First, there may be a degree of immunological tolerance induced. Secondly, there may be a gradual replacement of the cells lining the vessels of the kidney with host cells. Such replacement would very greatly diminish the antigenic disparity problem. Thirdly, enhancement may be playing a role. Finally, the small doses of immunosuppression could be dealing with whatever vestige of immune attack is still present.

Ethical, legal and logistic problems in transplantation

In this brief review, I have no chance to expose the various viewpoints that have been expressed about the ethical and moral aspects of organ transplantation. As regards cadaveric transplantations, most countries have now introduced legislation which defines under what circumstances organs may be used for transplantation. Morally I can see no harm, and only good, in an effective transplant from a cadaver, as the organ concerned is of no earthly use to the cadaver six feet under the ground! Of course, this statement is predicated on the organ donor being dead from every point of view. Fortunately in Australia this has traditionally meant cessation of all three vital activities—respiration, heart beat and brain function. The ethical problems relating to living volunteer donors are much greater. Here, one's judgment must obviously be coloured by the statistical likelihood of a successful outcome to the transplant. To take the extreme example, it would obviously be a moral and generous act for an identical twin to donate a kidney if the other twin were dying. On the other hand, this, and the general field of sibling transplantation in the presence of good genetic matching, places an enormous degree of pressure on the prospective donor, and obviously a thoroughly professional psychosocial work-up of the situation is vital. We have taken these problems of living volunteer transplantation very seriously in Melbourne, and at the Royal Melbourne Hospital have in fact relied exclusively on kidney cadaver donors.

With respect to the transplantation of organs such as heart, liver and lungs, there is of course an enormous moral problem for the doctor in that an unsuccessful operation will usually be irretrievable and may even hasten the patient's death. On the whole I believe that most people's fears on this score would be removed if they could see how very near to death nearly all of the cardiac transplant recipients were. However, clearly much more research needs to be done at both experimental animal and clinical levels before transplantation of any other organ becomes as successful as renal transplantation. My own feeling is that over the next few years heart transplantation should remain the province of a small number of highly specialised groups with a profound commitment to the area, rather than becoming a widely-hailed technical extravaganza. Fortunately there is every sign that this is happening spontaneously.

The organ transplantation field is here to stay, both as a branch of immunobiology and as an accepted method of treatment for some diseases. How far it will really revolutionize the fabric of medicine and surgery will depend on the research of the next decade.

FURTHER READING

Nossal, G. J. V. *Antibodies and Immunity* (New York: Basic Books, 1969; Harmondsworth: Penguin, 1971).

Nossal, G. J. V. and Ada, G. L. *Antigens, Lymphoid Cells, and the Immune Response* (New York: Academic Press, 1971).

Roitt, Ivan *Essential Immunology* (Oxford: Blackwell Scientific, 1971).

Population, Resources and Environment

A Multiple Crisis

by

Paul R. Ehrlich

Population, Resources and Environment: A Multiple Crisis

In mid-1971, there were 3,700 million people on Earth. If the current growth rate, 2% per year, continues, the world population will double in 35 years. Such enormous numbers and such a rapid growth rate boggle the imagination; both are completely without precedent in human history.

The stress put on the Earth's resources and life-support systems by this burgeoning population is only beginning to be widely realized. When such limiting factors as food supplies and non-renewable mineral and fossil fuel resources are considered, it becomes evident that this uncontrolled growth cannot be sustained for long. Our very efforts to keep increasing food production and to expand the industrial society put the Earth's ability to support such a huge population, let alone a much greater one, in serious jeopardy.

The growth rate of a population is the difference between the birth-rate and the death-rate (ignoring migration). The birth-rate is the number of individuals who are born, and the death-rate is the number who die, per thousand per year. As long as the birth-rate exceeds the death-rate, a population will grow; if the death-rate exceeds the birth-rate, it will decline. If they are equal, population size is stable.

Both our population growth and human impact on the environment have their beginnings in prehistoric time. The population grew extremely slowly, with numerous local and temporary setbacks, before the agricultural revolution that began about 10,000 years ago. At that time, there were perhaps 5 million people scattered around the world, fewer than now live in the metropolitan areas of Sydney and Melbourne combined. The development of agriculture provided more security than did the previous hunting and food-gathering existence, and a given amount of land generally could provide food for larger numbers of people. The expansion of food supplies and the establishment of permanent settlements meant that food could be stored in times of plenty as "insurance" against times of shortage. In response to these changes, death rates began to fall, while birth rates remained more or less unchanged. This trend continued as agriculture was adopted by nearly all human cultures. The population doubled about every 1500 years until the beginning of the modern era. It reached 500 million about 1650. Then population growth began to accelerate rapidly. Later improvements in agricultural practices, the industrial revolution in Europe and North America, and particularly the development of modern medicine led to further reductions in death rates, especially in Western countries. The population doubled to 1000 million in the 200 years between 1650 and 1850, and reached 2000 million 80 years later, in 1930. Since then, it has almost doubled again.

The rapidly dropping death rates in Europe and North America during the 18th and 19th centuries were followed by some reduction of birth rates over the last century, largely as a result of changing social and economic conditions accompanying industrialization. In agrarian societies, children are seen as a source of labor for farm work and a form of social security for their parents' old age. In industrial societies, on the other hand, children are expensive to raise and educate. It is generally thought that these new advantages to family limitation were the reason for the lower birth rates that followed the industrial revolution. This so-called "demographic transition" has partially compensated for the lower death rates and helped to slow growth in developed countries (DCs). Today, these populations are growing at an average rate of about 1% per year, which would double them every 70 years.

Following World War II, modern medical technology was introduced to underdeveloped countries (UDCs), without any serious effort to introduce birth control techniques. The result was a spectacular drop in death rates and practically no change in birth rates. The underdeveloped areas of the world—most of Africa, South America and Asia—are now growing at an average rate of about 2.5%, doubling every 30 years, sometimes even 20 years or less.

There is little hope for a spontaneous drop in the birth rates of these nations, because the socio-economic conditions that produced the industrial revolution and the demographic transition in the West are not present. In most UDCs the majority of people still support themselves by agriculture, often at the subsistence level.

Some consequences of this runaway population growth are obvious and horrifying. Perhaps as many as *half* the people in the world today are either undernourished (lacking calories) or, more commonly, malnourished (usually lacking in protein). Between 10 and 20 million die each year of starvation, most of them infants and children. Countless millions are doomed never to achieve their human potential because of nutritional and other deprivation. Food production is failing to keep up with population growth. If all the world's food were equitably distributed, there would be just barely enough calories to go around, but not enough protein. However, available food is very unevenly distributed, both within and between nations. The rich get the lion's share in both cases. There are also massive preventable losses to pests and spoilage between the harvest and the table. Publicized food surpluses are most often *economic* surpluses—food that no one buys. These surpluses create a misleading appearance of abundance where, in fact, shortages exist. Destitute people around the world (including in the richest countries) go hungry because they cannot afford to *buy* the food they need, even when it is available. The trend in food production per capita has been downward since the late 1950's. In 1969, there was no increase in total food production, while the world population grew by 2%.

If the human race is unable to feed itself adequately today, how can we hope to feed additional hundreds of millions a few years from now? The food that is being produced today, particularly in DCs, is gained at considerable expense to the environment through the use of powerful synthetic pesticides and inorganic fertilizers. Such "modern" farming practices may result in lowered soil fertility and less food in the long run.

Various schemes have been proposed to raise food production or to develop supplementary protein sources. The most likely of these to succeed, the Green Revolution, involves the introduction in UDCs of new, high-yield forms of traditional grains, wheat, rice or corn, combined with more efficient farming methods.

The Green Revolution, at best, may enable agricultural production to keep pace with population growth for the next decade or two. Even the men who were most involved in developing it, including Dr. Norman Borlaug, who won the Nobel Peace Prize for his development of new wheat varieties, have proclaimed that it can only buy us time to establish population control. Despite the hope the Green Revolution offers for increasing food production, there are numerous social, economic, and environmental obstacles to success. These new grains require large amounts of fertilizers and irrigation water to produce high yields. Without these, they are often inferior to traditional strains. Therefore, the use of the new varieties involves a considerable amount of related development to build fertilizer plants, distribute the fertilizer to farmers, and build dams or drill wells for irrigation projects. All this requires capital and the ability to take risks, so only the better-off farmers have been able to invest in the new seeds and methods. In many areas, increased harvests have resulted in lower prices. The small, poor farmer is caught in the squeeze. In other areas landlords have displaced their tenant farmers in their eagerness to get in on the Green Revolution bonanza. Such dislocations have led to a land-grab movement in India and considerable rural unrest in several countries.

Besides the need for fertilizers and irrigation, some new grain varieties may require much more protection from pests than naturally resistant traditional crops. Thus the Green Revolution will introduce into the UDCs the same fertilizing techniques and wide use of synthetic insecticides that have been causing so many environmental problems in DCs.

Essentially, the Green Revolution is an attempt to institute the same kind of fossil-fuel subsidized agriculture as is now practiced in most DCs, in place of traditional UDC agriculture. Since the future of the world's fossil fuel supplies is questionable and the ability of the UDCs to get their share is even more so, one may reasonably be apprehensive about many of the fundamental assumptions on which the "Revolution" is based.

It would be especially disastrous to mechanize UDC agriculture. Farm mechanization drives people off the land and into cities.

Most UDCs already suffer enormous problems of urban migration and unemployment, which would be catastrophically intensified by mechanizing agriculture. UDCs can avoid such upheavals by developing labor-intensive agricultures, such as are practiced in Japan and Taiwan.

Various other panaceas for solving the food crisis besides the Green Revolution have been proposed. Food from the sea particularly is promoted as a potential solution, especially in supplying needed protein. The sea now provides a significant fraction of the high quality protein in the world's food. If sea life were protected and carefully husbanded, it has been estimated that productivity might be doubled by around the end of the century. But since population size will also have doubled by then, obviously no improvement in the average diet can be expected from that source.

Unfortunately, far from protecting this vital source of food, we are polluting the oceans with perhaps a half million different substances, including deadly pesticides and poisonous mercury compounds. Moreover pollution is heaviest near continental shores, where sea life is most abundant. Rather than husbanding oceanic food resources, we are grossly overexploiting many fisheries. Some, including several whale fisheries (whales are mammals, but the harvesting of a species of whale is conventionally called a "fishery"), have been nearly destroyed. Our abuse of the oceans may be catching up with us already. In 1969, for the first time since 1950, there was a decline in fisheries production. Despite greater effort and increasingly sophisticated techniques, total production was 3% less than in 1968. In 1970, production rose again, but preliminary reports indicate another drop during the first half of 1971.

Besides food, many other resources are running short, particularly non-renewable ones such as copper and petroleum. The DCs consume the overwhelming bulk of these (the U.S. alone accounting for 30-40%), although increasing amounts of their raw materials must be obtained from the UDCs. Presently known reserves of petroleum and many important minerals will be exhausted within 50 to 150 years at the current rates of consumption, and those rates are rising.

The technology and way of life of Western countries demand the use of these resources, but their consumption in the present wasteful fashion plainly cannot continue much longer. The underdeveloped world expects to enjoy the blessings of Western style technology in the future. But the dream of industrializing the entire world can probably never be realized; the Earth simply does not have the resources. Even if they did exist, the environmental consequences of worldwide industrialization would turn the dream to a nightmare. Meanwhile, how long can we expect UDCs to allow DCs to extract their natural resources to support DC affluence? Not much longer, judging from the behavior of many UDC governments and the rise of civil unrest around the world. We can expect UDCs to drive increasingly hard bargains for these resources. There may well be more wars over resource pools—Vietnam was not the first.

Besides the frantic consumption of non-renewable resources, humanity is also depleting renewable ones such as forests, soils, and fresh water. In all three cases, consumption or destruction are far exceeding the rate of replenishment. Deforestation and soil erosion have been by-products of human activities for centuries, as the advancing Sahara Desert and overgrazed Mediterranean areas testify. America's demand for water is expected to exceed her domestic supply within 20 years. The U.S. is now beginning to develop water project schemes to utilize Canada's stores of fresh water. Needless to say, intelligent Canadians are less than enthusiastic about such proposals.

The crisis of population versus food and resources alone would be a massive challenge for humanity to face over the next few decades. But over this crisis hangs the huge problem of environmental deterioration. As a factor in human health and well-being, environmental pollution is a serious cause for alarm, especially in DCs where most of it is generated. Air pollution has led to an increase in respiratory diseases in heavily polluted cities; water pollution can contribute to increases in the incidence of other diseases; the ubiquity of chlorinated hydrocarbon insecticides and industrial compounds represent a dangerous long-term health risk to everyone in the world; incessant noise, crowded urban areas, and ugly surroundings pose unknown dangers to mental health.

However, man's impact on his environment has reached a state where "pollution" in the usual sense is only part of the problem. Man's activities now present a grave danger to the entire planet

Earth, which nourishes and supports us all. The poisonous substances we are heedlessly discharging into soils, waters, and atmosphere injure not only ourselves, but the ecological systems on which we all depend for oxygen to breathe, for food to eat, and for the recycling of our wastes.

Long-lived chlorinated hydrocarbons, such as DDT, are found in every living plant and animal around the world, on land and in the sea. DDT accumulates in the fat of organisms. It does not break down readily into harmless substances; rather it is concentrated by ecological systems. Herbivorous animals get it from eating plants and pass it on to the predators which devour them. Although only a fraction of the energy in the food an animal eats is incorporated into its body, almost 100% of the DDT is. Therefore concentration of DDT is greatest in such animals as predatory birds and fishes that feed at the upper end of food chains (eating sequences such as grass→cow→man→tiger). Most chlorinated hydrocarbons and some other pollutants have similar characteristics.

Numerous species of predatory birds and fishes are threatened with extinction today from the effects of chlorinated hydrocarbons alone. As the quantity of these poisons increases in the environment, species lower in the food chains will be affected. These organisms are often involved in the food chains which produce human food. The photosynthesis of tiny marine plants, phytoplankton, which form the base of oceanic food chains, is known to be affected by DDT. Some species of phytoplankton are affected more than others. Altering the composition of local marine flora could have drastic repercussions for the organisms that depend on it. This is one reason why food from the sea is likely in the future to contribute less rather than more to the average human diet than it does today.

Many other serious pollutants are produced primarily by agricultural activities, including inorganic fertilizers and atmospheric dust. Dust from agricultural activities influences the weather. Inorganic fertilizers not only pollute rivers and lakes, killing off useful fishes in the process of eutrophication, they also appear to interfere with the natural processes of the soil that refresh and renew it. Artificially high soil fertility thus may temporarily be achieved at the cost of lower soil fertility over the long term.

An increasingly common agricultural and industrial pollutant in water systems is mercury. A highly toxic, organic form, methyl mercury, circulates in food chains more or less as DDT does, and has also been shown to affect photosynthesis in marine phytoplankton. High concentrations of methyl mercury have been found in many North American river and lake fishes and in fishes in Europe and Japan. It has been estimated that enough mercury has already been deposited in North American waters to keep it circulating through the food chains for at least a hundred years, even if no more is allowed to escape. Recently it has turned up in deep sea tuna and swordfish. It appears that we are discovering a new environmental threat which may prove to be as grave as that posed by synthetic insecticides.

Air pollution damages trees and crops as well as human tissue, but, even more disturbing, it may induce changes in climate. Smog is no longer a local problem of large cities; the entire atmosphere is contaminated. Whether the long term effect of air pollution will be to warm or cool the planet remains uncertain, but at the moment it seems to be contributing to a general cooling trend. If air pollution continues to increase, there will be an acceleration of climatic change. Agricultural production would inevitably be profoundly and adversely affected by any rapid change in weather, regardless of what it was. Such a change is believed to have been responsible for the appearance of corn blight in the American midwest last year, where many farmers have responded by planting different crops this year.

Such local climate changes can be expected to increase in frequency if air pollution is not reduced. They may be further accelerated if the SSTs are developed and flown. That expensive, technological boondoggle, besides producing sonic booms, and an uneconomical, uncomfortable ride, may leave semi-permanent contrails in the stratosphere, thus reducing the amount of sunlight that can reach the surface. The results for the weather in areas where SSTs fly can only be guessed, but it is probable there will be results.

Another increasing human influence on climate is the heating which is generated by any use of power, whether it is the burning of gasoline in a car, the running of a refrigerator, or the generation of electricity by power plants. If power consumption continues to increase at current rates, what is now mainly a local problem of thermal pollution will soon escalate into a world-wide one. Heat due to man's activities may alter subtle balances in the global heat engine and cause rapid, perhaps catastrophic climatic change. Switching to nuclear power would add radiation dangers to the heat pollution, although it would reduce other forms of pollution produced by power plants. It is becoming clear that power consumption cannot continue to increase much further. Countries like the United States and Australia must curtail their per capita power consumption. The costs of power should be raised; power is too cheap, and cheap power leads to frivolous uses such as heating and airconditioning poorly insulated houses and providing lights where they are not needed.

Humanity now finds itself in a multiple bind. There are too many of us and our population is still rapidly increasing. We are finding it more and more difficult to feed ourselves and obtain other resources we need. Worse, we are wasting and mismanaging many of these resources. In our frantic efforts to provide more each year and to stoke the industrial society, we are seriously damaging the fragile life-support systems of the Earth, jeopardizing our future ability to provide for ourselves. Where is the way out? There are, of course, several ways in which the population explosion could be abruptly halted without conscious action on the part of men.

Medical propaganda to the contrary, man has not been entirely freed from the ravages of epidemic disease. From time to time new forms of old diseases appear or lethal diseases never before seen in man suddenly transfer from animals to humans. The human population today is the largest, weakest, and most densely packed that has ever existed; an ideal situation for an epidemic. For the majority of the world's people, medical care is essentially non-existent. A new virus could be carried around the world by modern transport within hours; an epidemic could spread out of control within days or weeks. A relatively lethal disease such as flu could produce millions of deaths directly and countless millions more through disruption of food distribution alone. In addition to the chance of a naturally occurring epidemic, biological warfare laboratories are industriously creating dangerous new disease agents. There have been accidental escapes before, and more are inevitable as long as research and development of biological weapons continue.

Thermonuclear war is another possible means of halting the population explosion. One important factor in most, if not all, modern wars is population pressure. This most frequently appears as pressure on resources; a need for more resources or a need to deny others access to them. The war in Vietnam is an example. The petroleum, tin, rubber and other resources of Southeast Asia have been determinants of policy in that area for a long time. Several major oil companies have already staked claims for offshore drilling in waters adjacent to Vietnam.

As the world's population grows, the potential for conflict will become even greater. War is always undesirable, but the advent of thermonuclear war has provided humanity with a new means to achieve self-extinction. It would not be necessary to annihilate the entire population directly. Even a limited nuclear exchange could sufficiently impair the life-support systems of the Earth and damage the structure of industry to prevent survival in the long run. At the very least, civilization would be utterly destroyed.

Even if war can be avoided (or better, abolished), what about the internal disturbances plaguing so many countries, including Australia and the United States? Most of the dissent and disruption in the U.S. can be traced to obvious social causes such as racial injustice, an exploitive economic system and a distant, costly, unjust war. Nevertheless, it seems reasonable to suggest that population pressure, especially in the malfunctioning cities, and the stress of rapid population growth may be contributing to these troubles. It has been shown that crime rates rise when people migrate rapidly into cities; mental disturbance may be more prevalent in areas of high density. On the other hand, people appear to have very high tolerances for crowding and great ability to adapt to adverse social conditions. Our adaptability cannot be limitless, however, and the stresses may be beginning to show among the more sensitive individuals.

A multiple crisis obviously cannot be met with a simple solution; fundamental changes will be required in several areas. One such area, of course, is the population explosion. It is clear that population growth must soon stop and ultimately be reversed. If humanity fails to regulate its own numbers, nature will do it for us, sooner or later.

Population regulation requires the setting of some sort of goal. Humanity must determine what its optimum population size is. The first step is to discover how many people the Earth can support. Before answering this question, we must decide what the average standard of living should be in terms of food, shelter, clothing, other material goods and services. At the average U.S. standard, it has been estimated that the world could temporarily support between 500 and 1000 million people ("temporarily" because this does not take into account the depletion of non-renewable resources—when they are gone the world will be able to support far less). By this optimistic measurement, there are already 3.5 to 7 times too many of us.

At a much lower standard of living—perhaps the average that prevails worldwide today—we probably could temporarily support more people than now exist, particularly if food and materials were more equitably distributed and less wastefully used. But we probably could not support so many for long; the environmental stress would be too great, even if we were far more careful in controlling pollution than we are today. It should be obvious that population-size decisions based on material living standards encompass far more than simply feeding ourselves. Fundamentally the choice is whether we will limit our numbers and preserve a decent life for all human beings or allow our continuing growth to reduce everyone to the barest subsistence level of misery—a choice between quality or quantity.

In addition to considering the problems of basic support and quality of life, thought must be given to the ideal number of people to keep each society functioning efficiently. It has been suggested, for example, that the United States could function very well with a population of 50 million (one quarter of the present population) or even less. Given the same level of technology, the pollution problems certainly would be diminished! Australia might need a somewhat smaller population than she has today. It seems unlikely that she could accommodate a great many more without severe environmental penalties. Such decisions must be based on the type of society that will prevail locally, regionally and nationally. A primarily industrial area will naturally require more people than an agricultural one, and some kinds of agriculture demand more people than others. Ideally, an optimum population would permit diverse life styles and the opportunity to choose among them. There should, for instance, be enough people that large cities can exist, few enough that there can be open space available for everyone.

Far more difficult than *determining* the optimum population size will be *achieving* it. The ancient, obsolete idea still prevails that among nations or political groups there is strength in numbers. If that were true, China and India would be the most powerful nations in the world, and Israel would long since have disappeared.

Many officials and economists in UDCs (and some DCs) still believe that more people are needed to carry out development. They cling to this belief, although high birth and population growth rates repeatedly have been shown to hinder economic progress and although most of these countries are suffering from severe unemployment problems. One of the consequences of a high birth rate is a very young population. In the UDCs some 45% of the population is under 15 years old. This large proportion of dependent children is an economic drag. Conversely, this group would proportionately be reduced first in a population control program, bringing immediate economic benefits. At the same time, the huge proportion of young people now living guarantees that, even with the most effective imaginable program of population control, these populations would continue to grow rapidly for *at least* another generation.

People in DCs also mistakenly believe that they have no population problem because they practice family planning and their populations are growing more slowly than those in UDCs. But these slowly growing populations are creating the bulk of the environmental deterioration that is endangering the entire world. They are the people who are consuming most of the world's renewable and non-renewable resources. They also take an unfair share of food, particularly protein foods in the form of fish, grains, and oilseeds. Most DCs are overpopulated in the sense that they cannot feed themselves. They may export meat and dairy products, but they must import huge quantities of vegetable and fish protein to feed their livestock. Most of this comes from UDCs, many of whose people are suffering from protein malnutrition.

Since DC population growth is therefore the greater threat to the future of man, it is most imperative that they, especially the superpowers, lead the way in population control. Otherwise, people in UDCs will quite reasonably consider recommendations of population control simply a new plot to keep them from gaining their share of the world's goods. Less powerful DCs, such as Australia, can also make important contributions by providing examples. Their motives are much less likely to be regarded with suspicion by UDCs.

It is important to remember that family planning is not population control. Population control is the regulation of population size by a society; family planning only serves to regulate the sizes of individual families. Unless a society at least attempts to influence the goals of family planning, there is no control of population. As long as people want and have an average of significantly more than two children per couple, a population will continue to grow.

Governmental family planning programs have been introduced into numerous UDCs since 1960. India's program was the first, having been started in 1952. In recent years it has become one of the most vigorous and one of the few actively promoting two-child families. But it has encountered severe difficulties, especially in rural areas. Even without overt resistance, which is not unusual, a family planning program in India would have formidable obstacles to overcome. In terms of lowering the birth rate significantly, it has not been successful; the 1952 birth rate was 40, in 1970 it was 42! Most of the programs in other UDCs operate as early birth control centers did in the U.S. and England. They open clinics, provide information and free or low cost contraceptives. Their propaganda chiefly consists of spelling out the personal advantages of planning and spacing children and of avoiding unwanted children. This is socially valuable, but it has had very little effect on birth rates. *Ways must be found to reduce the numbers of children desired by couples.*

An important prerequisite of any form of population control is that the means of birth control be readily available to all members of each society. The affluent U.S. is only now meeting this basic goal. Before now, these services were for practical purposes unavailable to the poor. Contraceptive information and/or devices are still illegal in many countries, both DCs and UDCs. This is largely but by no means entirely due to pressure from the Roman Catholic Church, which disapproves of "artificial" birth control.

When birth control fails, abortion can be used. There has been a great deal of controversy about the morality of abortion, but history, particularly in the past decade, shows that opinions can change very rapidly on this issue. Although it is illegal in the majority of countries, abortion is far the most commonly employed form of birth control in the world. At least one out of every four pregnancies the world over is believed to end in abortion. In some countries the ratio is much higher, especially where contraceptives are generally unavailable or illegal. Italy, a country which has effectively banned birth control devices, has an estimated one million illegal abortions annually—more abortions than births. In 1969, birth control pills were allowed in the country, which will probably reduce the abortion rate. There has recently been a trend to legalize abortion in many countries, including the United States. Sixteen states have liberalized their laws to some degree; four place virtually no restrictions on obtaining an abortion.

Complete availability of contraception and abortion will not alone produce population control, since people still want too many children. Individuals must be convinced of the need to limit their reproduction to an average of two children per family or less. In fully literate, relatively world-oriented areas such as North America, Europe, Japan, Australia and New Zealand, this might be achieved simply through an educational campaign to convince society of the consequences of continued population growth, and to foster new cultural attitudes. All human cultures today embody attitudes which are implicitly pronatalist. Childlessness and nonmarriage, conditions now commonly viewed with pity or disapproval, might instead be encouraged and approved. Women's roles as wives and mothers can be deemphasized, and employment and educational opportunities on an equal basis with men should be provided for them. Job security following marriage and pregnancy and low-cost or free day care for children are useful means of encouraging women to seek fulfilment outside the home. The very real advantages both to children and parents of very small families can be stressed by the educational campaign. First and only children generally do better in school and later life and (contrary to the myths about only children) are usually more independent than later children. Similarly, children from small families have been shown to be healthier, brighter and more successful on the average than those from large families, regardless of income level.

If family planning plus an educational campaign fail to reduce birthrates to the desired level, stronger socio-economic measures may be required. Tax laws which presently encourage reproduction can be changed, preferably in such a way that they do not penalize poor families (in poor families, it is the children, not irresponsible parents, who would suffer). There could be special bonuses or tax incentives for childless couples and single people, encouragement and subsidy for adoption, and bonuses for sterilization. Such measures, hopefully, would suffice to bring birthrates to the replacement level in most DCs.

The situation in UDCs may prove to be much more complex and difficult. Pronatalist attitudes there are typically much stronger than in DCs, where family planning is now a part of tradition. In most UDCs, high death rates among children and infants are a matter of recent memory; in most they are still considerably higher than in DCs. There is no tradition of family limitation, except perhaps in the small urban fraction of the population. For the rural majority, the old economic reasons for large families still hold sway. More stringent measures for population control may be required in most UDCs; these should be worked out in each area or country to fit the social, economic and political situation. But at the same time, DCs must stand ready to give whatever assistance is needed: manufactured contraceptives, trained personnel, educational materials, or logistic support.

Obviously, population control alone will not solve all human problems. Environmental deterioration, war and arms races, social injustice and race prejudice, food and resource needs and shortages, all will still be with us. Population control will, however, provide us with an *opportunity* to solve them. Without it, our efforts to solve our other problems are ultimately doomed to failure.

It means nothing less than that the present Western economic system and way of life must be fundamentally changed. The importance of material goods should diminish, and wasteful methods of production must be abandoned. All materials and goods should be manufactured with a view to durability and recycling. Pollution resulting from manufacturing must be controlled and minimized. Judgments must be made on alternative methods of manufacture that take pollution dangers into account as well as economics. For example, a manufacturing method that uses mercury might be abandoned in favor of a slightly more expensive method that does not.

Cities and other human habitations should be planned with human beings in mind, not automobiles. Although efforts are being made to produce a "smog-free" car, it is obvious that the larger population of the future will not be able to support such luxuries as private automobiles for everyone. The pollution, traffic congestion and comsumption of resources they represent will make them obsolete in many areas. The only reasonable choice is to move to efficient, convenient forms of mass transportation. Cities can be redesigned so that such systems would be far more convenient than what we have now.

Agriculture should be practiced in cooperation with nature, not as a war against it. Today in the U.S., livestock are raised in feedlots, and their excreta is dumped into the nearest river as sewage, thus creating a huge water pollution problem. Meanwhile, crops are fertilized with synthetic fertilizers, which are not retained by the soil, but run off and add even more to water pollution. The use of already centralized feedlot sewage as fertilizer appears not to have occurred to anyone.

Similarly, there are better ways of controlling agricultural pests than by indiscriminate use of powerful poisons which have damaging effects far beyond what is intended. These may involve such methods as enlisting natural enemies to help control pests and using only non-persistent pesticides as an occasional emergency control measure.

If tropical areas (where most UDCs are located) are to be successfully developed for greater food production, it will have to be done with much greater care than has been shown in the past.

It is clear that temperate zone agricultural techniques do not work with tropical biological communities and soils. A great deal of research is needed to find ways of supporting people in these areas without destroying what is there. Again, DCs must be ready to provide assistance to UDCs for developing ecologically sane agricultures. DCs can, for example, provide technical and material assistance and help establish local research and training institutions.

Societies around the world should be rated on the well-being of their citizens, rather than their gross national products. Much of our present difficulty stems from the fact that economic measures of affluence contain hidden costs in the form of pollution and the impairment of health and the quality of life it produces. Indeed, the higher costs of medical care are included in the GNP, as are the costs of pollution abatement. The GNP of the U.S. rose spectacularly during most of the decade 1960-1970. Yet the quality of life for the average American has considerably declined in a great many ways. How has the quality of life fared in Australia?

A solution to the 20th century's multiple crisis demands nothing less than a worldwide revolution in attitudes and values. Humanity must come to see itself as a single entity with many more common bonds than differences, each person dependent upon others and all dependent upon a fragile planet for continued life. Changed attitudes toward such basic elements as reproduction, marriage, families and the roles of women in society will be essential. The change of basic values must also include the rejection of war as a means of settling differences, the ending of racism, and the ending of exploitation of poor people and nations. So long as men in insulated groups compete for resources and power over other groups, consuming and destroying both natural and human resources in the process, we can never hope to solve our problems.

In generations past, each society had a vision of the future, and it built and planned for that future. Today we seem bent only on continuing to build on past goals, without evaluating their worth or feasibility. What vision of the future remains is out of alignment with reality. Resisting change is useless; our choice now is whether to control and direct the changes of the future toward new goals derived from the hard realities we face, or to continue drifting until the "solutions" find us.

Today's young people are already exploring new ideas and values, many of them in appropriate directions. If we are wise enough to work with them, rather than rejecting their apparently strange and radical views, perhaps together we can create a new vision for the future that can benefit and inspire all men.

FURTHER READING

Borgstrom, Georg *The Hungry Planet* (New York: Collier, 1967).

Borgstrom, Georg *Too Many, A Story of Earth's Biological Limitations* (New York: Macmillan, 1969).

Brown, Harrison (ed.) *The Next Ninety Years* (Pasadena: California Institute of Technology, 1967).

Brown, Lester *Seeds of Change: The Green Revolution and Development in the 1970s* (New York: Frederick A. Praeger, 1970).

Carson, Rachel *Silent Spring* (Boston: Houghton-Mifflin, 1962).

Cloud, Preston E., Jr. (ed.) *Resources and Man* (San Francisco: W. H. Freeman & Co., 1969).

Dasmann, Raymond F. *The Last Horizon* (New York: Macmillan, 1963).

Ehrlich, Paul R. *The Population Bomb*, 2nd ed. (New York: Ballantine, 1971).

Ehrlich, Paul R. and Harriman, R. L. *How To Be a Survivor: A Plan to Save Spaceship Earth* (New York: Ballantine Books, 1971).

Environment (formerly *Scientist and Citizen*) Official publication of the Scientists' Institute for Public Information, published monthly.

Fagley, R. M. *The Population Explosion and Christian Responsibility* (Oxford Univ. Press, 1960).

Falk, Richard *This Endangered Planet* (New York: Random House, 1971).

Food and Agriculture Organization of the United Nations, *Production Yearbook* (Rome: FAO-UN).

Food and Agriculture Organization of the United Nations. *The State of Food and Agriculture* (Rome: FAO-UN).

Gofman, J. W. and Tamplin, A. R. *Poisoned Power* (Emmaus, Pa.: Rodale Press, 1971).

Hardin, Garrett *Birth Control* (New York: Pegasus, 1970).

Hardin, Garrett (ed.) *Population, Evolution, and Birth Control* (San Francisco: W. H. Freeman & Co., 1969).

Holdren, J. P. and Ehrlich, P. R. (eds.) *Global Ecology* (New York: Harcourt Brace Jovanovich, 1971).

Holdren, J. P. and Herrera, P. *Energy* (New York: Sierra Club Books, 1972).

Hopcraft, Arthur *Born to Hunger* (Boston: Houghton Mifflin, 1968).

Illich, Ivan *Celebration of Awareness* (New York: Doubleday and Co., 1970).

Istock, Conrad E. "Modern environmental deterioration as a natural process." *International Journal of Environmental Studies*, 1971, Vol. 1, pp. 151–155.

Marshall, A. J. (ed.) *The Great Extermination* (London: Heinemann, 1966).

Marx, Wesley *The Frail Ocean* (New York: Coward-McCann, Inc., 1967).

Meadows, D. L. *The Limits to Growth: A Global Challenge* (Washington, DC.: Universe Books, 1972).

Mishan, E. J. *Technology and Growth: The Price we Pay* (New York: Praeger, 1970).

Morris, Desmond *The Naked Ape* (New York: McGraw-Hill, 1967).

Murdoch, William W. (ed.) *Environment: Resources, Pollution and Society* (Stamford, Connecticut: Sinauer Associates, 1971).

Myrdal, Gunnar *Challenge of World Poverty: A World Anti-Poverty Program in Outline* (New York: Pantheon, 1970).

Osborn, Fairfield *Our Plundered Planet* (Boston: Little, Brown & Co., 1948).

Paddock, W. and Paddock, P. *Famine—1975! America's Decision: Who Will Survive?* (Boston: Little, Brown & Co., 1967).

Park, Charles F., Jr. *Affluence in Jeopardy* (San Francisco: Freeman, Cooper & Co., 1968).

Population Reference Bureau, Inc., Washington, D.C. 200 publishes the *Population Bulletin*, *PRB Selections*, and the *World Population Data Sheet*.

Reinow, R. and Reinow, L. T. *Moment in the Sun* (New York: Dial, 1967).

Rudd, R. L. *Pesticides and the Living Landscape* (Madison: Univ. of Wisconsin Press, 1964).

Shepard, Paul and McKinely, Daniel (eds.) *The Subversive Science* (Boston: Houghton Mifflin Co., 1969).

Study of Critical Environmental Problems (SCEP) *Man's Impact on the Global Environment* (Cambridge: MIT Press, 1970).

Udall, Stewart *1976*, *Agenda for Tomorrow* (New York: Harcourt, Brace, and World, Inc., 1968).

United Nations, 1966–1967. *World Population Conference, 1965*. (Vol. 1, Summary Report; Vol. 2, Fertility, Family Planning, Mortality; Vol. 3, Projections, Measurement of Population Trends; Vol. 4, Migration, Urbanization, Economic Development). (New York: U.N.).

United Nations. *United Nations Statistical Yearbook*, published annually (New York: U.N.).

United Nations. *Demographic Year Book*, published annually (New York: U.N.).

Vogt, William *Road to Survival* (New York: Sloane, 1948).

The Interaction of Science with Society with Special Reference to the Media

by

Chapman Pincher

CHAPTER ONE

Science and Violence

My purpose in these three talks on the interaction of science and society with special reference to the media—meaning the newspapers, television and radio—is to promote discussion rather than to lecture you about my views. I shall therefore have no qualms about being controversial or even outrageous. And because there are so many points I want to raise, I make no apology for giving you a rather disjointed discourse or for posing more questions than I shall answer. Most of the problems are long-term and it is you and your generation which will have to resolve them.

I should perhaps first explain why Professor Messel chose me to deal with this diffuse and difficult issue. I must say I was surprised when he asked me but he quickly demolished my pride in his inimitable way by pointing out that his area of choice was narrow because there are not many people with scientific qualifications who have wide experience of writing about science on all its fronts in a "popular" newspaper, in serious journals and even in novels.

As I think you will agree when I define my terms, it is not only impossible for me to cover the field in three talks. It would be impossible for anybody to cover it in thirty talks. So in wondering what to select and what to omit I thought I might concentrate on three of the areas to which the media seem to devote most space and time—Violence, Sex and the various aspects of right and wrong which we can lump together as Social Conscience. They turned out —as might be expected if the media mirror life as they should—to cover the most urgent issues. This first talk then will be essentially about aspects of violence, with sex next and as is so common in life, with conscience last.

How should we define "society", which means something different for all of us because we each associate with a different part of the world's population? Unless you live in a highly regimented community, each person's experience of society is different, often very different.

A distinguished anthropologist has defined a society as composed of individuals and groups communicating with one another in three ways:—by messages, through the exchange of goods and services and through the exchange of women. I think he missed one out. They also communicate by means of aggression—by violence.

I suggest that as regards "society" we might as well adopt the attitude of the man who when asked to define an elephant said he was unable to do so but had no doubt about recognising one when he saw one. Meanwhile, we should be in no doubt about the dominating role of science in setting the course of society and of the fact that science, far from being an academic subject is a high-priority political issue and will remain so.

If society is hard to define, what about "science"? The term is very loosely used. To many it means just the mass of information discovered by people called scientists. If they think about it they realise it also involves the application of that knowledge. And of course to scientists themselves it is an attitude to life—the "scientific attitude".

Most scientists, who are usually specialists, also think of it in terms of their own branches of science. But to me as a popular journalist it means all the various disciplines—physics, chemistry, biology, astronomy and so on. It also includes medicine and military science—weapons and the study of strategy. It includes the social sciences like criminology and psychology. It also includes technology—that rather overused word about which there is also disagreement on definition. The best definition of technology I have heard is that technology comprises the skills, know-how and tools necessary for the efficient creation of wealth—which means that it includes engineering. The overlap between science and engineering is well shown by the remark of the engineer at the rocket launching station who said—"When the rocket works it's a scientific achievement. When it fails it's an engineering fault."

In my world "science" also involves dealing with scientists, their personal histories, their squabbles, sometimes their scandals, as when occasionally one of them turns out to be a Russian or commercial spy. Or when one discovers that some scientist who is complaining publicly that he is brain-draining to America because conditions for research are so bad in Britain, is really leaving because his wife is about to divorce him and cite the woman member of the research team that is going with him.

Since we are dealing with science in relation to society I think it will be most profitable to look at how most people view science and the scientist. To examine what the public relations men would call the image of the scientist and his work.

When I began to write about science in the *Daily Express* 25 years ago the mass of my readers visualised the scientist as living in a world apart—in what was referred to as an ivory tower. They pictured the scientist as a rather odd and paradoxical fish—myopic but having great foresight, inventive but absent-minded, brilliant mentally but naive in ordinary matters and definitely eccentric. (It has been suggested that some managers in business and industry still believe this is true.)

As a result of widely read novels and films about Frankenstein monsters and rocket ships, many people visualised the scientist as slightly crazy and irresponsible. I remember that at the first test of a nuclear device in the desert at Alamogordo in 1945 when the fireball was bigger than expected, one of the non-scientific observers shouted "My God the long-hairs have lost control." (Long hair and general untidiness were then also common features in the public image of the scientist as they now are of the young.) As late as 1955 I remember that the Association of Scientific Workers in Britain issued a solemn plea to the Press and Radio urging them "To stop portraying us as madmen who want to blow up the world."

In spite of this, "science" and "scientist" were largely good words so far as the public were concerned. Of course the scientists should stay in their labs and not poke their noses into business or politics. They should be on tap but not on top. But, by and large, their products like penicillin, radar and even the bombers that helped to fight the war were good things. The words "Science" and "Scientist" were even chosen for the title of a religion—Christian Science and the Church of Christ Scientist—and for movements like "Scientology". One can see why. To the mass of the public science meant certainty. There were facts which might turn out to be wrong but "scientific facts" never did.

But what of the image now? There is no doubt that over the past few years there has been a radical change in the attitude of the public to science and the men and women involved in it. There is in fact a growing disenchantment with science by society and many people see science as some kind of conspiracy in which scientists are forcing them to pay so that they can indulge in their wild ideas.

This disenchantment started with those aspects of violence which dominate the political scene—strategic nuclear weapons, H-bombs and the rockets and planes that deliver them; napalm, the terrible incendiary used in what is called "conventional" war and bacterial and chemical weapons like those used to destroy crops and defoliate whole forests in Vietnam. All these and many others are of course the product of modern science and technology and their impact has seriously corroded the image of the scientist in the minds of most people. If they are not mad like Dr. Strangelove, scientists are judged by many to be responsible for the uncertainty of life in the nuclear age. Their position is rather like that of the regular soldier in Kipling's day. You may remember Kipling's lines:—

"It's Tommy this and Tommy that, and throw him out the brute,
But it's 'Saviour of his country' when the guns begin to shoot."

During and immediately after the last war British and American scientists were hailed as the saviours of their respective countries. Now they are the brutes who could be responsible for the horror of World War Three and are already responsible for the fact that both the people of the East and West are living, in so-called peace-time, permanently under each other's guns—the intercontinental rockets which, as we sit here, are continuously targeted on the world's most populous cities.

This animosity is of course not confined to the relatively few scientists working on weapons. The disenchantment has spread with the increasing realisation that man is doing terrible violence to his environment and the creatures which live in it—by pollution, which is of course the product of science and technology. Clearly there is something radically wrong with the relationship between technology and the environment, and not only in the capitalist countries where pollution has been regarded by many as the result of selfish greed. The Soviet Union has a big pollution problem especially in its rivers. But there is nothing new in this. Britain's rivers have been open sewers fouled by industrial filth for half a century. Britain's power stations have been belching 9,000,000 tons of sulphuric acid into the air for many years. Smog may be a new word but the thickest smogs in history—the London "pea-soupers"—regularly brought the capital to a standstill during late Victorian times. Previously this was shrugged off by the politicians and the public on the grounds that "Where there's muck there's money" and you can't have one without the other. The media have been very much to blame for not campaigning against pollution much sooner but nobody much wanted to listen. There was more concern about continuing employment than continuing pollution. Recently that has changed, mainly because of developments in the U.S. in which constructive protests by scientists, journalists and young people have played an essential part. Pollution is now a political issue with the different parties competing to see which can offer the better clean-up programme, just as a few years ago they were competing on the grounds of which party was better fitted to advance new technology. Meanwhile, of course, the scientists get the criticism, even for developments once hailed as near miracles, like DDT, which has virtually eliminated malaria in many areas by wiping out the mosquito but is now in disrepute because of its effects on wild life and on human beings who all contain traces of it, though there is little evidence that it does much harm to us. The thalidomide disaster when hundreds of babies were born malformed because their mothers took the drug as a sedative was another savage knock because science was seen to have done the most terrible violence to children. The drug had in fact been very thoroughly tested but at that stage nobody had thought of investigating possible effects on the unborn child.

I will admit now that the media have been much to blame for promoting the concept of "wonder drugs" and "miracle drugs" without paying enough attention to the inevitable side-effects. What we should have done and must do in future is to make it clear to the public that every drug which promotes health carries its quota of sickness and even death because to be effective it must interfere with the natural working of the body and in a few sensitive people this interference may be damaging or even fatal. Our failure to get this message across has resulted in unnecessary set-backs to the use of certain drugs, of which perhaps the birth-control pill is the best example. The rare cases when the pill causes fatal blood clots have been widely publicised, usually without the information that though about 40 deaths in Britain a year are probably due to it—a sad loss for the families concerned—such a risk is no greater than the natural hazards of pregnancy, even with all the ante-natal care and medical aids available at childbirth.

This reaction to the over-stressing of the side-effects of oral contraceptives could set back world birth-control programmes simply because pharmaceutical companies will not go on with the costly research if their products are to be banned, as some have been in Britain. Research to devise oral contraceptives which carry less risk must of course be pushed ahead but an effective birth-control pill free from all risk is almost certainly a fantasy and the public should be made aware of this.

Another side-effect of a technological development is the supersonic airplane bang which is considered to be "ear-pollution". In the U.S. we have recently seen the Senate reject an application for money to develop a supersonic plane. This was a vote against technology. The designers of the Concorde supersonic airliner are similarly reviled as the perpetrators of something that will destroy still further the quality of life. (I must confess that when I first saw it fly I couldn't resist cheering it.)

Even the world epidemic of deaths by violence on the roads, which is mainly due to irresponsibility on the part of drivers, is being called a "science-based disaster". You might as well blame the Western-world epidemic of obesity on science for providing the means whereby so much rich food is available and so many people have the money to afford to over-eat.

For these and other reasons science is visualised as a juggernaut without a driver. We are said to be "riding the back of the tiger of technology". "Things" are in control and riding mankind.

I must say that one can see why such statements are current when one considers the history of the Concorde, with which I have been closely associated since its inception in 1962. Here was a machine which, basically neither the French nor British Governments wanted to build because of the cost; which the airlines did not want because they were already overloaded with machines; and which few of the public wanted. Yet the machine exists. It is almost as if the machine got itself made: that once the technology existed its application was inevitable. The Frankenstein stories and others have projected the idea of man being at the mercy of the machines he creates. The story of the Concorde is not all that far removed.

This attitude to science and the scientist seems to be peculiar to the West, at any rate at the moment. In the Soviet Union science and scientists still have an heroic aura. They are projected in the official newspapers and magazines as being dedicated to the welfare of the Soviet people. There has not been any outcry about the "Konkordski", the Tupolev 144 supersonic aircraft and the only criticism about nuclear weapons is aimed at the Americans and British. Everybody cheers when the gigantic SS9 missiles with the biggest and "dirtiest" nuclear warheads ever made are paraded in Red Square on May Day. Yet in the West nuclear weapons experts are derided as "merchants of death", and I am certain that neither the U.S. Government nor the British would dare to stage a parade of nuclear weapons because of the disturbance it would inevitably cause—much of it I might add aroused by pro-Russian sympathisers who never criticise their Kremlin idols for doing the very thing against which they are prepared to protest so violently in their own countries.

I am not suggesting that protests about supersonic bangs and other disturbances which reduce the "quality of life", to use that "in" phrase, are all politically motivated. On the contrary the underlying objections are felt by the great mass of the public who are beginning to realise that new technology is not necessarily worth what it costs: that "progress", as we have understood it in recent years, is not necessarily in the public interest or at least that the social costs of technological advances must be added to their apparent costs, which would rule some of them out by making them uneconomic. For example, if the estimated damage done by detergents to rivers and sewage plants had been put on the cost per packet some of them might never have been marketed.

Of course in many cases it is impossible to put a price on what is damaged or lost, when technology overtakes natural beauty for instance. I remember motoring through some delightful country in California with an American rocket scientist who told me that a certain area which was particularly beautiful was about to become industrialised. He said with pride that it was "the fastest growing area" in the U.S., as though that in itself was splendid news. We in Britain are perhaps the world's worst offenders in this respect because our reserves of natural beauty are so small. It required a terrible disaster about five years ago when more than 100 school children were suffocated by a slide of coal dust to secure public agreement that tipping industrial refuse in mountainous mounds near country villages was a crime against society.

It's sad to see that in the "developing" countries the lessons learned at such expense by the West are not being taken to heart in the rush to industrialise. Indeed the scars of science and technology are almost status symbols. Even the destructive weapons by which large sections of Western society are repelled, are symbols of prestige. The supersonic jet has replaced the Cadillac with results that are not always good for the nation concerned and may be bad for it. It was excellent for Britain's exports when the Saudi Arabian government ordered the very advanced Lightning supersonic fighter for its new air force. But the cost was enormous and it is now very doubtful if such a machine is really suited to a nation which is technologically backward and clearly has its priorities wrong.

It has been suggested that just as most countries now have panels of experts set up to advise on whether a new drug is sufficiently safe before it is released for general use, there should be similar advisory panels to examine the long-term effects of new technological processes and their end products on the environment and on society. I think the application of such advice is likely to prove very difficult in a free society. It would amount to the censorship of research and its applications and presumes the prosecution of those who might defy it. But we have recently seen one example of it in Britain. There was great pressure from the road haulage contractors for Government permission to raise the size and weight of lorries that may be used on British roads. They wanted 44 ton lorries, which are a technological development and are commonly used in the U.S. and in Europe. The Government ruled that with so many small villages on main roads in Britain such a development would decrease the quality of life for many thousands of citizens and banned it. I think that was a warranted interference by politicians given a simple and clear-cut choice. But in general I would defend the freedom of science as vehemently as I would defend the freedom of the Press.

Speaking of villages, in which most old countries abound, raises another way in which advancing science and technology is held by many—usually older people—to be doing violence to the "quality of life". I refer to the replacement of village craftsmanship by mass-production processes. There are many who believe that in this and other ways science is "dehumanising" our way of life. There is some truth in it but on the broad front I do not support the idea that technological progress is destroying human values. On the contrary, the great majority of people have a much fuller life in the technological age. Change has always been the essence of human development. Human values are not eternal and much of the nostalgia for the "good old days" is based on totally fallacious ideas of what those days were really like for most people. Further, the belief that old crafts are rapidly becoming extinct does not stand examination either. A few months ago I went to an exhibition of five sporting guns made and engraved by British craftsmen. They were as good as anything that has ever been made—as witnessed by the fact that within half an hour of their being on sale an American business man bought them for £23,000—no doubt with money he had made out of some technology-based business. In Britain there is a firm of craftsmen making harpsichords as good and as beautiful as any made for Bach. They are terribly expensive and nobody could afford them but for affluence based on technology. While walking through the back streets of Florence and Venice I notice that the craftsmen making the antiques there seem to be every bit as skilled as those who made the genuine article.

The commonly stated view that science does violence to taste and refinement by making society less aesthetic and more utilitarian is also untrue. Surely there has never been wider interest in classical music, literature and art—a laudable development almost entirely due to the science and technology which gave us the hi-fi record, the paperback book, the faithful colourprint, colour television—and cheap overseas travel.

It is of course no new thing that the scientist is imagined as a cold, calculating, dispassionate creature with little humanity or emotional feeling. I suspect that this image dates back to the conflicts between scientists and religion—to Galileo, Darwin and the rest. It is totally unfounded. I have met more scientists than most people and I can testify that they come in all shapes and sizes and with emotions, passions, prejudices, and intellectual capabilities as varied as in any other profession. While some of them are shy and retiring, others actively seek publicity. While some are restless, self-driving workers, others are lazy. (You may have heard the story of the director of a big research institute who was asked by a journalist "How many scientists work here?" He replied "About half.") While some scientists are highly stable, some are as volatile and as temperamental as prima donnas. As regards their passions I can testify from attending many international conferences that a man seen disappearing into the wrong hotel bedroom is just as likely to be a scientist as a salesman.

For these reasons in the novels I have written I have deliberately set out to show that scientists are ordinary human beings with all the human strengths and failings. It follows, if one takes this view, that scientists have no special right to put themselves forward—as a few do—as having special qualifications in the political field, of having more vision than ordinary mortals. Their views should carry no greater political weight on general issues than anybody else's, except in those fields where they have special qualifications and experience. The advice of scientists should certainly be heeded by Governments when scientific issues are involved. In this connection I have done what I can to campaign in the newspaper against those entrenched civil servants in Britain who object to the presence of the scientist at all in the administrative machinery: who think that even on matters concerning the procurement of highly technical weapons systems scientists should not be employed at such a level that they become involved in making decisions. But I disagree with those who maintain that the scientific attitude is necessarily applicable to all forms of politics. That way humanity may be driven out of the window.

Nor is the scientific attitude necessarily applicable to the running of businesses. A few scientists have done extremely well in business, even founding them, but recently in Britain there have been several major disasters when scientists and engineers have been promoted to positions where the acumen really required was in finance. Advanced projects devised with great skill had to be modified or abandoned because the costings were hopelessly wrong.

I think it can be said that so far as the impact of science on prosperity is concerned, in Britain there has of recent years been too much research and not enough commercial application. We could learn a valuable lesson from Japan in this respect and our newspapers should be more insistent in highlighting the gap between research and its application. In science-based businesses as in any other it is the creation of wealth per employee that determines the success of the business and, on the broad front, of the national standard of living. In Britain in 1967 a working population of 23 millions produced goods and services valued at £30,000 millions —an output of £1,300 per employee. In the United States a work force of 75 millions produced goods and services worth £350,000 millions, an output per employee of £5,000. The efficiency in creating wealth is three times higher in the U.S. and that accounts for the substantially higher standard of living there. Clearly the Americans are making much better use of technological possibilities than we are. So are the Japanese, the Germans and the French. One suspects that in all those countries the financial control of technical resources is more ably exercised than in my country.

It follows from what I have said that if scientists are just ordinary human beings there is nothing wrong when they speak out in more emotional terms than are normally associated with the disciplined mind. But this does not condone the use of violence or the approval of the use of violence by others in demonstration of their fears and demands. In Britain we have recently seen instances where highly qualified men, particularly social scientists, have associated themselves with this type of violence.

Student unrest is one of the major symptoms of the sickness of society throughout the world. In the last university year in the U.S.

there were nearly 1800 student demonstrations, sit-ins, arsons and building seizures with 8 people killed and 462 injured. The newspapers and other media have played a contributing role in giving student violence publicity but to what extent is *science* responsible? I will be very interested to hear your views on this. As you know there are many people, particularly those of my age who put the whole thing down to the absurdities of modern youth—to the process of being young in a society where adolescents get too much money and too much licence.

What do you think of this statement by an elderly and no doubt highly intelligent gentleman:—"The world is passing through troublesome times. The young people of today think of nothing but themselves. They have no reverence for parents or for old age. They are impatient of all restraint. They talk as if they alone know everything." That was written by a gentleman called Peter the Monk in A.D.1274.

I think the real truth of what the old have always thought about the young is best summarised in this statement "The universal trouble with the young is that whatever we teach them they will insist on behaving like we do." I can't remember the author. It may even have been me, though it sounds rather too good.

The impatience of youth is understandable in an age when because of scientific and medical advances the old are able to hang on to the plum jobs for so long. I certainly believe students should protest if the standard of teaching in their universities is as bad as it was in my student days. I had to put up with some brilliant researchers who were appalling lecturers and indeed gave the impression that the teaching for which they were paid was a regrettable interference with their research time. I always felt that I could have used the lecture time more profitably in the library with the set books but I never had the courage to say so. The professors and lecturers were not entirely to blame, for the golden rule for academic advancement was, and in many universities still is, "Keep on publishing research or perish", and the excellent teacher who was not a recognised authority on some Amazonian earwig or Himalayan moss—I was a biologist—did not get very far.

There were student demonstrations in my day mainly of a political nature, protests against Fascism and so on. There were "rags" which frequently led to violence and involved the police. And every generation has its share of non-conformists who would rather burn a city down than agree to live in it if they believe it conflicts with some principle they find overwhelming. But unrest among the young today is far more widespread and more violent than ever before and I suggest that there are at least five ways in which science may be a major cause of this:—

I have already touched on the first—the revolt against the development of the mass terror weapons and their effects which are rightly seen as being the result of the uncontrolled application of technology. There is unquestionably a close association between the soured image of the scientist and student unrest. Obviously it is right and necessary for the generation which inherits such a situation to protest about it and insist that something ought to be done and to put forward constructive suggestions. But why the violence? I suspect that fear may be at the root. As I said we are all living "under the guns"—the nuclear guns, with all the publicity given to their possible effects if they were ever used—as of course they were used in Japan. People are afraid if only subconsciously and fear breeds violence which on the surface appears to be motiveless.

This fear seems to have spread from the natural horror of weapons to other aspects of technology like computers and we have witnessed the extraordinary sacking and destruction of a university computer department in the United States, a modern counterpart of iconoclasm—the destruction of Church images—which was also possibly due to fear—the fear of God for disobeying the commandment against making graven images.

The second cause, which may be a direct offshoot of the first is the spiritual revolt against the material affluence resulting from technology—a revolt in which the newspapers and especially commercial TV have been a potent factor through persistently promoting the joy of material gains. We see this phenomenon most glaringly in the cult of the hippies and their rejection of all material possessions—even soap. These people, who are by no means all young, claim to express more interest in the spiritual values and as this requires contemplation they cannot let work eat into their time. While peacefully strumming the guitar or painting flowers on each other's bodies they are more interested in rights than in responsibilities; in innocence rather than in wisdom.

Again, this is no new phenomenon. To me the hippies seem to be the modern counterpart of the monks and lusty vagabonds of medieval days who withdrew from society to do what is now described "as their own thing". You may remember that when they became so numerous as to be a serious burden on society in Britain, Henry the Eighth sorted them out.

I have found it impossible to convince a hippy or near-hippy that by doing no work he is a burden on society. Yet they are because they usually draw public assistance or some other welfare benefit. I suppose there should be some tolerance for the occasional oddball who wants to drop out but drop-outs should appreciate that were it not for the great majority, who according to them have it all wrong, there would be no social structure for them to drop out from. And no structure can afford too many. So perhaps, unless the cult collapses—and such fashions do change—some modern Henry the Eighth may sort them out. (A well-known American businessman has been urging me to write a novel set in the near-future and involving a takeover of power in the U.S. by the military at the behest of a "great silent majority" determined to restore law and order and respect for property.) As some American statesman has put it "There is nothing wrong with the younger generation that becoming taxpayers won't cure."

Meanwhile the situation is dangerous. Many of the more extreme of the young rebels are consciously objecting to rational thought. They say "If this is where rational thought has brought us then let us be irrational." This is a return to mysticism, the contemplation of the novel rather than of what is really going on.

The third way in which science may be responsible is through the general teaching of science in schools. Science teaches you to question authority: to require authority to provide proof of its statements before you accept them. In the universities and schools today we see a questioning of authority which never existed in my youth. The situation is the same with the attitude of workers towards employers. The worker used to accept the boss's right to run things his way. Now everything is questioned.

The fourth way is through the sheer intellectual burden of the science courses. I thought it was tough in my day but the weight of material you young people now have to digest to complete the course and satisfy the examiners is far greater. To me it is no surprise that in many cases the brain just refuses to cope and I suspect that more than a few of society's young drop-outs are the result of inability to cope. Life is just too difficult for them and in an affluent society, particularly with welfare states, it is possible to drop out and not to starve.

I don't know whether science students or would-be science students tend to be any more involved in protest than any other kind. There don't seem to be any statistics.

The fifth, and possibly the most effective way in which science is responsible is through the sheer rate of change it is imposing on society. Man owes much of his evolutionary success to having a brain which can cope with change, which is another way of saying he is highly adaptable. But there must be a limit to the rate at which he can absorb change and it seems possible that for some individuals that rate is being exceeded. Nuclear weapons, man-on-the-moon, satellites, electronic eavesdropping, computerised and automated society, supersonic flight. Most of us can still take such things in our stride but maybe others can't. One would expect adolescents to be more vulnerable to this factor. It may be no coincidence that science and social unrest have advanced exponentially at the same time. I am not alone in thinking this. Psychiatrists have suggested it as a cause for the increasing incidence of mental illness. It may possibly explain why suicide rates are so much higher in affluent societies.

Some scientists have suggested a moratorium on research or at least a slowing down to give more time for the digestion of its results so that society can avoid some of the more obvious disadvantages—a suggestion I shall discuss in a later talk. But with the still increasing momentum of research and especially with the biological revolution which is the subject of my next lecture there might be a psychological benefit in providing society with a breathing space.

Of course the whole picture of student unrest has been distorted because it has been exploited by agitators, usually of the left, for political motives. Rudi Dutschke, the so-called "student" recently expelled from Britain was 30 years old. Mr. Tariq Ali, the Pakistani associated with student revolt in Britain is 27. I suspect it is this exploitation by professional agitators of a genuine disquiet on the part of young people, more than any other factor, which has led to the unforgivable excesses of the campus riots and other demonstrations involving burning and sniping even at firemen. Anybody who shot at a fireman trying to extinguish a blazing building in my youth would have been considered insane. One wonders how much such behaviour is the result of political despair because the far-left are not getting their way in the Western world quickly enough.

Such extremists can lead us back to the dark ages. We must remember that while science has always advanced through rebellion it has been an intellectual rebellion not violence. Einstein was noted for his lack of respect for authority but he didn't go round destroying laboratory equipment. The young must question authority but I suggest that they would be better advised to work *with* older people rather than against them.

Does what I have said so far constitute a "crisis" to use that overworked word? Is there a "crisis" in science, as so many headlines in the semi-technical journals insist there is? It depends what you mean by crisis. There has been a swing away from science as a popular course at the universities and that could be a dangerous sign for future progress. In the U.S. many highly qualified scientists are unemployed because of the cut-back in military and space research, which is partly due to the public revulsion at spending so much money on such projects. And there is the violence that seems to be associated with science. But according to my understanding of the word "crisis" there isn't one because a crisis is a peak of problems which are quickly resolved one way or the other. To say science is "in crisis" means that either its problems are going to be resolved in a reasonable time or it is going to collapse. Neither of these things will happen. The series of problems I have outlined and the others I will raise in my next talks are going to be with us for a long time, and science as a profession, as a discipline, as an attitude, as a way of life is not going to collapse. For in spite of all the noise created by the militants and in spite of all the space given to them by the media, reason is going to prevail. For surely it must be seen by all not blinded by prejudice that in spite of its unfortunate and dangerous side-effects, science is heavily in credit with society. Indeed, society cannot be what people of goodwill want it to be without the benefits of science, and that means further benefits as well as those on which civilised community life is already almost totally dependent.

CHAPTER TWO

Science and Sex

The relationship between man and woman is one of the dominating factors of society, and scientists, including applied scientists such as doctors and surgeons, are doing some extraordinary things to it. As this relationship changes, the social images of man and woman are changing too, with the newspapers and other media exerting their usual accelerating influence.

Technology has produced some obvious changes which we have all accepted, such as the effect of machines in lessening the advantage of man's physical strength and enabling women to work alongside and often in competition with men in factories, and the effect of the fractional horse-power motor, the electric heater and the thermostat in liberating woman from the household. With a much longer life expectancy, better health, better looks and a small family, the bride of 19 no longer necessarily dedicates herself to a whole lifetime in the nursery and the kitchen. She can look forward to 30 years of active life after her children have been reared and left home. This alone is having profound effects on society as seen in the number of women in the forties and early fifties who find new romances and end up in the divorce courts.

Like everything else sex itself, the physical relationship between men and women, which according to many authorities is the major driving force of all societies, is being revolutionised by technology, and the complete control of every facet of reproduction from sexual appetite to the deliberate moulding of the form and character of the resulting children is very near.

These particular changes which are widely, possibly too widely, publicised and discussed by the news media—even the staidest of the British papers are now said to have broken through the sex-barrier—are already posing serious problems for the law, as the issues of abortion, and artificial insemination have already shown. They are posing almost insoluble ethical questions for religions and for the medical profession through their erosion of society's attitude to the sanctity of human life.

Large areas of social custom are based on widely accepted agreements and taboos about sex and the roles of men and women in society, and anything which radically alters attitudes to marriage, family life and manners must have profound effects on the whole of society.

The undermining of some of the more pompous attitudes taken by man in the past can only improve the husband and wife relationship. I am thinking for instance of the Henry the Eighth complaint, quite widely held until recently, that it is a wife's fault if she produces a string of daughters and no sons. Biology has shown that the female normally plays no part in sex determination. Each of her eggs is the same so far as its sexual potential is concerned carrying one sex-chromosome of the type known as X. It is the male sex cells which are different, some sperms carrying an X chromosome, which in union with an egg produces the double-X female situation, others bearing a Y chromosome which on fertilising an egg gives the XY masculine composition. So if a monstrous regiment of women is anybody's fault it is certainly not the wife's. The same applies to man's previous attitude to the barren wife. We now know that in about two thirds of all cases of human infertility it is the man that is at fault. But it is no exaggeration to say, as I think you will agree, that the foundations of family life are being shaken when children are no longer considered an essential part of happy marriage but as a self-indulgence—something to be avoided in the national interest and when the one-parent family not only carries no stigma but is widely publicised as something rather splendid by those practising it.

Marriage is to some extent being actively subverted by female liberationists who claim that the family system has exploited women. But woman's gradually changing role in society is exerting a more potent and lasting effect. Thus, about 40 per cent of all American women now have jobs outside the home giving them a fuller, if more arduous life. It has been said that marriage can't hold two full human beings because it was designed for only one and a half and it is a fact that one in every four American marriages now ends in divorce. This is a major change in society and it could be dangerous. Learned anthropologists are warning that no society has ever survived after its family life deteriorated, pointing to Rome and Greece as examples.

The most obvious way in which technology is influencing sexual relationships is of course through birth control which has liberated many women from the annual child which made them old at 40, though of course in huge areas of the world where birth control has not been widely adopted for reasons of religion, politics or ignorance, women still suffer in this way. Indeed, many of the authorities who have been trying to induce Governments to introduce birth control as a means of reducing the rate at which the world is becoming desperately over-crowded, are deeply disappointed with their results. They tell me they are making negligible impact in India, South America and the other countries which matter most so far as numbers are concerned. But in those countries with what is called "Western culture" the "pill" and other methods of birth control are having increasing effect and there are many new developments in the pipeline. The most important of these, likely to have great influence are the natural substances, which can now be made synthetically, called *prosta-glandins*. These look like providing a safe and simple means of abortion — a drug which women could administer to themselves. In short women will be able to make their own decision as to whether to end a pregnancy or go on with it. The introduction of this development should also save many lives still being lost, especially in the developing countries as a result of illicit abortions by dangerous methods. The International Planned Parenthood Federation reckons that throughout the world more births are being prevented by abortion than by contraception.

There are of course millions of people who believe that any abortion except perhaps to save the life of the mother is a denial of the sanctity of human life. But you may care to consider whether general attitudes to that sanctity will not be more dangerously undermined by gross over-population than by birth-reduction. With so many Roman Catholic readers this is a difficult problem for the journalist to tackle. We need to present the facts, and we do, yet no newspaper can afford to alienate a large section of its readers by campaigning for birth control when these customers consider it an affront to their sincerely held principles. An article criticising the Pope for his opposition to the birth-control pill for instance could result in all editions of that issue being banned from entry into Ireland, where British newspapers sell a lot of copies.

"Birth control" will soon mean far more than the prevention of pregnancies and spacing them. It will include the deciding of the future sex—boy or girl. It will include *genetic control,* the diagnosis of defects in a developing child with the offer of abortion to the parents or, in some cases, repair of the defect before birth. We are also nearing the stage when we shall have what I call "procreation by proxy" which is already partly with us and is producing ethical and legal considerations which cannot continue to be ignored much longer.

I am sure you have all heard of artificial insemination which is widely used with cattle. It involves collecting the semen from a bull, diluting it and injecting several cows with what would normally have been used to fertilise only one cow. The technological problems of this gross interference with nature have all been solved and greatly extended by the discovery that the sperms can be stored for long periods in the deep freeze. In this way bulls have continued to father calves long after they have died.

About 14 years ago I wrote a report in the *Daily Express* which began "An Englishwoman has given birth to four fine children of whom she has never met the father. The children were all conceived by AID—artificial insemination by an anonymous donor. They will never know their true father and he will not even know that they are alive." This was regarded as something of a sensational disclosure at the time but AID is now so common that it seems certain that in excess of a million people are alive today because of it. (In the U.S.—and I suspect occasionally in Britain too—donors are paid up to £100 for a sample of sperm proved to be fertile.) In Britain and I believe elsewhere such children are officially illegitimate if their status is ever challenged, though their birth certificates illegally list the man they call father as the true father. The lawyers and the Church dignitaries have closed their eyes to this problem but it will have to be faced one day.

With artificial insemination a bull selected for the valuable characters he can pass to his offspring, can sire 1000 times as many calves as by natural mating. Similarly one man could father several thousand children if enough women were willing to bear them. (It is more than likely that some regular donors have each fathered more than 50 children.) By storing the sperms in deep freeze his progeny could be tested for intelligence and physical fitness before they were used on a large scale. In a totalitarian community a few men could be selected to become "Fathers of the Nation", even continuing to produce children after their death. This process, which would eliminate the accepted ideals of love and family life may sound far-fetched but the Nazis got near it in principle and one suspects that if the technology had been available they might have made use of it.

Frozen sperm has been successfully used in the human female so it is already possible for men to father children after they are dead. And this may already have happened to human donors killed in accidents since supplying a sperm-bank. It has been suggested that in future men might build up sperm-banks when they are young and healthy against the day when they might become infertile through ill-health or accident. There might even be national sperm-banks set up to counter the mass sterilising effects of nuclear war.

Even more remarkable and with more far-reaching consequences are the developments in the transplantation of female reproductive tissue. Transplants of complete ovaries have been successfully made in animals and there seems to be little doubt that this will become a standard operation in humans. It would enable a woman who has defective ovaries to produce children but of course, genetically, they would not be hers. She would simply be serving as an incubator for fertilised eggs belonging genetically to another woman, who was probably dead when her ovarian tissue was removed to supply the graft.

The same applies to individual eggs transferred from one female to another. The technology of this has been fairly fully worked out in animals. Eggs shed from the ovary are collected, fertilised in the test-tube and then inserted into the uterus of a female which then acts as an incubator producing offspring which are not her own. This has been repeatedly achieved in rabbits and to demonstrate the possibilities, fertilised eggs have been transported by air from the U.S. and inserted into does in Cambridge, England. As a result a black English rabbit has given birth to two white rabbits which were 100 per cent American. Rabbits have been used to export fertilised sheep eggs, merely serving to keep them at the right temperature during the journey. In the U.S. scientists have even successfully transplanted sheep eggs into goats so that a goat gave birth to lambs. In my childhood there was a riddle that asked "What is it that a cat has that no other animal has?" The answer used to be "Kittens" but that is no longer true.

This process of egg transplantation is obviously applicable to women, and research to that end is being actively pursued at Cambridge University and elsewhere. Human eggs have been fertilised in the test tube and cultured in blood serum until they have begun to divide. The next step is to introduce such an embryo into a woman who wants a child but cannot otherwise have one. The egg could be one of her own or one provided by an unknown donor. There has been considerable publicity about this in Britain and several women have already volunteered to serve as "guinea-pigs".

Again the question of legitimacy arises. Perhaps there is already need for legislation like that existing in Sweden where there is now no such thing as an illegitimate child. But of course the issue still arises when questions of inheritance are involved.

I have mentioned the awful possibility of Fathers of the Nation. Clearly egg transplantation and culture offer selected women equal opportunity of becoming Mothers of the Nation, with the rest acting as incubators. When such possibilities become a little closer in time the media will have a duty to campaign either against or for them or at least to point out the social consequences.

There is already one benefit which could quickly accrue from the latest work on the culture of human eggs in the laboratory. The eggs could be developed to a stage where examination showed them to be free from genetical defects of an easily discernible type, and they could then be reintroduced to the original mother.

By examining a few cells floating in the fluid cushioning the embryo in the womb, scientists can detect peculiarities in the chromosomes. These give early warning that a child will develop into a mongol, an intersex or some other potential human tragedy. Thus it is known that babies developing with only one sex chromosome or an extra sex chromosome may fail to mature sexually. Females with three X-chromosomes instead of two often have reduced intelligence while males with a double dose of the Y-chromosome may not only be of substandard intelligence but often violent and criminal. This last discovery has been confirmed by studies in many countries and it seems that the psychopath—the unstable person with little social conscience likely to commit a crime of violence on impulse—was born with an extra Y chromosome. What should be done about such people? What is their liability for punishment? It would certainly seem that when they come to court their extra Y chromosome must be taken into account in their defence and it has been so used successfully in murder cases, including at least one here in Australia. But should they be segregated from the community because they have this inherited defect *in case* they commit a murder? The trouble is that not all double-Y men are violent. Some seem to be quite normal.

The extent of these abnormalities in natural pregnancies is indicated by the fact that studies show that in about a third of all spontaneous abortions—the events usually referred to as miscarriages—the developing infants have an abnormal number of chromosomes. Were it not for spontaneous abortions the number of live children born with defects and deformities would be much greater than it is.

Clearly the examination of fertilised eggs as a service to parents who could then decide whether to allow them to complete their development or not would offer a way of reducing the incidence of inherited defects with which society is already overloaded because of the general acceptance that it is not morally right to compel people with faulty genetic make-up to be sterilised or even remain unmarried. If this seems far-fetched I would point out that already under British law if a woman believes she may have conceived a defective child she can opt for a legal abortion. So "genetic screening" as this process of unnatural selection is called may already be happening to a limited extent and its potentialities for the future are very great, with, of course consequent changes in social attitudes.

Now that embryos can be grown to a certain stage of their development in culture, it seems inevitable that the full process will be achieved and with animals at least the complete test-tube offspring so often a subject of science-fiction will become a reality. At first sight it would seem to be almost impossible to replace the highly complex functions of the placenta—the organ which links the embryo with the mother and is the vehicle for the passage of food, water and oxygen, besides serving many other purposes. But the human egg is a versatile object. It can, for instance, form a placenta after attachment to the oviduct or even the intestine, instead of to the wall of the uterus—a capability which sometimes results in what is called an ectopic pregnancy and has in rare instances led to the birth of a normal child. Some kind of placenta might therefore be formed in a culture medium in an artificial womb. It has been suggested that women could then give birth like kangaroos. Delivery of an embryo conceived in the normal way could be induced when it was about three months old and still small. The embryo would then be cultured in the artificial womb rather as the baby kangaroo which is only about an inch long when born, is nourished in the pouch. The advantages of this system would be easier and safer births plus the opportunity to examine the embryo for defects at any stage right up to its complete development.

One rather far-sighted scientist has suggested that this would enable babies to be developed with bigger brains than is possible now. The size of the brain is conditioned by the size of the head and this is in turn limited by the size of the birth canal.

The technical possibility of removing human eggs and replacing them also offers what may be the first really practicable way of choosing the sex of one's children. I have already mentioned the chromosome differences by which the future sex of a fertilised egg could be detected under the microscope but other methods which promise to be easier and quicker are being developed. By giving fertility drugs it is easy to induce a female to shed several eggs at once instead of only one as the numerous recent instances of quintuplets and other multiple births have shown. The eggs can be moved, examined and one of the desired sex replaced in the mother to complete development.

There are other possible ways of choosing sex before birth. Thus there is some evidence from experiments with rabbits that alcohol in the blood stream may increase the chances of producing male offspring—which might account for the belief that more boys are born when their fathers are intoxicated at the time of conception. This suggests that it might be possible for men to take some drug that would at least increase the chances of male births. If this becomes available as yet another "right" under the welfare State or privately, an important new factor will be introduced into society. Clearly the process could quickly affect the balance between the sexes. In view of the common preference for boys it has been calculated that in 50 years time there could be two men for every woman. This could conceivably lead to polyandrous marriages, a wife being permitted to have two husbands—a situation which seems common enough now in practice, if you read the more sensational newspapers, but is not yet legal in the Western world.

All these developments are likely to revolutionise the relationship between men and women in the future but there is one area of new knowledge resulting from the scientific study of sex which, in my opinion is likely to have wider repercussions much sooner. I refer to the growing realisation that men and women are not anything like as different in structure or nature as generally believed. Studies, anatomical, genetical, physiological and psychological have shown that there is no such thing as the totally masculine male or the totally feminine female. Each man or woman is a mixture of the attributes commonly considered to be male or female. In between the so-called "normal" man and "normal" woman there are many gradations of diminishing maleness and increasing femininity with types in the centre who are 50/50 mixtures not only temperamentally but anatomically. These intersexes, as they are called are by no means rare, and they are frequently the result of a reversal of sex occurring before birth. In the view of specialists in this field, probably 10,000 men alive in Britain today were conceived as females.

It seems that the human body is basically female, the male being the result of male hormones acting in this basic structure. Left to itself, the body would develop in the direction of femininity and the extent to which it becomes male depends on the strength of the male hormones. If the organs which produce the masculine hormone are inactive the body tends towards femaleness even though it is genetically male so far as its chromosomes are concerned. Thus if male rabbit embryos are castrated by surgery inside the uterus they develop into females. Of course this basic femininity has been appreciated for a long time in the case of eunuchs. It is now known to be true of several other medical conditions.

Until recently it was believed that the sex of the child was fixed at the moment of conception by the chromosome mechanism I have described. An egg fertilised by a Y-sperm should produce a boy, while an X-sperm should produce a girl. But as research almost always shows if it is pursued deeply enough, the situation is rather more complicated because it is now known that this mechanism can be over-ruled by the placenta—the pad of tissue fixed to the wall of the womb which nourishes the embryo through the umbilical cord.

What happens is that up to the sixth week of its development in the womb, whether it is XX or XY the human embryo is potentially bisexual. Prototype sex organs are already formed but they can develop into either male or female or something in between.

The reason for this is that in *all* embryos the placenta secretes a mixture of both male and female hormones. If the mixture contains a preponderance of male hormones, the organs will develop on the side of maleness and the child will be a "normal" boy at birth. With the right amount of female hormones the child will be a girl. But because both masculinising and feminising hormones are always present, every baby is bound to be a compromise showing some male and female features. This even shows up in the reproductive organs. There is a counterpart, however small, of all the male organs in the female and vice versa. In the normal man there is even a minute womb—the uterus masculinus. If an embryo has the XY chromosome set and therefore should develop into a male, it will still be predominantly female if the placental hormones are predominantly female. And vice versa. Examination of the skin-cells of boys has shown that many of them contain two X chromosomes showing that they began life as females and changed their sex.

When I was a student I was taught that males had male hormones and females female hormones but not only is it now known that a mixture is present in both but this persists throughout life. Every child produces a mixture of male and female hormones throughout adolescence and beyond and neither of these ever achieves total victory. The hormone responsible for sexual desire in females is now believed to be the same that motivates it in males—testosterone, so called because it was once supposed to be exclusively masculine. Even the hormones present in pregnant women also occur in man. Throughout life the body of each sex remains a compromise compounded of the features of both, a reasonable balance being essential for psychological stability.

In practice this turns out to be a sensible arrangement for an essentially social creature like man. Without the leavening of kindness and gentility associated with femaleness men would be aggressive boors. Without a dash of the aggressiveness which engenders a spirit of independence, critical faculty and drive, women would be impassive and dull. I think we have all noticed that the extreme masculine man and the extreme feminine woman are often not well balanced individuals.

As the proportions of male and female hormones in the mixture can vary so much, they can produce an infinite variety of effects. That is why some "normal females" have deep voices whilst some highly fertile men have soft, high-pitched voices and may never need to shave: why there are probably as many men with sizeable breasts as there are flat-chested women: and probably why there are as many aggressive dominating women as there are passive hen-pecked men.

It is also the reason why some children are born with sex organs that are such a mixture that their sex is mistakenly recorded at birth and may never be definitely established throughout their lives. These intersexes lie in the centre of what doctors term the "spectrum of sex"—a range of individuals varying gradually from the most masculine man to the most feminine woman. Latest estimates suggest that as many as one in 1000 of all people are 'intersexes' to the extent that they are anatomically abnormal and usually sterile.

Microscopic examination shows that some of these intersexes result from faults occurring at conception. They receive an abnormal number of 'chromosomes'. So the *"genetic sex"* which should be settled at birth is faulty from the beginning.

In other intersexes the genetic sex is normal but has been reversed or altered by an imbalance of sex-hormones during development.

For millions of people, not just the few unfortunates who hit the headlines, this compromise creates personal problems which can be overwhelming. Thus it is believed that perhaps one in every 5000 males is conceived as female but reverses sex before birth. Many of these men look perfectly normal and produce children. (In theory such children should all be female as they carry only X chromosomes and this may account for the fact that some men produce daughters only.)

The sex that is apparent at birth is called the *clinical sex*. But even if the genetic sex and the clinical sex are absolutely normal they can still be over-ridden by psychological factors. Psychologists are now convinced that mentally, a baby is not born boyish or girlish, whatever its sex. The 'gender role'—the sex a child will consider itself to be—is gradually established during the first three years and then remains indelibly imprinted. So if through a mistaken diagnosis at birth or through a quirk of the parents, a male child is brought up as a girl, he will have the urge to continue in that sex irrespective of the true genetic or clinical sex, while a girl may be forever obsessed with the belief that she was really intended to be a man.

One man who had always been reared as a boy and was 'happily married', proved to be essentially a woman when doctors operated on him for suspected appendicitis. They found that with fairly simple surgery they could convert him to a woman who could almost certainly bear children but there was no possibility that he could ever father them.

When told this, the 'man' and his wife decided that he should continue as a man because his outlook was entirely masculine. Among 113 American patients with sex disorders studied by one team, 30 had been brought up as one sex but were biologically of the other. And of these 30, 27 behaved according to the accepted social standards of their assigned sex and not of their anatomical or chromosomal sex. Rarely in the past have such cases had any impact on society as a whole but more recently the apparent woman who is really more masculine than feminine has become an issue in the world of sport. We have seen several instances where women athletes—usually from the Iron Curtain countries—with outstanding prowess at jumping, running or discus-throwing have withdrawn from international sport rather than face the medical examination which was obviously becoming a necessary qualification for entry.

There are thousands of case-records which show that 'psychological sex' can be the over-riding factor so far as sexual behaviour in adult life in concerned. A woman who is normal in every physical sense may be attracted only to other women while it is estimated that one in every 20 physically normal men is homosexual. Just as there is a spectrum of anatomical sex so there is a spectrum of sexual behaviour with a scale of activities ranging from complete heterosexuality to complete homosexuality, with bisexuals in between.

The general acceptance of these facts due largely to the work of social scientists and psychiatrists has led to legislation in Britain and other countries removing such behaviour from the list of crimes so far as consenting adults are concerned—an important effect of science on society in which the newspapers and other media played a prominent part.

So far the law has done very little to improve the tragic lot of the trans-sexual, in spite of great publicity given to certain cases.

The trans-sexual male is convinced that he is really a woman trapped in a man's body, the reverse being true of the trans-sexual female. This seems to be an entirely psychological phenomenon, for anatomically trans-sexuals are usually completely normal and fertile, and on every other issue they are normal mentally. It is estimated that perhaps one in every 100,000 people in the world is a trans-sexual. In the U.S. there are "Gender Identity Units" to help them to adjust to society. A group of consultants disturbed by these cases has formed a Gender Identity Unit in London.

Some of the men involved are content to settle for permission to dress and live as a woman and there are several in Britain who have been permitted to change their sex on paper by the Registry Office at Somerset House. This is possible because under the Births and Deaths Registration Act of 1953 the sex of a person can be changed if a doctor makes a statutory declaration that subsequent events have shown that an error was made when the birth was registered. The doctor does not have to produce his evidence and he can take the view that a psychological change is sufficient. Such men who have been registered as women may not have undergone any surgical or hormone treatment yet they have the legal facilities of ordinary women such as exemption from front-line service in war and old age pensions at 60 instead of 65.

Other trans-sexuals are so repelled by being males that they insist on 'conversion surgery' to change them into something as anatomically like a woman as possible. The late Sir Harold Gillies, the plastic surgery pioneer, reported the case of two sisters who served as women volunteers in the London Fire Brigade and were both turned into "men" by plastic surgery. He also revealed that a woman athletics champion who was really a man disappeared after he had operated to convert her more fully to that sex because he told her she would be liable to military service.

As the understanding of such cases becomes more widespread through the work of scientists and the publicity given by the media there is greater public sympathy for them but as usual the law is lagging way behind. As the recent and tragic case of the British model April Ashley showed, in British law there is no legal definition of what is male and what is female. April Ashley started life as a male but underwent surgery to secure a status as near that of a woman as possible. She went through a form of marriage and then the husband sued for nullity. The judge ruled that no amount of surgery or hormonal treatment on someone who started out as a male could produce a person capable of performing the essential role of a woman in marriage, which he considered to be child-bearing. But is child-bearing the essential role? Thousands of women who have undergone hysterectomy have later been married. Could their husbands sue for nullity? And what about women who marry after they have passed the child-bearing age? Are they not women?

Miss Ashley found herself in the no-man's land of sex so far as the doctors were concerned and in the no-woman's land in the eyes of the law. The law ruled that in such a case the body is more important than the mind but many psychologists would dispute that, for reasons I have already mentioned.

Research on chimpanzees and other primates is even undermining the notion that maternal behaviour is something essentially feminine. Caring for children may be largely learned behaviour rather than instinctive. The female chimpanzee reared in isolation from other females does not know how to care for her infant. The accepted instinct of motherhood is now suspect as a myth—perpetuated of course by masculine society—and we may live to see a demand for greater sharing of child-care between wives and husbands.

The final stage—if anything ever is final in science—may be the baby produced without the involvement of sex at all.

When artificial culture of the fertilised egg is achieved, it should be technically possible to produce a baby from any cell cut from a body—even from the body of a newly-dead person—as plants are propagated from cuttings.

You might not be able to take your wealth with you, but you could leave it to someone who should turn out to be a near replica of yourself.

To recap let us consider what could be the fate of a person born in the not too distant future. First he could be the product of an egg produced by means of a fertility drug from a woman who was not the person he knew as his mother. Second, the person he called father, if he had one at all, could be some anonymous donor not even known to the woman he calls mother. He could have been sexed like a chick before birth and examined for possible defects. Having escaped abortion he could have outwitted the genetic screeners by changing his sex before birth and if he is dissatisfied with that he could go at least part of the way to changing it back again as an adult.

This is in the future but already the understanding that the sexes are far more alike than they are different is having an impact on the attitude of women to men. It is not a coincidence that we are seeing an upsurge of women's liberation movements determined to secure what they believe to be a more rightful place for women in society at a time when this new knowledge is becoming widespread. The movements are in fact making use of it in their propaganda writings. The plea for equality of pay and opportunity for women in careers is of course old. It was expressed violently 60 years ago when British suffragettes, as they were called, were fighting for the right to vote. What is interesting about the new movements is that the accent is on man's attitude to women as human beings. They are calling their movement "a slave revolt".

Instead of racism as their social target they have "sexism". They are opposed to any symptoms of society—and they are numerous—which indicate that woman's major role is sexual, and that in men's eyes women are something of a commodity. They are resenting, for instance, the Miss World competition, which in their eyes helps to perpetuate the role of woman as a sex symbol and little else.

It is a movement which is gathering strength especially in the U.S. and in Holland where one gimmick is that the girls wolf-whistle the men to embarrass them or open doors for them. Some women are preaching violence and learning karate. A few have practised it. Some are urging chastity because love is counter-revolutionary and men are sexual racists.

In Britain there is one organisation called SCUM—the Society for Cutting Up Men. In the U.S. there are WITCH—Women's International Terrorist Conspiracy from Hell, the Redstockings and others, as well as the more reasoned and more influential National Organisation for Women.

Naturally, all this makes splendid copy so the movements are getting plenty of publicity in the newspapers.

Though these militants are something of a nuisance, adding to what is already a surfeit of militancy in the world, they have much justice and scientific accuracy on their side. There is no escaping the fact that though women constitute the majority in almost every country they face the problems of a minority in professions, wages and political openings.

Though technology has helped to create opportunities for women in industry they are mainly of a menial type. Equality of opportunity for women in education and employment may be legally a fact but it is factually a fiction because in daily life long-established social attitudes prove to be more restrictive than the law. Woman's role in society is still largely governed by the established concept of femininity which steers her away from competing with men in many areas. Women can also call on a welter of anthropological evidence from tribes where males and females behave differently from us to show that it is attitude and upbringing which make the sexes behave in different ways socially. For the social image of man and woman is undoubtedly a product of the specific culture in which they were born. The more women realise that masculine dominance in society is due more to custom than to constitution, the more their role in society is likely to change.

We cannot discuss the impact of sex-science on society without looking at the appalling prospect of over-population—the population explosion—but I would rather reserve this for my next talk on Social Conscience. But to put it in perspective at this stage just consider this question—"If the world had the capacity to send its daily increase in population to the moon how long would it be before the moon became as densely populated as the earth is now?"

The answer is only about 15 years! The world's population is rising by about 170,000 a day. In 15 years that adds up to about 975 millions which would make the moon's 15 million square miles as overcrowded as the earth's land surface. So even this heroic and expensive effort would only delay slightly the day when the world becomes dangerously overstocked—which could be early in the next century. There are already enough young people alive to ensure a further increase of 1000 millions. As I have pointed out, there is gloomy disappointment so far with the effects of birth control in even delaying this disaster and governments may have to take recourse to purely social rather than scientific methods which I will mention in my next talk. Suffice it here to say that if they are adopted we shall have to stop looking on parenthood as a human right and treat it only as a privilege. That, of course, would be a major change in attitude. So far, civilised society has not even taken the step of denying the right to breed to those people who are known to be defective and pretty certain to pass on their defects to further generations. Such a change is nevertheless possible under the pressures which over-population will bring.

It is astonishing to me how far man and woman has gone in the last 30 years as regards attitude changes. I remember that when Lord Horder suggested the possibility of oral contraception in a conversation with me I said that women would never be prepared to accept such a gross interference with their metabolism. How wrong I was! As for artificial insemination, egg transplantation and so on, such things seemed interesting and possibly useful for domestic animals but academic so far as human beings were concerned. I was wrong again. Whilst in my youth sex was publicly regarded as a means of procreation it is now openly accepted as a means of recreation. Even the accepted symbols distinguishing the sexes are changing and we now have the cult of unisex—both sexes with long hair, trousers and smoking cigars.

Well, man's dignity may be outraged by scientific and medical developments and sex-scientists will no doubt be blamed for the unfortunate side-effects of their discoveries, but one thing is certain—the research and its applications will continue. As one of the pioneers of what we journalists call "test-tube baby" research says "We are well aware that this work presents challenges to a number of established social and ethical concepts. But in our opinion the emphasis should be on the rewards that the work promises in fundamental knowledge and in medicine."

That is also my opinion and we must hope the challenges will be met without eliminating too many of the complementary differences between the sexes which give social life so much of its grace and savour. Someone has said that the battle between the sexes will never be won because there will always be too much fraternising with the enemy. I, for one, hope that turns out to be true.

CHAPTER THREE

Science and Conscience

To what extent should those scientists who are involved in research hold themselves to be responsible for its misuse? There is a growing belief that they should and in Britain there is now a Society for Social Responsibility in Science, which like science itself is to a large extent dependent on the media for putting its message across.

The nature of this message depends on the degree of militancy of the scientists concerned and I am in no doubt that this new society has been quickly seized on as a useful vehicle for left-wing propaganda—a platform for castigation of the United States for manufacturing nerve-gas with no comments whatever about the Soviet Union which is doing exactly the same things.

It was almost inevitable that just as the Communists cornered the word "peace" so that most organisations with Peace in their titles are Communist front outfits, so they would latch on to "social responsibility". I am not suggesting that all militants are left-wing or that scientists should not protest militantly if they genuinely feel their results are being misused. But it is wise to bear in mind that agitators are at work for reasons which have little to do with science or social responsibility.

The more detached of the scientists who are genuinely troubled about their image and their consciences seem to agree that there can be no moral or ethical issues involved in pure research, by which I mean the study of nature's laws for itself alone. It is only in the application of such knowledge—when attempts are made to use it to modify Nature in some way—that moral questions arise. So it is in the realms of applied science—engineering, medicine, agriculture and, of course, defence—where the watch must be kept.

But where is the line to be drawn? Recently a Swedish nuclear scientist who is working on hot plasmas wrote to a scientific journal seeking advice as to whether he had the moral right to publish his "pure" research findings because they might eventually be misused for military purposes. This of course raises a further ethical question—has any scientist the right to suppress his findings to satisfy the doubts of his conscience? Or do those findings belong to the world whatever they are? I suppose to some extent it depends on how the research was financed. A scientist financing his own ideas possibly has the right to suppress. Another who has been taking money from a university, government or a private source so that the sum of human knowledge can be extended may not. I shall be interested to hear your views.

There are of course many fields in which thousands of scientists work knowing that their results will be kept secret—at least for several years and who may be in serious trouble if they do anything to break that secrecy. The scientists working on new weapons and defence equipments like radar are the best examples and this secrecy sometimes invades the pure research laboratories. You may know that when Frisch and Meitner realised that a chain reaction in uranium 235 was almost certainly possible in 1939 they kept the information secret from the Germans but alerted the British and Americans.

Then there are the large numbers of scientists and technologists bound by commercial secrecy—the undertaking that their results will not be published until the firm financing them has patented its processes and ensured a return on its investment. This has even led to criminality among scientists. There have been several cases in which scientists have been jailed for selling commercial secrets and even smuggling out mould cultures used for the production of new antibiotics—which supports my contention that by nature scientists as a race are little different from any other section of the community.

There are a few areas in which organisations have deliberately refrained from undertaking research because the dangers of the results were foreseen. The best example I can think of occurred some years ago when the Scottish fisheries authority decided against some interesting marine research to discover the migration routes and feeding grounds of the salmon in the sea. They realised that once these were known commercial fishermen would exploit them and the salmon would be netted in thousands when immature in the sea instead of back in the rivers. How right their fears were! The Greenlanders and others have since discovered these feeding grounds by accident and the salmon is threatened with extinction through savage over-fishing.

In the realm of social conscience the scientists for whom I have always had deepest sympathy are those working in Government organisations on micro-organisms. One of them was the head of what is popularly known in Britain as the germ-defence establishment and he suffered agonies of anxiety about the ethics of his job, finally leaving on this account. Yet all his work was aimed at saving lives by developing mass-production methods of making antidotes to bacterial weapons and nearly all of it was published so that doctors could make immediate use of it for their patients. But regrettably it is not possible to work in the field of virulent bacteria without discovering something that can be of value on the offensive side.

Anyone so deeply troubled by his conscience should leave that type of work. It should be possible for a scientist to be a conscientious objector as regards certain areas of research without disgrace. But it is not permissible, I believe, for a scientist who has accepted the secrecy conditions of military research to remain in it and, because his political conscience pricks him, to reveal those secrets to another nation. I am of course referring to the notorious cases of Klaus Fuchs and others who betrayed their countries for what they called reasons of conscience. Currently I note a move on the part of some eminent scientists to do a rehabilitation job on these men. I believe that they were rightly convicted. Being dedicated to the pursuit and dissemination of knowledge does not set any scientist above the laws of a country which he is free to leave if he disapproves of those laws.

In my view every scientist not only has a moral right to use his talents to help defend his country in time of danger, but has a moral responsibility to do so. This, of course, was the view of almost all scientists in all countries during the last war but since then the nuclear and bacterial weapons have exacerbated the problem beyond the limits acceptable to some. Nevertheless I think we should realise that if all the high-grade scientists in a particular country became conscientious objectors so far as defence research is concerned, then that country would run a grave risk not only of being defeated in war but of being taken over under the threat of war. Of course if all scientists throughout the world agreed not to work on weapons that might be a different matter. But that will not happen in our lifetime and in any case nations would still be at the mercy of Genghis Khan type hordes. It is instructive to note that while President Nixon has ordered the destruction of all bacterial weapons as a result of public pressure there is no evidence that the Russians have done the same or have any intention of doing so.

Personally I can see no difference in principle between working on what are called conventional explosives and nuclear explosives. The nuclear bomb is essentially a huge incendiary and ordinary incendiaries killed more people in Tokyo than nuclear weapons killed in Hiroshima.

There is little doubt that because of the ballistic missile with the huge nuclear warhead, casualties in a future global war would be much higher than in previous conflicts but nobody seems to take into account that even without nuclear weapons a global war in future would be almost certainly catastrophic because the technology of conventional explosives and their delivery systems would advance rapidly if the research effort were put into them. If one plots the casualties in major wars starting with the first really big one, the American Civil War, the curve rises exponentially through about 20 million dead in World War Two to a projection now of about 60 millions in a conventional war fought over several years. The figures for Europe alone are indicative. Taking account of the Russian Revolution, the Spanish Civil War, the Russia-Finland war, and the two World Wars, the total of dead and injured for this century approaches the appalling figure of 100 millions. And, I repeat, we are speaking only of so-called conventional war.

Before leaving the military considerations I should mention the recent arguments—much featured in the media—about the rights and wrongs of university departments accepting research contracts paid for by government defence departments. If scientists wish to undertake a certain type of research for which any government department is prepared to pay I can see no objections so long as they do it willingly. This seems to me another valuable way of getting extra money out of governments for research that might not otherwise be done. It is surprising what defence departments can be induced to finance, like work on sleep, alertness and the effects of stress. Such work, which is usually published, is perfectly permissible for a university. So, in my opinion, is research on the chemistry and physics of explosives.

Another special aspect of loyalty affecting scientists has arisen as a result of the so-called "brain-drain". The 65,000 scientists and engineers who have left Britain and Europe for the U.S., Australia and elsewhere have been referred to scathingly as "scientific mercenaries". The previous Prime Minister of Israel called those who left his country after being educated there "gypsies", implying that they lived on the community then put nothing in. One Israeli Minister went so far as to call them "traitors". This is totally unreasonable in my view. Why should a scientist who finds better facilities or even just a more attractive life abroad be subjected to criticism any more than an advertising agent or business manager who does the same thing?

It has been suggested that one answer to the conscience problem is for scientists to take a kind of Hippocratic Oath, as doctors are supposed to, though in fact, few do so. One version of this reads "I vow to strive to work towards the co-existence of all human beings in peace and dignity with all the necessities for a self-fulfilling life and freedom from fear, stress, ugliness, pollution and noise."

That reads very well but I suggest it would turn out to be pretty meaningless and highly restrictive in practice. For instance if you believe that nuclear deterrents keep the peace and help prevent a spread of Communism into the West—and many highly respected scientists do—then this gives you carte blanche to work on nuclear weapons. In particular this stressing of the "necessities for a self-fulfilling life" is extremely stultifying. Necessity is rarely the mother of invention. Curiosity is the real mother of invention. So many developments, now considered as necessities of life were discovered by scientists while they were looking for something else. If scientists restrict themselves to social necessity, invention will surely come to a crawl and science will cease to attract inventive minds. Some people, of course, would consider that a good thing!

No. If you are troubled by conscience or feel you ought to be then you must treat each offer of research on its merits. But however much you try to think ahead you are bound to miss the unforeseeables which give research so much of its appeal and excitement. It is crazy to blame scientists for the application of research results they could not possibly foresee, especially when there is usually so much on the credit side for their work. This has been true right down the ages. Was the man who invented the wooden roller which raised water from the well to be held blameworthy because it was later used to torture people on the rack?

If the scientist begins to look too far ahead and begins to wonder about all the possibilities of misuse he will end up doing nothing. Let us consider for example what is probably the gravest problem faced by man—the problem of over-population.

The rate of population increase is now about 2 per cent for the world as a whole which means that the population will double in about 35 years if the rate remains unchanged, meaning that there will be 6000 million people by the end of this century. This growth-rate of course is greatest in the less-developed countries where production of food is being steadily outpaced by production of mouths, so the prospects for really dangerous overcrowding there leading to famine and strife are grim indeed. Two world wars and various natural disasters have shown that population growth cannot be effectively curbed by violence, accidents or acts of God. The 500,000 people killed in the Pakistan tidal wave in November, 1970 were replaced by world births in three days.

Population is not a pressing issue just for the developing countries. By 1980 the odds are that the population density in Greater London will be around 14,000 per square mile—the sort of overcrowding which in animal species leads to aggression. We in the West think of the population problem as being something affecting the Asiatic and South American nations but in Britain we could soon be facing a national emergency if only from the increased food imports that will create an impossible balance of payments situation. And the greed for land for housing and industrial development could challenge the whole concept of private property.

Globally it is not just food that is a problem. The rate at which man is using up reserves of minerals, fuel and water will become catastrophic long before the teeming millions of the underdeveloped countries reach Western living standards. There just isn't enough to go round to give everybody the same living standards enjoyed by you and me. That alone is sufficient need for rapid restriction of population growth. A big population and individual affluence is a dangerous combination in any country and almost certainly fatal world-wide.

Clearly all the possibilities of scientific research and technology should be applied to ameliorating this problem. I say ameliorating because there is now no hope of preventing it. As I told you previously there are already enough young people in the world to ensure a further increase of 1000 millions. Science has already produced an answer in the form of the birth-control pill but those doctors and social scientists involved in administering it are very pessimistic about inducing sufficient women to take it. Progress is extremely slow even in those countries like India where the governments favour it.

Scientists are also making impressive strides in developing new foods to nourish the growing multitudes—the so-called "Green Revolution". The new types of rice now being grown have doubled yields. But again, I find that most authorities are pessimistic about the long-run results if the world population doubles.

While hundreds of scientists work on birth control, thousands are specialising in death-control—the prevention of premature deaths and even the prolonging of old age by methods which might be called "medicated survival". Of course, every advance in death-control makes the population problem worse. Should scientists therefore desist from the saving of lives which already exist to leave living space for those yet to be born? Obviously not. The whole idea of limiting death-control is contrary to the concept of the sanctity of human life on which most of the great civilisations have been based. But I do suggest to you that scientists and everybody else might have to have second thoughts about this as the population problem worsens. We have already seen that they have done so to some extent in the general acceptance of abortions. How long before married couples are being bribed to avoid having children, which I think many people would regard as rather immoral? Not long, according to social scientists who have already worked out possible schemes such as tax relief for men if they agree to be sterilised; annual awards or tax relief to couples who avoid breeding; national savings certificates to married women who remain non-pregnant for five years and Government bonds payable at the end of fertile life to couples who accept a specified limit to their families.

In India men are already being bribed into accepting sterilisation by the offer of free transistor radios. Surgeons are taking the minor operation to the men by using mobile surgeries in buses standing at railway stations and other places where men congregate.

I find this rather distasteful but social niceties will have to be discarded if there is to be any impact on the population problem. Methods like selective taxation may be more refined but they are still definite inroads into the concept of the sanctity of human life and one wonders what will happen to social conscience in a few years' time if—I think I can say when—the increasing child population is regarded as part of the deterioration of the environment? When the birth of quintuplets is regarded not as a matter for local celebration but for dismay? Will people be upset when they read of tidal waves destroying thousands of human beings or will they console themselves that from the viewpoint of the world as a whole it is not such a bad thing? As population increases, the dangerous concept that life is cheap, which existed in former times for different reasons, may well return.

Obviously any scientist who strives to prolong life, cure the sick, increase fertility or prevent accidents is making the global problem worse. But these are the very areas which are regarded as ethically the most laudable.

I think most people would agree that a pill that would prolong the human life-span to say 110 and at the same time preserve the intellectual and physical faculties would be a "good thing", and there are scientists working patiently towards this end. But in a novel I published a few years ago I set out to examine the social consequences of such a development and they turned out to be uniformly disastrous for society as a whole, particularly if the elixir arrived suddenly and was demanded as a right, as it almost certainly would be.

Like novelists and newspapermen, scientists should speak out when they foresee some real danger through misapplication of their findings. Silence can be an expression of irresponsibility in such circumstances.

I can't leave this subject without warning you that scientists who are most vocal about "social responsibility" are often highly emotional and unobjective about issues which interest them. When Professor Jensen of Berkeley, California put forward evidence that the intelligence of negroes must inevitably remain inferior on average because it is inherited, he was abused as being irresponsible for making such a terrible suggestion. But surely it was his critics who were being irresponsible. The way to rebut Professor Jensen's research findings is by contrary evidence from further research. It would surely have been socially irresponsible for Professor Jensen to suppress his findings because they offended a political view. I am convinced that making science subservient to social values which happen to be approved by some faction or other must undermine the very foundations of science. Remember Galileo and remember Lysenko, the Russian charlatan who set back Soviet genetics for 20 years by trimming his scientific theories to suit the political dogma of the Communist Party.

Making science "subservient to social values" is of course another way of saying that science should be controlled. There is already considerable control through the money which research requires. Governments control pure as well as applied research—as for example the huge atom-smashing machine proposed for CERN in Switzerland which is so costly that several countries must join forces to provide enough money. This Science Foundation here, can control its research by withholding or providing money for projects.

There just has to be some control over the choice of technological possibilities because there are too many of them either for the available money or available talent.

In the totalitarian states like the Soviet Union control is applied so that research can be planned systematically "for the benefit of the State". I think we can see what happens when this is done.

Russia has devoted immense resources to medical research with very little to show for it. I am told that the same is true in biology and to a considerable extent in chemistry and physics. The Nobel awards alone show that free unfettered science will almost always beat regimented science. This was also demonstrated, I suspect, in the prestige race to put the first man on the moon. But it is especially true of pure research and invention. For any government, foundation or university to believe that it has the foresight to realise which research is going to be beneficial and which might have dangerous side-effects is an illusion. Lord Rutherford was quite convinced that his atomic researches would never have any social application beyond advancing human knowledge of the structure of matter. Faraday had pretty well the same view about electricity. Who could have foreseen that the cathode ray tube would lead to colour television? Or that Fleming's interest in a mould would lead not only to penicillin but to a huge antibiotics industry?

I don't see how we can control curiosity and that, as I said before, is the mainspring of science. In human evolution curiosity has been as great a driving force as sex and both are here to stay. There is little hope of curbing it even if over-rapid change may be responsible for some of the sickness of society.

We can't turn our backs on technological change because millions will starve to death if we do. The critics who say we must stress the quality of life rather than the quantity of production fail to realise that for most people in the world the two are intimately connected. Without the production of *cheap* food and *cheap* clothes, which means mass-production, the majority go without. As man increases in numbers the more he becomes a slave to his machines.

We're stuck with a technological age whether we like it or not and it's up to the elected politicians—with goading from the public and the media—to make use of what is best in it and ensure that what is obviously worst is not used. We want *more* rationality, not less. For, looking round the world one is left in little doubt that it is emotions that are more urgently in need of control than science is. In the final analysis passions still rule the world and ideological passion is probably a bigger danger than the bomb.

As a newsman I believe I should devote more space to complaining about the gap between discovery and application than about the over-rapid pace of research.

We can't discuss social conscience without considering the conscience, if any, of the media themselves. I say "if any" because there are occasions as for instance when a newspaper pays huge sums for the memoirs of a convicted criminal or sponsors a series on sexual perversion on the phony grounds that it is a "social document", when the existence of social conscience is in grave doubt. But with respect to science I think that on the whole newspapers and TV do not come out too badly. I should perhaps first ask you to consider what the prime purpose of a newspaper should be. Most people answer this by saying "To bring the news to the public". That is not correct. The prime function of a newspaper is to *sell* and this is true of the *Times* of London, the *Sydney Morning Herald,* the *New York Times,* the *Asahi Shimbun* and every other so-called quality paper. If you can't sell enough papers to make expenses and reach a sizeable or influential public—preferably both—then you don't begin to put over science or anything else. Bringing the news to the public is the second priority function and of course the two are linked because it is the content of the paper that sells it. But when we look at that content and compare it with sales what a depressing picture confronts us. In Britain the only newspaper steadily gaining circulation has been the *Sun.* It has achieved this by a deliberate policy of printing material that can only be described as salacious and, as a result, other papers have followed its example—a sad decline in standards.

It has to be admitted, I fear, that we are all to a considerable extent part of the entertainment industry. Even the *London Financial Times* feels it necessary to devote many of its columns to articles which are purely entertainment. But then I take some consolation in realising that writing about science can be entertaining. And if it is not most people will just not read it. The extremes to which this entertainment requirement is taken depend on the newspaper. One friend of mine got himself appointed science reporter to one of the more lurid Sunday papers in Fleet Street and he lasted three weeks. During that time he produced three articles entitled "Can Butterflies Slim?", "The Kinky Life of the Kinkajou" and "Indian Satellite TV to be powered by Cow Manure".

When the next story to appear under his name was about some film starlet of doubtful virtue I asked him what had happened. "Oh we've done science," he explained. The editor had decided that it was taking up valuable space.

The sad truth is that this is what so many people want. We are told that as more people are given what is called a proper education this requirement for trash will disappear. I fear I do not believe it. In Britain almost everyone has had very considerable sums spent by the State on their education yet when any random section of the public is stopped by TV interviews and questioned on any topic the result is usually deplorable—little capacity to make any sensible judgement or even to complete a sentence.

I fear we have had a rather pathetic trust in the fruits of education. It has long been argued that if only people were properly educated crime would disappear. Yet it increases.

It is the same with gullibility. One of the disservices most newspapers do to science is to promote superstition through the publication of horoscopes based on astrology and other methods of divination. None of the editors believes it. They do it because there is a demand for it—probably a bigger demand than ever before in history. It is an extraordinary phenomenon but I suspect that in addition to being the great age of science this is also the great age of superstition. Consider for example the cult of the Flying Saucer. A Gallup poll suggested that about one half of the American population thinks flying saucers are real while about one-third of these is convinced they are visitors from other planets. Those who claimed they had seen them were said to be no different on average from the sceptics in education or age. I am sure from the numerous readers' letters I get that the situation is similar in Britain. There we also have a widespread belief among highly intelligent people in a device called the Black Box with which its practitioners claim they can diagnose and cure disease by means of rays capable of operating over ranges of thousands of miles on people the operator has never met. One practitioner using the Box in Oxford claimed to have photographed the inside of a sick cow in Australia and then cured it.

The number of highly successful businessmen wearing copper bracelets to ward off rheumatism is quite astonishing.

I suspect the reason for this credulity is that science generates superstition. For every naive fallacy that science explodes it provides the breeding ground for a dozen more. By performing feats which seem magical and beyond the understanding of ordinary people scientists have created a credulity for anything with the imprint of science about it. When robot satellites and men are orbiting the earth it is not difficult to believe in the flying saucer. When radar can detect an object millions of miles away the idea of a box which can send out healing rays 12,000 miles is not so ridiculous. For millions today the use of words like "cosmic", "radionic" and "astral" is enough to cover up a multitude of logical discrepancies. As a result the situation is in many respects comparable with medieval times when credulity was equally honest and equally misguided. Today instead of witches on broomsticks there are bug-eyed monsters in flying saucers. Instead of the wizard's hat the sincere practitioner wears the white laboratory jacket.

The uncertainty created by nuclear weapons may be a factor in stimulating such beliefs but I suspect that for millions a belief for which no rational explanation is required, must satisfy some deep-seated psychological need and only the outward forms by which this need is fulfilled seem to have changed since primeval times.

It is a pity that newspapers encourage it but horoscopes do help sell them. A pity because one of the first things the media should do is bring the "truth" about science in all its forms, however unpalatable or frightening to the public. Right away, of course, we run up against the problem of what is truth? It is often very hard for a journalist to find out what the truth is and when you do you find there are several variants of it. I have for instance lunched with a senior civil servant who attended a Cabinet meeting where an important issue concerning atomic energy was discussed and dined on the same day with another equally eminent civil servant who was also there. Each gave me an entirely different account of what went on.

Sometimes one is driven to give wide publicity to some development or theory which turns out to be wrong. I remember being very impressed by the early evidence for the theory that coronary thrombosis may be caused by an excessive amount of animal fat in the diet. Yet this is now considered very doubtful. Excessive sugar intake and smoking are now regarded as likelier causes. Readers sometimes write to me pointing out that I must have been wrong, which of course I have to admit.

Those of us associated professionally with science know that uncertainty is the very essence of it. If geography's about maps and history's about chaps, then science is about "perhaps" or as we would say "probabilities". But this is not the public's view of science. To the mass of the public science means certainty. There are ordinary facts which turn out to be wrong but "scientific" facts should not.

All we in the media can do is to press on—against the clock as always—and do the best we can. This domination of daily journalism by the clock is, of course, the major cause of inaccuracy in science reporting and on this subject of social conscience I would like to say here that we are not inaccurate deliberately. We err because we are misinformed or misinterpret. But the old jibe that reporters never allow the facts to stand in the way of a good story is not true. What we call "stories" cannot be held up for checking as long as we would like because of the competition from other papers who may print them earlier.

There is a common complaint by scientists that newspapers concentrate too much on the new discovery, the ingenious invention and the exciting innovation and spend too little of their space explaining the methods whereby these were achieved. This trend has been called "the gee-whiz syndrome". I agree that we do err by making it appear too easy, by giving the impression that all you have to do is give the scientist a laboratory and the money and the answers will surely follow. They often do but sometimes they don't and people then ask, plaintively, "Why can't *they* do something about it?" "they" being the scientists and "it" any unbelievably difficult problem from finding a cure for cancer to producing a silent aeroplane.

There is now a move away from the gee-whiz approach but so long as science goes on providing wonders like moon landings it will always be with us. For there is the over-riding factor that editors believe it is the discoveries rather than the scientific methods in which their readers are most interested.

Whether scientists like it or not newspapers and TV are the

major means whereby *they* communicate with the public and that communication is now essential to them. Of course not all scientists want to communicate so much. Some of them suspect that one of the reasons science is in disrepute is because it has been "over-exposed", as public relations men say. In the old days the scientist could hold himself aloof and play with his string and sealing wax in his ivory tower. Some splendid work came out of ivory towers but those days are over. Most science now needs big money and it is the public which provides it and they not only have a right to know how their money is being spent—including occasions when it may be being wasted—but can't be expected to go on forking out through taxes unless they see it is worthwhile. Professor Messel was one of the first scientists I know to realise this fully and it is largely because of that realisation that he was able to raise funds, create this Science Foundation and do so much to put Australian science in the forefront where it now is.

I must tell you one story about Professor Messel. He invited me to join him at lunch in London to celebrate the fact that some American concern had just given the Foundation a huge quantity of dollars. He didn't eat much because he had just paid an emergency visit to a dentist. He'd never met the dentist before but when he emerged from the chair he had done such a good public relations job for Australian science that instead of giving him a bill the dentist gave him a cheque for £150.

The media can even play some role in bridging the communications gap between scientists working in different fields. There is a sad chasm between the different disciplines and though a paper like the *Daily Express* cannot give a scientist sufficient technical information it can make him aware of new developments by giving the bare bones of the story that let him know it exists so that he can find it. I can recall at least three occasions, which gave me great pleasure, where young scientists were given very substantial grants by industrialists who heard of their work through my columns.

New knowledge is interesting and entertaining for its own sake but the determining issue so far as the impact of science on society is concerned is the use which man makes of that knowledge. In deciding this the media can—and sometimes do—exert great influence. We can exert influence through conditioning the public to provide the money for research—through private means or through the Government. We can help the public to understand scientific problems which only the politicians can resolve—such as the pros and cons of building the Concorde aircraft or the huge atom-smashing machine at CERN in Switzerland. They can then play their part in elections in supporting or rejecting policies. In short, we can do a public relations job for science by creating the climate of opinion in which its fruits will be accepted. If we only help to bring about general acceptance of the idea of *change*—that with burgeoning technology rapid change is inevitable—we shall be doing an essential service. That is especially important with the trade unions when they take a reactionary attitude to technological change. (In this context I should like to pay tribute to science fiction for the effect it has had on making society receptive to new ideas. In some cases they have done the job too well. Some children watching the first moon landing live on television thought it had been done much more excitingly in TV fiction films.)

We can also intervene when we believe that the wrong things are being promoted—as happened in Britain and other countries recently over heart transplants. I think there is little doubt that by promoting heart transplants before they had solved the problems of tissue rejection, some surgeons were running before they could walk and the press campaign against what was admitted to be a degree of experimentation on human beings brought it to a halt.

Many people believe that we promote the wrong things, as when we champion space research, as my newspaper has done, and we hear loud cries that the money for missiles ought to be switched to cancer research. There is of course no evidence that if the money were denied to moon exploration, it would necessarily go to cancer research. In any case I suspect that the limitation on cancer research has been in ideas and talent rather than finance. The money devoted world-wide to cancer over the last 50 years has been enormous.

The media should protest, I think, when scarce resources are being dissipated in keeping the aged and terminally sick in a state of medicated survival. We have reached a situation when the State is prepared to spend thousands in keeping an ailing kidney patient alive but is reluctant to give a poor, healthy man a roof over his head. We can of course do harm by promoting the *wrong* things ourselves and we may well have done this in the past. The idea, for instance, that high density living in cities is desirable, when it means that people live rather like battery hens, is a possible example. The Concorde may be another. Perhaps we have been much to blame for promoting the concept, against which there is now revulsion, that technology can solve all.

We certainly do not do enough to point out the contribution which women can and do make to science and technology. The successful woman scientist is usually projected by the newspapers as something of a freak yet seven women have won the Nobel Prize, and one has orbited the earth.

We can help to destroy the charlatan, when we can detect him. This was done in the West with Lysenko. Had there been a free press in Russia he surely could never have survived so long.

At a lower level we can puncture the pomposity and arrogance of well-respected scientists if they try to foist some nonsensical notion on the world. I remember having a hand in this when a solemn conclave of doctors and scientists issued a definition of "health" on behalf of the World Health Organisation. They defined health as "complete physical, mental and social well-being." This of course meant that practically nobody in the world was healthy. Similar nonsense about race was issued by UNESCO some years ago without proper evidence.

On a popular paper one gets bombarded with ideas from crack-pots, which can usually be ignored but occasionally have to be exposed. People still come in with perpetual motion machines and usually there is some crook hovering in the background in the hope of making money out of it.

I hope you will agree that we have a very important function in seeing that in scientific matters, as elsewhere, the loud voice of a destructive minority does not drown the more tempered voice of the responsible majority.

Finally the media must guard against the inroads of technology into the humanity of man and the quality of his life. A few months ago I was alerted to the fact that the government was secretly carrying out a national survey of health in which doctors from selected practices were sending full medical details of all their patients to a computer centre together with their names. All this was being done without the patients' permission. This seemed such an infringement of the confidentiality between patient and doctor that I exposed it. Previously the Census Office statisticians had insisted that the patients' names were essential for accuracy but after the publicity they quickly changed their tune and dispensed with them.

In this connection I heartily approve of the move made by the managing director of a British computer firm who wants rules to make it a professional offence for any manufacturer to sell a computer to a government or private organisation if it knows the computer is to be used to store details of people's personal lives which could be made available to others. If governments start using computers surreptitiously for this purpose, then the manufacturers should apply sanctions by refusing to service them.

Yet we must be careful not to exaggerate the side-effects of science. We could easily exaggerate pollution to the detriment of society through its excessive impact on industry—if for example it led to unemployment because industries were shut down because they were unable to cope with new laws. I understand that three volcanic eruptions—in Krakatoa 1883, Alaska 1912 and Iceland 1947—polluted the atmosphere with more dust, ash and gas than all the activities of man down the ages.

The media surely have a very important role in convincing governments and the people they represent that science is the slave of man and not its master. The prevalent view that science and technology have somehow failed mankind and even betrayed it is quite ludicrous.

Over the centuries science has not been a dehumanising influence. On the contrary, the scientific spirit has been intimately involved in man's struggle to throw off the shackles of tyranny, dogma and injustice.

In regretting the passing of what may have seemed to be a more idyllic way of life let us remember that when it existed for a few, most men and women lived like brutes. Above all we must go on asserting the general supremacy of reason over emotion in overcoming social problems and help scientists to convince the public that its fear of science is irrational.

While recognising the dangers I remain optimistic for the future in the belief that reason will prevail.

I shall be most interested to hear your views because, as I said when I opened these lectures, while you and your generation benefit enormously from the scientists who have gone before you, you have also inherited the problems they created.

Index